Postgraduate Symposium in Civil and Environmental Engineering (PSCEE 2019)

IOP Conference Series: Materials Science and Engineering
Volume 601

Batu Pahat, Malaysia
31 March 2019

Editors:

Norashidah Abd Rahman
Nor Hazurina Othman
Noorwirdawati Ali
Siti Nazahiyah Rahmat
Alvin John Lim Meng Siang

ISBN: 978-1-5108-9479-2
ISSN: 1757-8981

Printed from e-media with permission by:

Curran Associates, Inc.
57 Morehouse Lane
Red Hook, NY 12571

CURRAN ASSOCIATES INC.
proceedings.com

Some format issues inherent in the e-media version may also appear in this print version.

Institute of Physics
Dirac House, Temple Back
Bristol BS1 6BE UK

Phone: 44 1 17 929 7481
Fax: 44 1 17 920 0979

techtracking@iop.org

Additional copies of this publication are available from:

Curran Associates, Inc.
57 Morehouse Lane
Red Hook, NY 12571 USA
Phone: 845-758-0400
Fax: 845-758-2633
Email: curran@proceedings.com
Web: www.proceedings.com

Table of contents

Volume 601

Postgraduate Symposium in Civil and Environmental Engineering 2019 (PSCEE 2019)

31 March 2019, Universiti Tun Hussein Onn Malaysia

Accepted papers received: 31 July 2019
Published online: 3 September 2019

Preface

Papers

Sustainable Infrastructure Engineering

Sustainable Environment and Water Research

IOP Conf. Series: Materials Science and Engineering **601** (2019) 011001 doi:10.1088/1757-899X/601/1/011001

PREFACE

On behalf of the conference organising committee, it is my pleasure to welcome researcher to the "Postgraduate Symposium in Civil and Environmental Engineering 2019". We must always acknowledge that this event continues to be offered to FKAAS postgraduate researchers through the ongoing support of our university, the commitment of individual professional and academic staff and the participation of postgraduate candidates and their supervisors.

The theme of this conference is "Sustainable Engineering for a Better Future", which expected to be a pioneer in focusing sustainable civil and environmental engineering to ensure the conservation of natural resources, renewable resources and sustainable development.

This conference is a forum which offers the opportunity to experience and practice a range of skills that will be useful as progress within student degree and into their career. To the participating candidates, thank you for accepting this opportunity to: prepare and present their research in a friendly conference setting; share their research across the university; hear how researchers from other disciplines are approaching their research and; how research practices are carried out in different contexts. It is worth noting that their preparation and presentation could well provide an opportunity to target attendance at an international conference and possibly be the precursor to the creation of another publication.

Like all worthwhile challenges, research requires dedication, conviction, perseverance and specific training. Also in common with many significant life challenges, research requires the will to succeed in a task that necessarily has an unknown result and outcome. This conference provides an excellent opportunity to explore and share the skills and professional and personal attributes that student can develop in order to succeed.

The training and experience gained through a higher degree provides those involved with specialist skills and very often specific knowledge on a particular subject. Opportunities to follow the specific areas focused on during higher degree research may or may not present themselves on completion of a degree. Within this reality it is important that higher degree researchers appreciate and recognize the toolbox of skills that they develop and then use this as a utility to explore the broad and perhaps unimagined opportunities that open up.

On behalf of the organising committee I would like to thank you for your support and contribution, acknowledging that without it we would not have a conference at all.

Dr. Nor Hazurina Othman
Chairman of PSCEE2019

Symposium Committee

Chairman
Dr. Nor Hazurina Binti Othman

Asst. Chairman 1
Dr. Noor Azlina Binti Abdul Hamid

Asst. Chairman 2
Dr. Mohd. Ariff Bin Ahmad Nazri

Secretary
Dr. Sharifah Salwa Binti Mohd. Zuki

Publication Committee
Ir. Dr. Zainorizuan Bin Mohd Jaini
Dr. Sabariah Binti Musa
Dr. Nur Adila Binti Ab Aziz

Editor
Dr. Norashidah Abd Rahman
Dr. Nor Hazurina Othman
Dr. Noorwirdawati Ali
Dr. Siti Nazahiyah Rahmat
Ts. Dr. Alvin John Lim Meng Siang

Symposium Committee
Dr. Siti Radziah Binti Abdullah
Dr. Nor Amani Filzah Binti Mohamed
Prof. Madya Dr. Razfidah Binti Hamdan
Pn. Rosnawati Binti Buhari

Organising Committee (Students)

Registrations
Nur Amira Afiza Binti Saiful Bahari

Emcees And Time Keeper
Akhtar Izzaty Binti Riwayat
Nurul Nabilah Binti Ismail
Muhammad Shabery Bin Sainudin
Nor Syafiqah Binti Ghadzali

Food Preparations
Muhamad Afif Bin Iman

Certificates
Muhammad Afiq Bin Tambichik
Siti Khatijah Binti Abu Bakar

Websites & Promotions
Alif Syazani Bin Leman
Normarliana Binti Zulkiflee

IOP Conf. Series: Materials Science and Engineering **601** (2019) 011003 doi:10.1088/1757-899X/601/1/011003

Peer review statement

All papers published in this volume of *IOP Conference Series: Materials Science and Engineering* have been peer reviewed through processes administered by the proceedings Editors. Reviews were conducted by expert referees to the professional and scientific standards expected of a proceedings journal published by IOP Publishing.

Driver Parking Behaviour: An Observational and Experimental Intervention Study

Nurul Najihah Mad Rosni[1,*], Ahmad Fadzil Jobli[1] and Maureen Neging[1]

[1] Faculty of Civil Engineering, Universiti Teknologi MARA, 94300 Kota Samarahan, Sarawak, Malaysia.
*Corresponding author:ahmad151@sarawak.uitm.edu.my

Abstract. Illegal parking has become a major problem in Malaysia especially in urban area due to the increased traffic volumes and limited parking space. When the parking facilities are limited, driver factor thus has a significant effect on the formation of traffic congestion. Extensive work has been conducted to address this issue, however most studies did not incorporate the impact of driver behaviour at limited parking condition whereas it is understood that "cruising for parking" at limited parking facilities has a very strong influence on the road traffic jams. Thus, objectives of this study are to analyse drivers' personality traits which focus on consideration and selfishness traits; and to investigate drivers' personality traits and their impact on driver parking behaviour in Kuching, Sarawak. Analyzation of driver parking behaviour was based on observational study of driver, vehicle, and situational variables. Experimental intervention then conducted to different age groups of drivers. They were asked to respond to questions that related to the driver parking behaviour characteristic. Analyses of the data reported in this study revealed that most of the drivers are considerate enough but they become a selfish person when confront with difficult situation at the parking space and their action sometimes are susceptible to illegal parking. The findings significantly beneficial to the traffic engineer, policy maker and researcher in enhancing parking control and management system.

1. Introduction

Illegal parking has become a serious issue especially in dense area. Driving etiquettes is often ignored when some of drivers deliberately park their vehicle inappropriate way. Some of drivers tend to double parking, improvise their own parking, and others when supply and demand ratio of parking space is not balance. Land use has an impact on parking demand [1]. If the land use comprised of government service offices, retail stores, banks and others, these will generate a significant numbers of people to come to this area every day and they will demand for a parking space. Illegal parking occurs when the demand of parking space is high while the supply is limited in dense area due to unplan urban growth. This situation will lead to selfish driving behaviour when drivers tend to disobey road and traffic regulation and park their vehicle in appropriate way due to limited parking space.

Illegal parking may cause traffic congestion as claimed by Morillo and Campos [2]. Road will become narrow due to on-street illegal parking and may create a bottleneck. The blockage of lane due to illegal parking may reduce the road capacity and cannot accommodate high demand of traffic. According to Haslinda et al. [3], illegal parking occurs because lack of enforcement, insufficient warning signage, lack of parking spaces especially in areas where activities are overwhelming and bad driving attitude. Driving behaviour will reflect to driving attitude.

Lin et al. [4] noted that each driver behaves differently depending to their ages, genders, driving experiences, situations, emotions and so on. Hence, their personality traits will influence their behaviour. Therefore, driver parking behaviour can be related as a reason of traffic congestion. Previous studies found that most of the drivers have experienced some form of parking violation and road congestion. Some of drivers who are the victim of parking violation and traffic congestion will respond with horn honking and rude gesture. Although the victims did not commit any offence, but their respond show the bad side of them too. Hence, it is important to look at the drivers' personality traits because it is vital to understand the possible causes leading to such act.

2. Illegal Parking Behaviour

Each vehicle needs a parking space for every trip that they are making. Less supply of parking space will trigger driver to park illegally. Hence, illegal parking is one of the critical issues that cause traffic congestion. Illegal parking will cause obstruction of roadway which will eventually making the road become narrow for traffic movement. In order to overcome traffic congestion, the governments has invested heavily in transportation and infrastructure as stated in the 11th Malaysian Plan (RMK 11). The government has drafted plans to increase the level of transport service to reduce traffic congestion. Meanwhile, under Government Transformation Programme (GTP) agenda, government tries to manage travel demand by enhancing parking control and management especially in urban area. Therefore, in order to overcome and find solution for illegal parking that occur throughout the world, abundance research on parking enforcement have been carried out such as parking enforcement policies for commercial vehicles [5], legalizing the illegal parking, a solution for parking scarcity in developing countries [6], enforcement of illegal parking under the new parking legislation [7] and others. However, it was found across several studies [8-9], level of enforcement towards illegal parked vehicles was low, although regulation and parking policy already exist. In both of their studies, drivers prefer to park illegally near the destination. This is very similar to the finding by Conway et al. as cited in [5], which concluded that drivers commonly park on the bike lane which results in conflict with cyclist.

My Thanh & Friedrich [6] conduct a research in Hanoi and found that illegal parking occurs in 80% where they tend to park their vehicles near the destination for short period and leave immediately after settle their business. They also found that illegal parking occurs more frequently near the sidewalk as it is wide enough for parking. Khan and Lourenco, as cited in [9] claimed that illegal parking creates obstacle for the pedestrian who feel discomfort to walk as the walkway become discontinue.

The different use of land also has significant impacts on driver parking behaviour. When the parking turnover rate is low, drivers tend to park illegally. Yin et al. [10] noted that the traffic condition, land use and economic factor should be considered in arrange parking facilities. For example, Spiliopoulou & Antoniou [8] found that the presence of the train station raises the demand for parking, and generally the illegal parking occurs when the legal spaces have been saturated.

Overall, illegal parking is a topic that has been thoroughly studied. However further study of driver parking behaviour will lead to better understanding of why some drivers behave in such a way that cause traffic violation. Hence, in this study we address the issue of illegal parking that trigger traffic congestion with respect to human behaviour.

3. Methodology

Self-administrated questionnaire was used to measure driver parking behaviour. The questionnaire was design to collect the required information from the respondents for research purposes. The questionnaire was divided into three main aspects. The subjects contained in the questionnaire are as follows:

i. PART A: Background respondents

ii. PART B: Personality traits (which measure consideration and selfishness traits)
iii. PART C: Roadside illegal parking behaviour (which consist of 3 situations)

Before the data collection activity was carried out, a pre-test of questionnaire was done to ensure that the questions forms are easy for the respondents to understand and full cooperation can be expected from respondents. According to Dikko [11], a preliminary study serves as a tool of helping to detect any mistake or error in the measurement. Hence, pre-testing the instrument on 30 samples of respondents having the same characteristic as those in the main study was executed to get the information and response on the questionnaire. It was found out that the overall respondents had understood the questions presented in the questionnaire.

In part A, five questions on background respondent such as gender, age groups, driving experiences, involvement in road accident and number of parking summonses received was asked as it may help researcher to relate the factors may influence a respondent's choice of answer. In part B, participants were asked to complete the questionnaires which assess their personality traits on parking behaviour. The questionnaire is conducted to measure selfishness (concern only for own interest regardless for others) and consideration (kindness and thoughtful regard for others). These two personality traits were selected in this study as, consideration trait can lead the driver to park their vehicle in designated parking space while selfishness trait will lead the driver to park their illegally. Seven questions are given which represent each consideration situation and selfishness situation.

Whereas, in part C, participants were evaluate through questions after watching a video that contains roadside illegal parking in three main roads within Kuching city. The video acts as a tool to trigger participant's reaction regarding the scenario. The participants were asked three questions which represent situation 1, situation 2 and situation 3.Situation 1 indicates that drivers will double park their vehicle which is parked legally if designated parking spaces have full. Situation 2 explains about drivers will park their vehicles illegally when valid parking space is full. Whereas situation 3 explains that the drivers do not think that they are selfish when they park their vehicles illegally on the roadside because other people do that also.

3.1. Sample and Data Collection

Data collection from respondents was conducted through questionnaires and distributed to local people who live in Kuching and has driving license. The majority of the respondents who participated in this study are those who always drive around the study area; Jalan Tun Abang Haji Openg, Jalan Main Bazaar and Jalan Gambir as shown in Figure 1 [12]. In this study, the questionnaire is conducted to measure selfishness (concern only for own interest regardless for others) and consideration (kindness and thoughtful regard for others). The questionnaire consists of a combination of questions which are developed specifically after reviewing process. Questionnaire with both version of English and Malay language is used. In this study, we had used random sampling method in the approach to obtain a sample of respondents. The reliability analysis of the instrument had a Cronbach Alpha value of 0.813 which indicated that the closer the Cronbach alpha value to 1, the greater the internal consistency of the items in the scale [13]. With a population in Kuching City between 325,135 to 1 million at 95% confidence interval for the estimation to be within ± 5% error, the required sample size was around 384. A minimum sample size 384 were needed to reduce biases and sufficient to represent the general characteristic of the population in Kuching. Only 402 questionnaires manage to be collected and being considered for further action.

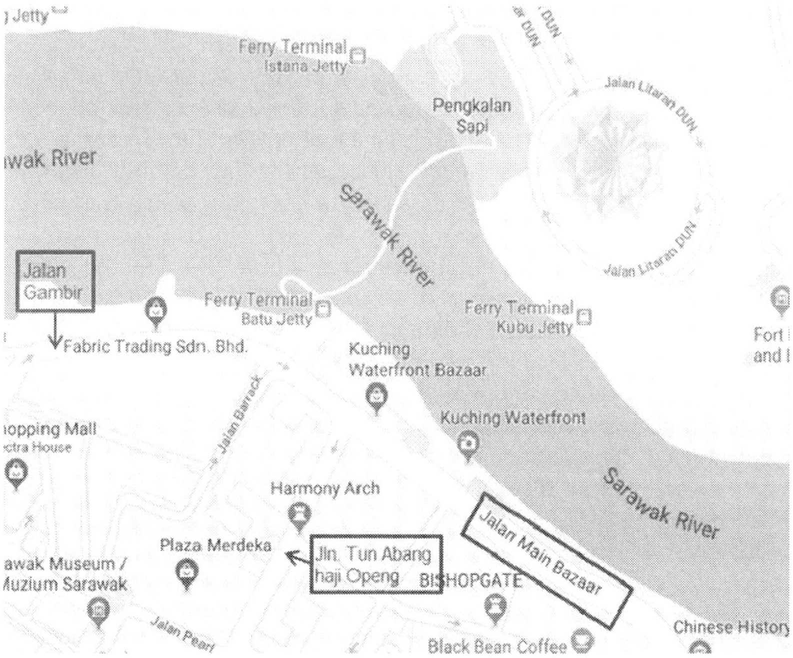

Figure 1. Three main roads for study area are labelled with red box [12].

4. Results

In this study, data analysis was conducted through SPSS using descriptive analysis, which includes measure of central tendency such as mean, standard deviation, and measure of frequency values. Once the data have been appropriately described, then inferential analysis was used to make reasonable analyses on driver parking behaviour. Non-parametric test such as Kolmogorov-Smirnov and the Shapiro-Wilk for normality test and Spearman rank correlation test were employed.

4.1. Demographic Result.

Sample of respondents consisted of 402 participants where 58.7% are males while 41.3% are females. There were four age groups in this survey specifically: 30 and below, 31 to 40 years, 41 to 50 years and 51 and above. Most of the drivers had 11-20 years of driving experiences. The summon result shows that 31.3% of respondents had received one and over traffic parking summon while the rest of them were not received any for the last 12 months. The majority of respondents 36.1% were in age group of 30 and below while the rest of them were older. Age of respondents can be interpreted as age group below than 50 years were dominated as they drive more than older drivers. All age groups were male dominated with the higher percentage of driver involved in collisions in the last 12 months as shown in Figure 2. However, there were more young male drivers involved in collision than older drivers. Tabibi [14] noted that rate of accident can be relate to driving behaviour where drivers with high risk behaviour and unsafe road practices often involved in road accidents. Therefore, accident involvement also can be relate to parking behaviour as well. The drivers with risky driving style and unsafe road practices also have tendency to park their vehicle illegally as they being selfish and refuse to obey traffic rule and regulation. This is compatible with the finding of Hu et al. [17].

IOP Conf. Series: Materials Science and Engineering **601** (2019) 012002 doi:10.1088/1757-899X/601/1/012002

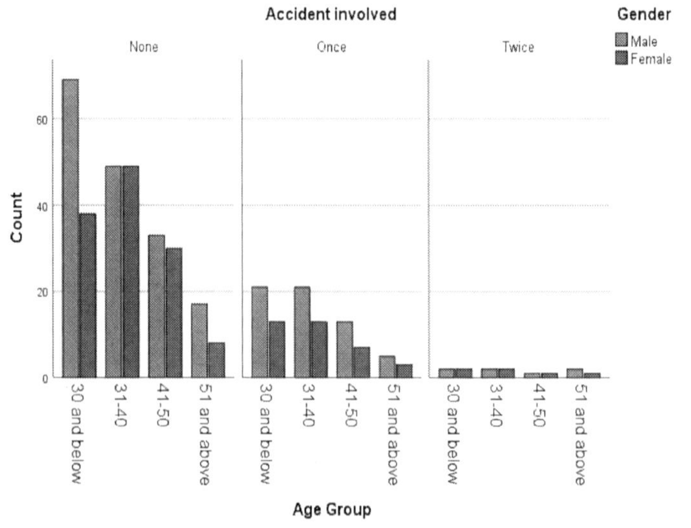

Figure 2. Respondent's age group involved in road accident in last 12 months.

4.2. Descriptive Analysis

The results of the survey have been tabulated based on mean and standard deviation of important survey factors. Respondent's personality traits was measured using 14-items which are divided into two subscales: consideration and selfishness. Consideration traits was measured using 7-items representing situation of kindness and thoughtful regard for other road users. Whereas selfishness traits also were measure using 7-items where seven potentially selfishness situations were asked and respondents were required to rate the statement of consideration and selfishness traits based on four-point Likert scale. (1= none of the description fits me), (2= only some of this fit me), (3= half of the description fits me) and (4=the statements describe me perfectly). Their responses were scored to provide overall measure of their propensity to become consider and selfish. The questions were designed to evaluate the consideration and selfishness traits among drivers. General situation related to driving behaviour on the road (not restricted to parking situation only) were asked. The analyses result from consideration and selfishness traits will be used to relate with the situation 1, situation 2 and situation 3 in part C.

The mean and standard deviation score for each item are listed in Table 1 and Table 2. Result of C1, C2, C3, C4 and C5 shows the score ranged between 2.88 to 3.16 suggesting that description of those consideration situations are half fits the respondents' behaviour. In addition, C6 showed that most of the respondents will never use their horn to indicate their annoyance to other road users. Meanwhile, C7 showed none of respondents claim that they are not considerate the needs of other driver. Most of respondents always consider and showing their empathy towards other drivers. Hence, based on the responses, it can be concluded that the average mean score for consideration traits was 2.85 out of 4 suggesting that drivers sometimes kindness enough and thoughtful regard for other road users.

Table 1. Means and standard deviation of consideration responses.

Item	Description	Frequency distribution of responses				Mean	SD
		1	2	3	4		
	Consideration (Mean =2.85)						
C1	You will give consideration to vehicles which have indicate their intention to change lanes	34 (8.5%)	57 (14.2%)	179 (44.5%)	132 (32.8%)	3.02	0.900
C2	When driving on the road, you will give consideration to other vehicles entering the road from the slip road	19 (4.7%)	94 (23.4%)	164 (40.8%)	125 (31.1%)	2.98	0.858
C3	You will give way to buses to move out from stops	27 (6.7%)	77 (19.2%)	215 (53.5%)	83 (20.6%)	2.88	0.809
C4	You will give way to pedestrians who crossing the road even there is no marked crossing	6 (1.5%)	98 (24.4%)	204 (50.7%)	94 (23.4%)	2.96	0.733
C5	You will slow down and move to the side and let the emergency vehicles overtake you	0	84 (20.9%)	169 (42%)	149 (37.1%)	3.16	0.745
C6	You will use your horn to indicate your annoyance to another road user	139 (34.6%)	131 (32.6%)	75 (18.7%)	57 (14.2%)	2.12	1.042
C7	I will always consider the needs of other drivers when driving	0	118 (29.4%)	228 (56.7%)	56 (13.9%)	2.85	0.645

The result of S3 declares that most of the respondents at least have tried to take chance and go through lights that have turned red. S5 is another proof that most of the drivers would become selfish when they attempted to overtake slow moving vehicle when they were queuing. In addition S2 prove that some of the respondents were intentionally drive in the emergency lane while most of them will not drive in the emergency lane. Based on all of these justifications, the mean score for selfishness traits was 1.97 indicated that most of the drivers have at least being selfish while driving depending on situation.

Table 2. Means and standard deviation of selfishness responses.

Item	Description	Frequency distribution of responses				Mean	SD
		1	2	3	4		
	Selfishness (Mean =1.97)						
S1	If you accidentally scratched, dented or damaged another car, no matter how minor the damage you will leave without admit your fault	156 (38.8%)	148 (36.8)	90 (22.4%)	8 (2%)	1.88	0.823
S2	You deliberately drive in the emergency lane	184 (45.8%)	137 (34.1%)	63 (15.7%)	18 (4.5%)	1.79	0.866
S3	You always take a chance and go through lights that have turned red	142 (35.3%)	203 (50.5%)	57 (14.2%)	0	1.79	0.672
S4	When you are crawling along in rush hour traffic, you will let someone in from a side road	116 (28.9%)	177 (44%)	85 (21.1%)	24 (6%)	2.04	0.860
S5	You attempted to overtake slow moving vehicle when they were queuing	96 (23.9%)	106 (26.4%)	122 (30.3%)	78 (19.4%)	2.45	1.057
S6	You always park where it is not allowed and risk a fine.	219 (54.5%)	156 (38.8%)	27 (6.7%)	0	1.52	0.620
S7	I am a selfish person when driving on the road	66 (16.4%)	177 (44%)	135 (33.6%)	24 (6%)	2.29	0.810

4.3. Correlation Analysis

Normality test were carried out to calculate the frequency of selfishness and consideration behaviours in other drivers. Result of Kolmogorov-Smirnov and the Shapiro-Wilk normality test for consideration and selfishness behaviour indicate that both tests have significant values less than 0.05. This value indicates that the frequency distribution of value for both behaviours did not show normal distribution. Therefore the correlation analysis using spearman rank correlation method is the most appropriate method for this study.

Spearman Rank correlation was used to test the correlation between consideration and selfishness personality with roadside illegal parking behaviour. The results showed that there was a significant strong relationship between selfishness trait and roadside illegal parking behaviour for situation 1, situation 2 and situation 3 with correlation coefficient 0.682, 0.751 and 0.745 respectively at the level 0.01. However as showed in the Table 3, the correlation between consideration trait and roadside illegal parking behaviour for situation 1, situation 2 and situation 3 is 0.089, 0.081 and 0.058 respectively. It shown that the correlation is not significant and the two variables are not linearly related as the value is approaching to 0 and $p>0.05$.

Table 3. Correlation between personality trait and roadside illegal parking behaviour.

Correlations						
Variables		**consideration trait**	**selfishness trait**	**Roadside illegal parking behaviour (situation 1)**	**Roadside illegal parking behaviour (situation 2)**	**Roadside illegal parking behaviour (situation 3)**
consideration trait	Correlation Coefficient	1.000	.032	.089	.081	.058
	Sig. (2-tailed)	.	.517	.075	.104	.246
selfishness trait	Correlation Coefficient		1.000	.682**	.751**	.745**
	Sig. (2-tailed)			.000	.000	.000
Roadside illegal parking behaviour (situation 1)	Correlation Coefficient			1.000	.671**	.688**
	Sig. (2-tailed)				.000	.000
Roadside illegal parking behaviour (situation 2)	Correlation Coefficient				1.000	.727**
	Sig. (2-tailed)					.000
Roadside illegal parking behaviour (situation 3)	Correlation Coefficient					1.000
	Sig. (2-tailed)					

**. Correlation is significant at the 0.01 level (2-tailed).

5. Discussions

The current study investigated the drivers' personality traits and their impact on driver parking behaviour in Kuching, Sarawak. There were two kinds of personality traits being focus in this study which were consideration and selfishness traits. Overall, majority of respondents claim that they are more considerate rather than being selfish while driving on the road. The result from Table 1 shows about 13.9% of respondents admit that they always consider the needs of other drivers when driving on the road. However, when they were being asked whether they are a selfish person when driving on the road, only 6% of them admitted. Majority of them claims that they will become selfish depending on the situation. The respondents refuse to admit that they are more selfish than considerate when being asked. However, results from correlation analysis prove that they are more selfish as the result show positive relationship with the given situation.

On the other hand, the result on consideration traits did not correlated with roadside illegal parking behaviour on the road. No significant correlations are shown with the frequency of illegal parking behaviour for all situations 1, 2 and 3. However, the findings contradict to the finding by Ismail et al. [15], as they found that low empathy traits correlate significantly with driving behaviour. Although the result showed that most of respondents think that they show enough empathy or considerate towards others, but there are situational variables such as mood, feeling, stress at the moment that play a role and may cause consideration trait change into selfishness trait when drivers confront with real situation. This finding is in line with the finding of Sucha et al. [16].

Meanwhile, the results from correlation analysis selfishness trait significantly correlated with roadside illegal parking behaviour for all situations 1, 2 and 3. When these three situations being asked to respondents, their respond towards the questions represent that they were selfish when come into this kind of situation. This is in line with the finding of Hu et al. [17], which mentioned that unsafe driving behaviour will lead to illegal parking. On top of that, traffic speed will be reduced while traffic density will be increased with the increasing number of unsafe driving behaviour. This finding also was supported by numerous studies [15-17], which found that personality traits were related to driving behaviour. Roadside illegal parking will reduce traffic speed, hence volume of traffic will increase and this scenario may lead to traffic congestion. Hu et al. [17] also claimed that roadside parking will cause subsequent vehicles waiting in line or deceleration while waiting the respective vehicle from drive way to park their vehicle.

Driver's behaviour is a very complex matter that is influence by human factors such as people's ability, skill, knowledge and personality traits (such as selfish, empathy, consideration and anger). This was supported by previous study [16, 18]. Personal characteristic such as gender, age, driving experience correlated to driving behaviour. The reason why driver are prone to getting selfish not only due to personal characteristic but can also be relate with situational factors such as road congestion, time constraint, competitiveness and others [19].

6. Conclusions

This study has highlighted the consideration and selfishness trait in affecting roadside illegal parking behaviours. Major conclusions drawn from this study are as follows;

i. Number of accident involvement by driver can be relate to parking behaviour as driver with risky driving style and unsafe road practices have tendency to park their vehicle illegally.
ii. Most drivers are considerate when driving on the road however they will turn into selfish person when they confront with difficult situation in parking space.
iii. The results suggest that those drivers who become selfish when driving on the road are susceptible to illegal parking behaviour.

Hence, these findings can provide useful information for public transport providers in the efforts of improving road safety and traffic congestion. It is recommended to increase the supply of transportation facilities and strengthen the traffic management in order to properly guide and control the transportation demand. Besides that, driver's evaluation on personality trait should be proposed to be included in the process of taking license course at driving school.

7. References

[1] Wenneman A, Habib K N, and Roorda M J 2005 Disaggregate analysis of relationships between commercial vehicle parking citations, parking supply, and parking demand. *Transportation Research record : Journal of the Transportation Research Board* **2478**, 28-34.
[2] Morillo C and Campos J M 2014 On-street Illegal Parking Costs in Urban Areas *Procedia - Social and Behavioral Sciences* **160** 342–351
[3] Haslinda A, Ng R and Mohamed R 2016 Factors Influencing Haphazard Parking in the Klang Valley, Malaysia. *International Journal of Academic Research in Business and Social Sciences* **6**(9) 159–181
[4] Lin N, Zong C, Tomizuka M, Song P, Zhang Z and Li G 2014 An Overview on Study of Identification of Driver Behavior Characteristics for Automotive Control. *Mathematical Problems in Engineering* **2014**, article ID 569109 15 pages
[5] Nourinejad M and Roorda M J 2017 Parking enforcement policies for commercial vehicles. *Transportation Research Part A: Policy and Practice* **102** 33–50
[6] My Thanh T T and Friedrich H 2017 Legalizing the illegal parking, a solution for parking scarcity in developing countries. *Transportation Research Procedia* **25** 4954–4969

[7] Ishida T 2006 Enforcement of Illegal Parking under the New Parking Legislation. *IATSS Research* **30**(2) 120–125
[8] Spiliopoulou C and Antoniou C 2012 Analysis of Illegal Parking Behavior in Greece. *Procedia - Social and Behavioral Sciences* **48** 1622–1631
[9] Aljoufie M 2016 Analysis of Illegal Parking Behavior in Jeddah. *Current Urban Studies* **4** 393–408
[10] Yin J, He Y and Sun X 2016 Area Classification of Surrounding Parking Facility Based on Land Use Functionality. *Open Journal of Applied Sciences* **6**(7) 380–385
[11] Dikko M 2016 Establishing Construct Validity and Reliability: Pilot Testing of a Qualitative Interview for Research in Takaful (Islamic Insurance). *The Qualitative Report* **21**(3) 521–528
[12] Jalan Main Bazaar Map. Google Map. Retrieved on November 10, 2018. http://maps.google.com/
[13] Gliem J A and Gliem R R 2003 Calculating, Interpreting, and Reporting Cronbach's Alpha Reliability Coefficient for Likert-Type Scales. *Midwest Research to Practice Conference in Adult, Continuing, and Community Education* Columbus pp 82–88
[14] Tabibi Z 2011 Risky Driving Behavior Relating to accident Involvement : Investigating Some Identified Factors in An Iranian Sample. *International Journal of Psychology* **5**(1) 51-72
[15] Ismail R, Voon N L, Saad M H, Saleem M and Ibrahim M 2016 Aberrant Driving Among Young Malaysian Drivers *Jurnal Teknologi (Science & Engineering)* **78**(6–10) 55–56
[16] Sucha M, Sramkova L and Risser R 2014 The Manchester driver behaviour questionnaire: Self-reports of aberrant behaviour among Czech drivers. *European Transport Research Review* **6**(4) 493–502
[17] Hu L, Xue G, Wang M, Chen Z, Zhang T, Li L and Qin L 2017 Influence of typical drivers' unsafe driving behaviors to traffic operation: An exploratory study in Kunming, China. *Advances in Mechanical Engineering* **9**(9) 1-9
[18] Targosinski T 2017 Preliminary simulation research of driver behaviour in response to outdoor advertisements. *MATEC Web Conf.* **122** 3009
[19] Rakotonirainy A, Schroeter R and Soro A, 2014 Three social car visions to improve driver behaviour. *Pervasive and Mobile Computing* **14** 147-160

Acknowledgement
The research is supported by Universiti Teknologi MARA under grant number 600-RMU/DANA /SS 5/3 (7/2017). We are very thankful for this support. Sincere thanks are extended to those who have contributed directly or indirectly for all their valuable comments, guidance, assistance in laboratory work and suggestion during preparation of this research.

IOP Conf. Series: Materials Science and Engineering **601** (2019) 012003 doi:10.1088/1757-899X/601/1/012003

Shear Strength Characteristics of Coastal Soil Treated with Sodium Silicate

Adriana Erica Amaludin[1], Melvin Teo Twan Ann[1], Mohammad Radzif Taharin[1] and Salinah Dullah[1]

[1] Civil Engineering Programme, Faculty of Engineering, Universiti Malaysia Sabah, 88400 Kota Kinabalu, Sabah, Malaysia
*Corresponding author: adriana@ums.edu.my

Abstract. In recent times, rapid developments near coastal areas are gaining attraction and attention from industry players. This necessitates further understanding of soil behavior found in these areas. Soil stabilization can further enhance the physical and engineering characteristics of soils used for development purposes. For this study, a series of laboratory tests were conducted to determine the potential of Sodium Silicate (TX-85), a liquid-type chemical soil stabilizer, to improve the properties of coastal soil obtained in Kota Kinabalu, Sabah. The soil samples were subjected to a series of laboratory tests, which includes the pH and Unconfined Compression Strength (UCS). The dosages of Sodium Silicate mixed with the soil were 4, 5, and 7% by soil sample weight, with curing intervals of 3 hours, 24 hours and 48 hours, respectively. The optimal dosage of Sodium Silicate observed in this study is 4%, at 48 hours curing period. This combination of stabilizer dosage and curing period produced the highest strength increment, where the UCS value increased by 90.3%, from 262.1 kPa to 498.8 kPa.

1. Introduction

Development projects near coastal areas are currently in demand within the construction industry. Therefore, the characterization and enhancement of coastal soil properties is an emerging research area to be explored by researchers in the geotechnical engineering field. Improvement of soil properties can be achieved through soil stabilization, where the increased resistance to softening by water is possible via water proofing the soil particles, better bonding of soil particles, or a combination of both [1].

The stabilization of naturally occurring or native soil has been in practice for centuries [2]. In order to solve the problem of weak soils, high costs might be incurred if good quality fill soils are needed to replace the weak soils for land development purposes. However, soil stabilization is as an economic solution to this weak soil problem. For this study, the soil stabilization method chosen to fulfil the research objectives is chemical stabilization. More specifically, the chemical stabilization technique involves adding a chemical compound in order to promote improved soil particle bonding through the formation of a stronger soil mixture when the stabilizer is mechanically mixed and compacted with the natural soil [3].

According to Eisazadeh [4], chemical soil stabilizers are either categorized as traditional or non-traditional stabilizers. While the use of traditional stabilizers i.e. cement, lime and fly ash have been well documented, this is not the case for non-traditional stabilizers i.e. enzymes, liquid polymers, and silicates [5]. Sodium Silicate, or its commercial name, TX-85 is a liquid-type of non-traditional stabilizer produced by Probase Manufacturing Sdn. Bhd. and is typically marketed as a material used to stabilize plantation road soils [6]. However, several studies conducted to study the TX-85 performance when combined with other types of soils, i.e. laterite [5, 7, 8], marine clay [9], and soft clay [10] have shown encouraging results in terms of improvement of soil engineering properties. In general, the trend for soil strength improvement for the abovementioned soils is about three to four times the strength increment compared to the untreated soil [5, 9, 10].

Therefore, the aim of this study is to improve coastal soil by determining the optimum dosage (percentage by weight) of soil stabilizer (TX-85) and optimum curing period to obtain the highest unconfined compression strength (UCS) value; where the stabilized soil strength characteristics was compared to the untreated soil.

2. Methodology

2.1. Coastal Soil

In this study, the soil sample was collected at a coastal area at a depth of 1.5 meters in Sepanggar Bay, located about 13 kilometres from Kota Kinabalu, Sabah. The color of the soil is dark brown. The samples were air-dried and broken into individual particles using a rubber mallet before testing. Table 1 shows the physical and engineering properties of the untreated soil sample, in comparison with soils tested by other researchers using the TX-85 soil stabilizer. From the tabulated data, the coastal soil sample was classified as low plasticity clay and silt soil (CL+ML) according to Unified Soil Classification System (USCS) [11].

Table 1. Characteristics of the untreated soil sample and comparison with past studies [5, 9- 10]

Physical and Engineering Properties	**Values**			
	This Study	**Laterite Soil [5]**	**Marine Clay [9]**	**Soft Clay [10]**
Liquid Limit, LL (%)	28	75	58	73
Plastic Limit, PL (%)	22	41	23	29
Plasticity Index, PI (%)	6	34	35	44
pH Value	5	5.35	-	-
Specific Gravity, Gs	2.51	2.69	-	-
Maximum Dry Density ($kg\text{-}m^{-3}$)	1640	1310	1600	1343
Optimum Moisture Content (%)	18	34	21	30
Unconfined Compressive Strength (kPa)	262.1	270	23	33.6
Soil Classification (USCS) [11]	CL+ML	MH	MH	MH
Type of Soil	Low Plasticity Clay and Silt	High Plasticity Silt	High Plasticity Silt	High Plasticity Silt

2.2. Sodium Silicate

The Sodium Silicate (TX-85) stabilizer material used in this study is a proprietary chemical compound and its chemical composition is not known to the public. Pakir [9] conducted an Inductively Coupled Plasma Mass Spectrometry (ICP-MS) on the TX-85 stabilizer to discover that the material largely comprises of Sodium, Aluminium, Silicone and Ferum; while Latifi [5] revealed that the stabilizer has a pH value of 12.54. Table 2 shows the average pH values for untreated soil and soils stabilized with

TX-85, tested based on BS 1377 [12]. The alkalinity of the stabilizer has significantly influenced the final pH value of the stabilized soil since the silty soil tested in this study is highly acidic in nature (pH 5). According to Eisazadeh [13], the pH of soil mixture reflects the changes occurring in the medium, where it influences the solubility of clay materials and the distribution of charge on clay lattice. After the initial mixing, the chemical reaction takes place almost immediately and no further changes to the pH value is observed after a short period of curing, similar to the findings by Latifi [5].
In Table 2, the pH value of the soil mixture stopped increasing after 9% and the value plateaued at pH 10.1. The selected dosage of TX-85 added to the soil sample in this study (by sample weight) were 4%, 5% and 7%, with the consideration of limited amount availability of TX-85 to complete this study.

Table 2. Average pH values for soils stabilized with Sodium Silicate based on British Standard [12]

Percentage of Stabilizer Added (%)	pH Value
0	5.0
1	8.4
2	8.7
3	9.0
4	9.2
5	9.5
6	9.6
7	9.8
8	9.9
9	10.1

2.3. Testing Programme
Before commencing the Unconfined Compression Strength (UCS) tests, the soil sample compaction characteristics were determined. A series of Standard Proctor Compaction (SPC) tests was conducted based on BS 1377 [14]. The obtained SPC results were used to plot the compaction curve of the soil which is a graph of dry density versus moisture content. Subsequently, the optimum moisture content (OMC) can be obtained by ascertaining the maximum dry density (MDD) on the graph. As previously shown in Table 1, the maximum dry density was valued at 1640 kg-m^{-3}. Subsequently, the MDD value for soil mixed with 7% of sodium silicate was valued at 1528 kg-m^{-3}. Although the results of the Standard Proctor Compaction (SPC) test results of soils mixed with TX-85 is not discussed in depth in this paper, the observed trend of the MDD value was found to have decreased with an increasing TX-85 content.

In order to prepare the silt sample for the UCS test, the soil was mixed with each proportion of TX-85 and the optimum water content. The sample was then compacted in a mould, before three (3) specimens are extruded using thin-walled sampling tubes for each compaction [14]. The specimens were trimmed to produce specimen sizes of 38 mm in diameter and 76 mm in height. Then, the samples were wrapped in several runs of cling film [5] before being labelled correctly, stored in polythene bottles and placed in a well-ventilated room [9]. The curing period chosen for this study is 3 hours, 24 hours and 48 hours. Even though other researchers had chosen curing periods of 3 days, 7 days, and 28 days [5,9,10], the short curing period for this study was chosen to see the short-term chemical reaction between the sodium silicate and the weak soil. The specimens wrapped with cling film and were cured at a temperature-controlled room with a temperature of 27° ± 2°C, before the specimens were subjected to the Unconfined Compression Strength (UCS) test. This specimen preparation method was applied for the untreated silt soil specimens and for the stabilized silt soil specimens.

After the curing process, the specimen was taken out of the polythene bottle and was placed carefully on the lower platen on the Unconfined Compression Strength (UCS) machine. The platen was adjusted manually until the sample just makes contact with the top platen, where this was indicated by a fractional movement of the load dial gauge.

Next, the load was applied so that the equipment produces a compression rate of 0.5% to 2.0% per minute [15]. In this experiment, a compression rate of 1.0% was chosen. Then, the load and deformation dial readings were recorded on a data sheet at every 0.2 mm to 0.5 mm divisions on the deformation dial. Subsequently, the testing continued until peak load was achieved to cause the sample to fail, and the last loading applied was recorded. Lastly, the sample was removed from the equipment, and all recorded readings were tabulated.

3. Results and Discussion

The original and stabilized soil specimens with varying percentages of TX-85 were subjected to a series of UCS Tests. Table 3 shows the summary of the unconfined compression strength test results, while Figure 1 shows the UCS values of stabilized soil specimens plotted against varying curing periods.

Table 3. Summary of UCS test results for silt soil treated with Sodium Silicate (SS)

Soil Sample Type	Unconfined Compressive Strength (kPa) at given curing period		
	3 hours	24 hours	48 hours
Untreated Soil (US)	262.1	262.1	262.1
Soil + 4% SS	322.4	448.4	498.8
Soil + 5% SS	330.4	444.4	468.2
Soil + 7% SS	385.5	448.7	462.7

Based on Figure 1, when the soil was added with TX- 85 and cured for 3 hours, the compressive strength of each soil sample had increased from 262.1 kPa to 322.4 kPa (4% TX-85), 330.4 kPa (5% TX-85), and 385.5 kPa (7% TX-85), respectively. The highest compressive strength cured for 3 hours was contributed by the soil added with 7% TX-85, with an increment of about 45% compared to the original soil sample.

Meanwhile, for the compressive strength of specimens cured for 1 day does not show a marked improvement compared to the soils cured for 3 hours. Soil added with 4% TX-85 gained strength from 322.4 kPa to 448.4 kPa, while soil added with 5% and 7% TX-85 gained strength from 330.4 kPa to 444.4 kPa, and from 385.5 kPa to 448.7 kPa, respectively.

Lastly, when the modified soil was cured for 2 days, the soil added with 4% TX-85 gained the highest strength of all specimens, which had a value of 498.8 kPa. The soil added with 5% and 7% TX-85 does not show significant strength gain with only 468.2 kPa and 462.7 kPa respectively, which is about only less than 2% increment compared to soil sample cured for 1 day.

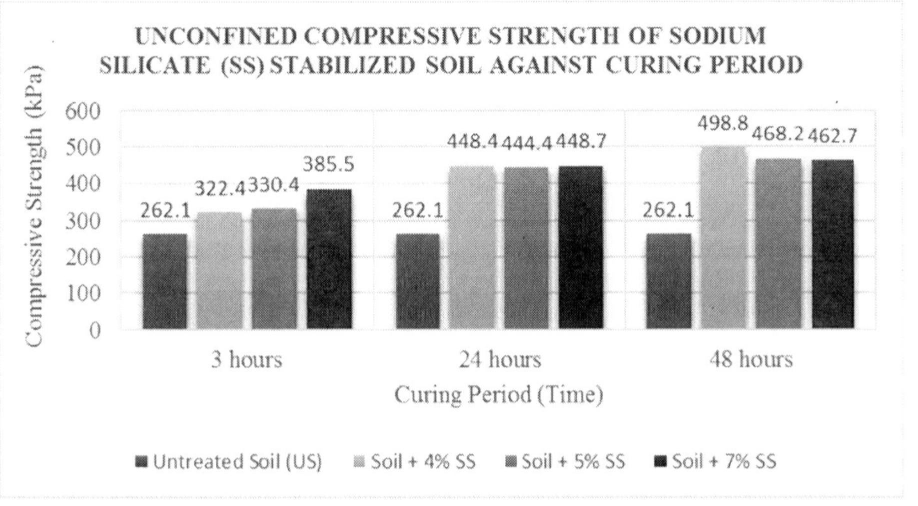

Figure 1. Unconfined Compressive Strength values of soil specimens stabilized with varying percentages of Sodium Silicate (SS) against curing period

In general, the soil gained more strength with an increasing curing period. According to Latifi [7] and Eisazadeh [13], this is due to an increase in the positive surcharge, causing the repulsion of soil particles inside the soil stabilized with Sodium Silicate - this soil particle repulsion phenomenon improves the soil-stabilizer reaction, but this process is time-dependent. On the other hand, the strength gain during curing was enhanced through the presence of cementitious material formed via soil-stabilizer, similar to that reported by Pakir [9].

For this study, it was found that the major improvement of soil strength was observed in the first day of curing, and only minor increments of strength improvement were seen after that. Due to the limitations of this study, where the maximum curing duration was set at 48 hours, a direct comparison to the other researchers' findings of soil improvement for curing duration of more than 48 hours cannot be directly compared [16, 17]. This is because in comparison with the other types of treated with Sodium Silicate i.e. laterite [6], marine clay [9] and soft clay [10], the silt soil used in this current study has lower plasticity index compared to the other soils. Since TX-85 is an ionic type of soil stabilizer (which causes cation exchange inside the soil and influences the surfaces of soil charge area) this increases the plasticity index of the soil. Hence, it is suitable to be used with the silt soil in this study since it is not an expansive soil [5] and has been shown to improve the soil strength in a shorter curing period.

Meanwhile, Figure 2 shows the UCS values of stabilized soil specimens plotted against varying percentages of Sodium Silica. The addition of 4% TX-85 in the soil sample produced a marked improvement in the soil strength characteristics, especially after it was cured for 2 days (in comparison with the addition of 5% and 7% TX-85 in the soil sample). As seen in Figure 2, the compressive soil strength value after 48 hours curing period reached its peak value at 4% TX-85, and the UCS value drops after an addition of 5% of TX-85. This finding is in line with the observation made by Latifi [5], which saw a similar phenomenon in laterite soils treated with Sodium Silica – whereby an addition of more than 6% of Sodium Silicate had reduced the strength of the laterite soil due to an increase of the soil moisture content and caused the soil to weaken. Furthermore, as seen in Section 2.2, the pH value for silt soil treated with 5% and 7% of Sodium Silicate is pH 9.5 and 9.8, respectively – which is close to the highest pH value obtained for a Sodium Silicate stabilized soil (pH 10.1). Sukmak [18] suggests that the increased amounts of the highly alkaline Sodium Silicate (pH value of 12.54) may have exceeded the requirement for chemical reaction in the soil samples, hence causing a reduction in the soil strength value. Therefore, it appropriate to conclude that the optimum dosage of TX-85 to improve the strength of the selected soil sample in this study is 4%.

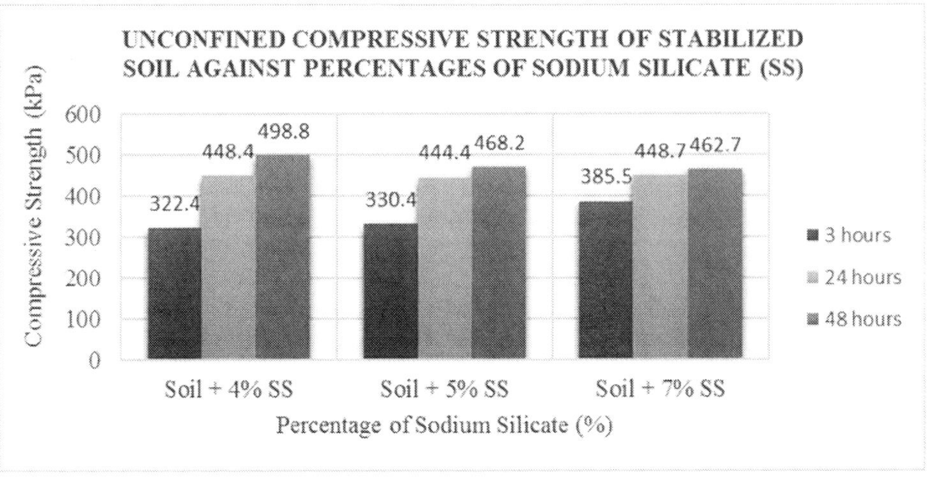

Figure 2. Unconfined Compressive Strength values of soil specimens stabilized with curing period against varying percentages of Sodium Silica (SS)

4. Conclusion

Based on the findings, the soil was classified as CL+ML (Low Plasticity Clay and Silt) according to the USCS standard. The tabulated data of each test shows that the addition of soil stabilizer TX-85 in silt soil for this study has enhanced its unconfined compressive strength characteristics. In this study, the optimum curing period for the modified soil to obtain its optimal strength was found to be 48 hours. However, if the curing time is extended further beyond 48 hours, as reported by [9] a further increase in strength (up to three to four times its original strength) is possible with seven (7) days curing. The general trend of unconfined compressive strength improvement is directly proportional to the curing period, whereby longer curing periods produces soil specimens with high compressive strength values. As such, the recommended curing period for the soil used in this study is 48 hours. However, the optimum dosage of TX-85 chemical stabilizer to improve strength and compression characteristics of the silt soil found to be 4% (90% improvement compared to original soil). At 48 hours curing period, the silt soil compressive strength was found to have reduced with the increment of TX-85 after 4%. Consequently, Sodium Silicate was found to be a viable product to be used to enhance the physical and engineering characteristics of coastal soil, especially for land development projects. As a part of an on-going research project, further experimental works are in progress to investigate the effectiveness of Sodium Silicate as a soil stabilizer for other types of soils e.g. marine clay and peat soil.

5. References

[1] Sherwood P T 1993 *Soil stabilization with cement and lime* (London: Transport Research Laboratory, HMSO) pp 38-87
[2] Kulkarni R P 1975 Soil stabilization by early Indian methods *Indian J. History Sci.*, **10**(1) 9–15
[3] Liu C and Evett J B 2008 *Soils and Foundations (7th Edition)* (New Jersey: Prentice Hall) pp 85-90
[4] Eisazadeh A, Kassim, K A and Nur H 2011 Characterization of phosphoric acid-and lime-stabilized tropical lateritic clay, *Environ. Earth Sci.* **63**(5) 1057-1066
[5] Latifi N, Marto A and Eisazadeh A 2013a Structural characteristics of laterite soil treated by SH-85 and TX-85 (non-traditional) stabilizers, *Electron J. Geotech Eng.* **18** H 1707–1718
[6] Lim S M, Wijeyesekera D C, Lim A J M S and Bakar I B H 2014 Critical review of innovative soil road stabilization techniques *Int. J. Eng. Adv Technol.* **5** 2249–8958

[7] Latifi N, Marto A and Eisazadeh A 2013b Strength behavior and microstructural characteristics of tropical laterite soil treated with sodium silicate- based liquid stabilizer *Environ. Earth Sci.* **72**(1) 91–98
[8] Marto A, Latifi N and Eisazadeh A 2014 Effect of non-traditional additives on engineering and microstructural characteristics of laterite soil *Arab J. Sci. Eng.* **39**(10) 6949-6958
[9] Pakir F, Marto A, Mohd Yunus N Z, Ahmad Tajudin S A and Tan, C S 2015 Effect of sodium silicate as liquid based stablizer on shear strength of marine clay *J. Teknologi* **76**(2) 45–50
[10] Mat Nor A H, Pakir F, Arifin A and Sanik M E 2015 Stabilization of Batu Pahat soft clay by combination between TX-85 and SH-85 stabilizers *Akademia Baru: J Adv. Res. Mater. Sci.* **14**(1) 1–7
[11] ASTM D2487-17 2017 *Standard Practice for Classification of Soils for Engineering Purposes (Unified Soil Classification System)* (Pennsylvania: ASTM International) pp 2-4
[12] BS 1377 1990 *British Standard methods of test for soils for civil engineering purposes: Part 3, Chemical and electro-chemical tests* (London: British Standards Institution) pp 20-21
[13] Eisazadeh, A 2010 *Physicochemical behavior of lime and phosphoric acid stabilized clayey soil* PhD Thesis (Johor Bahru: Universiti Teknologi Malaysia) pp 1-282
[14] BS 1377 1990 *British Standard methods of test for soils for civil engineering purposes: Part 4, Compaction-related tests* (London: British Standards Institution) pp 5-8
[15] BS 1377 1990 *British Standard methods of test for soils for civil engineering purposes: Part 7, Shear strength tests (total stress)* (London: British Standards Institution) pp 20-23
[16] Latifi N, Marto A and Eisazadeh A 2014 Analysis of strength development in non-traditional liquid additive-stabilized laterite soil from macro- and micro-structural considerations, *Environ. Earth Sci.* **73**(3) 1133–1141
[17] Latifi N, Meehan C L, Majid M Z A and Horpibulsuk, S 2016 Strengthening montmorillonitic and kaolinitic clays using a calcium-based non-traditional additive: A micro- level study, *Appl. Clay Sci.* **132** 182– 193
[18] Sukmak P, Horpibulsuk S and Shen S L 2013 Strength development in clay–fly ash geopolymer *Constr. Build. Mater.* **40** 566-574

IOP Conf. Series: Materials Science and Engineering **601** (2019) 012004 doi:10.1088/1757-899X/601/1/012004

Effectiveness of Non-monitored Green Roofs for Stormwater Management

Hartini Kasmin[1*], Siti Khalisha Khalidan[1], Wan Afnizan Wan Mohamed[1], Nur Syafira Razak[1] and Mohamad Norfekry Md. Yacob[1]

[1]Micropollutant Research Centre, Department of Water and Environment Engineering, Universiti Tun Hussein Onn Malaysia, 86400 Parit Raja, Batu Pahat, Johor, MALAYSIA *Corresponding author:hartini@uthm.edu.my

Abstract. A green roof is an alternative roofing system of any building or house that provides a better environment for urban areas. It has the potential to mitigate urban water problems. This study aims to investigate the rainfall runoff performance of two non-monitored green roofs with two different slopes. The experiment uses two types of non-monitored green roofs, namely flat roofs and sloped roofs that were exposed to climatic conditions in Parit Raja. The physical characteristics of the roofs were observed and rainfall runoff data was collected. The results showed that when 4 L of water was manually poured onto the roofs, an average of 60.4% of water was retained on sloped roofs whereas 48.4% was retained on flat roofs. Based on the 5-minute interval data, the retention ranges between 55.6% and 71.2% for both roofs. In terms of daily rainfall for a week, the average percentage was 69.2% for sloped roofs and 70.6% for flat roofs. All three approaches show that the green roofs produced less water runoff compared to normal roofs with both slopes retained water efficiently. Therefore, rainfall runoff performance for two types of green roof slopes was investigated and the available storage for the green roofs was recorded. In conclusion, with little maintenance, non-monitored green roofs can be efficiently used as a water mitigation system for urban areas

1. Introduction

The hydrological cycle is the process of water changes from the oceans, rivers, lakes and any water bodies, in the form of precipitation, transpiration, percolation, infiltration, run-offs and its return to the source areas [1]. The most readily available source of water with the highest volume is groundwater, followed by lakes, reservoirs, rivers and wetlands [1-2].

Urban areas are mostly covered with concrete and solid surfaces that are impermeable and impervious surfaces have been increasing. Urban development has also reduced the natural spaces that normally retain water before it turns into runoff and significantly changes the natural hydrological cycle. Rainwater will turn into runoff faster in urban areas compared to natural areas. Decreasing permeable areas had reduced infiltration volume and increased runoff production [3]. Increasing runoff means increasing the probability of peak flow and volume of flood in urban areas, where higher volume of rainfall becomes runoff compared to the volume of rainfall infiltrating the soil layer.

The urban drainage system has the potential to be improved by sustainable development known as Sustainable Urban Drainage System (SuDS) where the aim of this system is to design, manage and provide storage for water quantity (flooding), water quality (pollution), biodiversity (wildlife and plants) and the amenity of urban areas [4]. One of SuDS structure is green roof. Green roofs have been an increasingly popular structure used globally for urban stormwater management in order to promote SuDS [5-8]. Green roofs have the potential to reduce peak flows and runoff volume which help to reduce flooding events [5-8]. Vegetation planting on wasted roof areas has the potential to manage stormwater at source by reducing a small amount of the volume per roof area and delaying the velocity flow of stormwater runoff before reaching downstream.

Green roofs act as a storage structure for retaining water in the green roof layer. A green roof is also known as a vegetative roof that consists of plants and soil layers on the roof surface which is normally constructed for its aesthetic values. Green roofs have increasingly been chosen over conventional roofs due to its benefits towards the environment [9]. However, [10] it has been reported that the dynamic state of living roof systems that experience growth and changes due to weather, plant species competition and insect disease may reach its maturity due to the aging process and maintenance regime. The preliminary work by [11] also found that older extensive green roofs are correlated with substrate properties (pH, organic matters, depth and nutrients) and vegetation (cover abundance and species diversity) where lower soil pH, less biomass, less substrate depth and greater soil organic content were observed compared to the younger roofs [11]. Fortunately, [12] also reported that the hydrological performance of their green roofs has not declined after 5 years of operation. In fact, they found that the aged substrates have properties that enhanced hydrological performance.

Although this paper has not extensively explored the properties of aged substrates and their compositions, the preliminary performance of rainfall runoff and the physical characteristics of unmonitored green roofs were examined.

2. Materials and Methods

2.1 Data collection

Four green roof test beds with a dimension of 60 cm (L) × 60 cm (W) × 15 cm (H) were constructed in 2015 in Universiti Tun Hussein Onn, Malaysia, Parit Raja, Batu Pahat Johor. The substrate media consists of perlite, vermiculite and peat soil at a weight ratio of 3:5:2 and a depth of 80 mm. Based on its attractive flowers, easy and rapid grow, drought tolerance and its ability to survive under bright sunlight, the low-cost, succulent flowering plant *Portulaca Grandiflora* was used for the vegetation layer.

The test beds were used to simulate the typical roof housing/building with the top roof area for green plantation and the gutter for runoff water collection, as well as its downspout system for rainfall runoff measurements. The four green roof test beds were constructed based on two types of slope gradients; two sloped roofs at a 30° gradient (Figure 1a) and two flat roofs at a 10° gradient (Figure 1b). Different roof slopes indicate the typical types of roof slopes in Malaysian housing developments (the steep roofs), where flat roofing resembles the top roof of high buildings. In this experiment, one of the roofs functioned as a control roof for each gradient (empty with no substrate and vegetation layer) to simulate rainfall data collection. Meanwhile, the other roof contained the substrate and vegetation layer for runoff data collection. Comparisons were made between the control roof (rainfall) and the green roof (runoff) for each slope gradient.

Figure 1. Green roof test beds; (a) sloped roof with a 30° gradient, and (b) flat roof with a 10° gradient

2.2 Data Collection

Since the construction of these test beds in 2015, no maintenance has been conducted. After a number of initial studies, these test beds were left for a year in 2017 without any monitoring. Therefore, before any rainfall runoff data collection was conducted, these test beds were examined and verified based on two main viewpoints: i) Maintenance of roofing structure – in terms of the growth of the vegetation layer and the leaking structure of the roof layer; ii) Verification of the data collection after maintenance. The monitoring of data collection was based on the following approaches:

i. Manual data collection using 4 litres of water was used as a preliminary experiment for early verification. Four litres of water was manually poured onto the roofs (using a 5 L mineral bottle with a few holes underneath to create rainfall) in 5 minutes with a cumulative 30-second interval (using a measuring cylinder). The experiment was carried out four times with intervals of one hour, two hours and four hours between each experiment that also represented dry days.
ii. Manual monitoring with a 5-minute interval during local rainfall events. Two rainfall events were recorded on 22/4/2018 and 1/5/2018. Rainfall and runoff data was collected from the start of the rainfall until the runoff from all the roofs stopped for at least 15 minutes. The cumulative rainfall runoff volume was collected using a measuring cylinder under the roofs.
iii. Manual monitoring with a daily interval of local rainfall events between 2/5/2018 and 8/5/2018. The observations were conducted daily at 8.00 am and 5.00 pm for all roofs. The rainfall runoff data was collected cumulatively using a measuring cylinder.

3. Results and Discussion

As mentioned above, for the purpose of examining and verifying the test beds for further monitoring activities, maintenance activities were done for four experimental roofs which are i) onsite inspection on the roof's physical characteristics; and ii) examination and verification of the rainfall runoff data performance.

3.1 Physical characteristics of the test beds

i. Vegetation layer: Figure 2a – 2b shows the growth and health progress of the vegetation layer. Figure 3a shows healthy *Portulaca G.* when they were first planted on the roofs. All the vegetation was in good condition during the first year in 2015-2016, with good nutrition from fertilisers and watering. Since December 2016, no monitoring was conducted on these roofs, and the condition of the vegetation after one year without any monitoring and fertilisation is shown in Figure 2b(i). The substrate media was almost full of weeds, and a few of the *Portulaca G.* were still trying to survive together with the weeds. It also seemed that they had grown longer. Later, the weeds were removed and the vegetation layers were watered, fertilised and replanted (the same long *Portulaca G.* were used and replanted on the cleaned soil areas). It seemed that the vegetation can grow healthily and flower again as seen in Figure 2b(ii).

Figure 2a. Vegetation layer condition during test bed development – after one week of plantation (September 2015)

(i) (ii)

Figure 2b. (i) Vegetation layer condition after 1 year (November 2016) (ii) Vegetation layer condition after some maintenance (March 2017)

ii. Roof structure: The main purpose for the roof structure examination is to check on the leaking points. The roofs were poured with water from a pipe until it was saturated and the wet areas (that should not be wetted) due to leaks were observed. All the leaking points spotted were then repaired by filling the leaking gaps with silicone fillers.

3.2 Verification results – Manual collection using 4 litres of water

After a few weeks of maintenance, the first examination and verification of the data were done manually. The first preliminary experiment was conducted on 20/3/2018 (Figure 3). The average percentage of retention for the sloped green roofs had a water retention of 60.5% while the flat green roof had a water retention of 48.1%. Table 1 shows the summary of the experiment results on 20/3/2018 and 25/3/2018. The data shows the time to start runoff, time to start peak flow, peak flow reduction (between rainfall and runoff) and rainfall retention. The highest rainfall retention of flat roofs occurred during the experiment on 20/3/2018, was the first experiment (after few days of dry days) at 9.30 am with water retention percentages of 77.3% and 73.3% for flat roofs and sloped roofs, respectively.

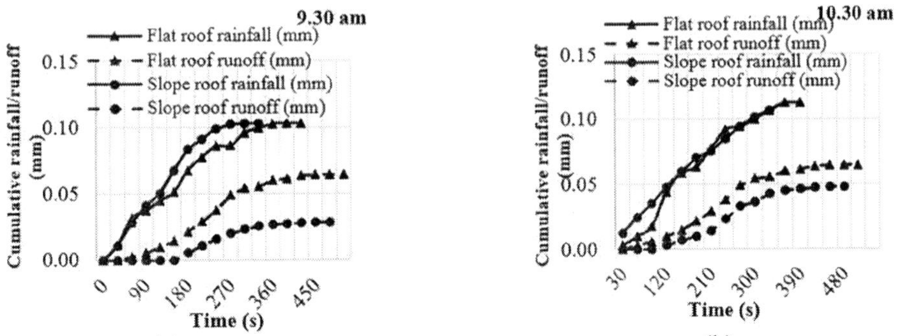

(a) (b)

Figure 3. Examples of cumulative rainfall runoff with different dry days (during short intervals) on 20/3/18 at (a) 9.30 am; and (b) 10.30 am

The second preliminary experiment on 25/3/2018 (Table 1) resulted in average water retention percentages of 60.4% for sloped green roofs and 48.7% for flat green roofs. After 5 days of dry days, the highest rainfall retention of flat roofs during the experiment occurred at 9.00 am with water retention percentages 74.8% and 72.9% for flat roofs and sloped roofs, respectively. Both experiments show that slope roofs retained more most of rainfalls than the flat roofs, but peak flow reduction were varies between roofs. It seems that both roofs can retained not more than 80% of the rainfall (of 4 L volume), and the variety values found in peak flow reduction, may be due to the technical issues on the experiment itself. Slope roofs also delays runoff more than 120 s, while flat roof delays runoff at minimum of 60 s. The roofs also delay the time for runoff to start peak flow compared to peak flow of rainfall more than 30 s.

Table 1. Summary of runoff results from the preliminary examination test on 20/3/2018 and 25/3/2018

Date & Time	Time to start runoff (s)		Time to start peak flow (s)		Peak flow reduction (%)		Rainfall retention (%)	
	Flat	Slope	Flat	Slope	Flat	Slope	Flat	Slope
20/3/18 9.30 am	210	180	210	30	65.7	66.7	73.9	72.2
20/3/18 10.30 am	60	120	150	150	57.8	22.7	42.4	54.9
20/3/18 12.30 pm	60	150	120	120	41.4	51.9	39.5	54.5
20/3/18 4.30 pm	120	150	120	120	50.0	26.9	36.5	60.4
25/3/18 9.00 am	210	180	90	90	66.1	65.6	74.8	72.9
25/3/18 10.00 am	60	120	60	90	33.3	49.2	41.0	54.0
25/3/18 12.00 pm	60	180	30	120	35.9	47.1	40.1	57.3
25/3/18 4.00 pm	120	180	60	210	38.3	50.9	38.8	57.4

3.3 Verification results-Manual monitoring at 5-minute intervals and daily intervals.

For the monitoring process with a 5-minute interval, only two rainfall events were collected on 22/4/2018 and 1/5/2018 (Figure 4). All the roofs were monitored simultaneously whenever rainfalls occurred, with empty roofs representing rainfall data and vegetated roofs representing the runoff data. Table 2 presents the summary of rainfall runoff performance for these rainfall events. It seems that for local rainfalls, it takes at least 5 minutes for both roofs to start runoff and the green roofs delays the peak flow at minimum of 5 minutes. Peak flow and rainfall retention values for the flat roofs seems less than the slope roofs for both events with differences not more than 10%.

Table 2. Summary of results for rainfall event on 22/4/2018 and 1/5/2018

Date & Time	Time to start runoff (min)		Time to start peak flow (min)		Peak flow reduction (%)		Rainfall retention (%)	
	Flat	Slope	Flat	Slope	Flat	Slope	Flat	Slope
22/4/18	5	10	10	5	37.7	45.6	55.6	59.6
1/5/18	10	5	20	15	76.4	78.3	65.0	71.2

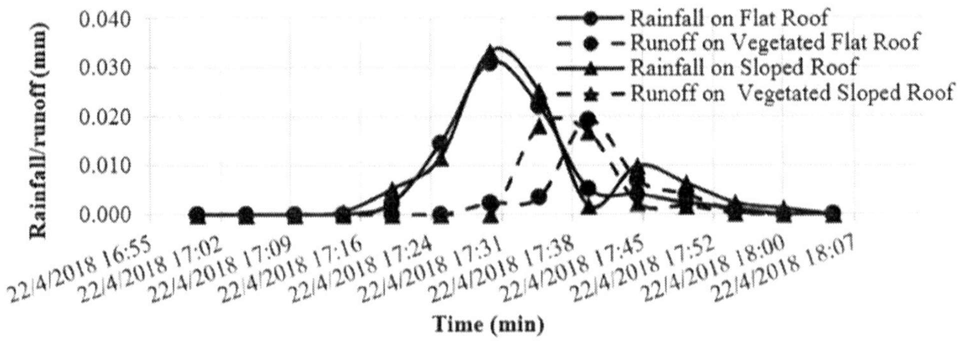

(a)

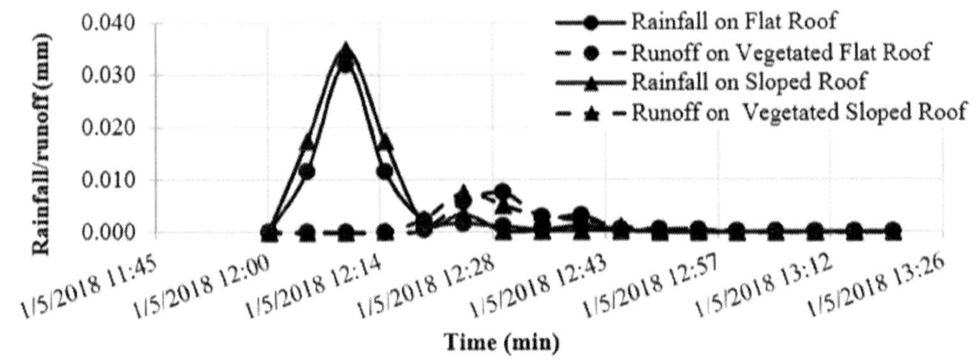

(b)

Figure 4. Hydrograph of rainfall runoff data on (a) 22/4/18; (b) 1/5/18

For daily rainfall events, the data was collected between 2/5/2018 and 8/5/2018. The observations were conducted daily at 8.00 am and 5.00 pm for all roofs and the values of these runoffs were totalled as the rainfall runoff of the day. During seven days of observation, there was rainfall on 4 days and 3 days without rainfall. Out of 4 rainy days, only one day had more than 1 rainfall. Three consecutive rainfalls occurred on 6/5/2018 at 2.20 - 3.00 pm, 6.50 - 7.20 pm and 9.00 - 9.30 pm.

Figure 5 shows the daily performance of the roofs for a week. The total rainfall on 2/5/2018 was 0.0544 mm and the water retention percentages for sloped roofs and flat roofs were 83.3% and 84.4%, respectively. On 3/5/2018 and 4/5/2018, lesser rainfall was recorded, namely 0.0234 mm and 0.0321 mm, respectively. The water retention percentages were still between the range of 80% – 83% for both roofs, namely 80.0% and 81.8% for sloped roofs and 82.5% and 82.7% for flat roofs on 3/5/2018 and 4/5/2018, respectively.

Due to the occurrence of three consecutive rainfalls, the highest total rainfall event on 6/5/2018 was observed at 0.1087 mm. A reduction in water retention was observed at 31.7% for the sloped roof and 32.8% for the flat roof. It seems that the water retention for both green roof slopes was quite similar but there was a slightly high retention observed in the flat roofs. The difference was only 1% for every rainfall event. Both roofs can also efficiently retain water during moderate and light rainfall as both roofs retained more than 50%. For higher total rainfall, less water retention was observed as only 31% - 33% of water was retained by both roofs. However, more data is needed in order to measure the efficiency of the roofs in retaining water.

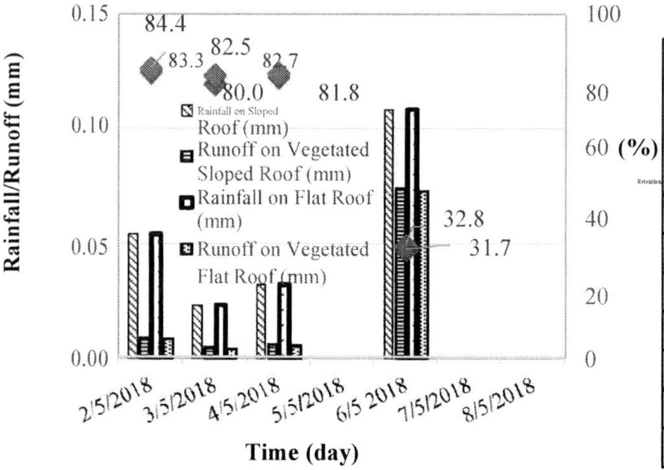

Date	% of retention on sloped roof	% of retention on flat roof
2/5/18	83.3	84.4
3/5/18	80.0	82.5
4/5/18	81.8	82.7
5/5/18	0	0
6/5/18	31.7	32.8
7/5/18	0	0
8/5/18	0	0

Figure 5: Daily rainfall runoff performance for a week

All roofs were tested simultaneously so that they experience the same rainfall runoff condition. Vegetated roofs produce less runoff compared to conventional roofs. This proves that the green roofs are able to provide storage capacity and have a good ability to retain rainfall in the soil layers even though the dry time between two rainfall events was less than 6 hours. The green roof also acts as sustainable drainage that releases the remaining rainfall water gradually to downstream in term of time and volume. Thus, it will give time for rainfall to disperse from the upstream and prevent flash flooding at downstream. Overall, both slopes exhibit the ability and the efficiency of green roofs in retaining rainfall.

4. Summary

Through the experiments, the efficiency of rainfall runoff performance of green roofs was measured by indicating the percentage of water retained. Based on the results, an average water retention of 60.4% was obtained for sloped roofs whereas an average water retention of 48.4% was obtained for flat roofs when 4 L of water was used. During the rainfall with 5-minute intervals, there was a similar ranges of retention between 55.6% and 71.2% for both roofs. An average retention of 69.2% was obtained for sloped roofs and an average retention of 70.6% was obtained for flat roofs from daily rainfall for a week. All three approaches show that the green roofs produced less water runoff compared to normal roofs with both slopes retained water efficiently. However, further observation on the substrate and vegetation properties may be needed. The results also prove the potential of green roofs to be used in urban stormwater management for long term operation, especially for flash flood problems.

5. References

[1] Nagarajan R 2006 *Water: Conservation, Use and Management for Semi-arid Region.* (New Delhi: Capital Publishing Company) 352

[2] Rattan L and Stewart B A 2012 *Soil water and agronomic productivity* (CRC Press: Taylor & Francis Group) 594

[3] Environment Protection Authority 2006 *Urban Stormwater: Best Practice Environmental Management Guidelines* (CSIRO Publishing) 268

[4] Susdrain 2018 *Sustainable Drainage* Retrieved on November 10, 2018 from https://www.susdrain.org/delivering-suds/using-suds/background/sustainable-drainage.html

[5] Berndtsson J C 2010 Green roof performance towards management of runoff water quantity and quality: A review, *Ecological Engineering* **36** 351–360

[6] Stovin V, Vesuviano G and Kasmin H 2012 The hydrological performance of a green roof test bed under UK climatic conditions *Journal of Hydrology.* **414–415** 148–161
[7] Qin X, Wu X , Chiew Y M and Li Y 2013 A green roof test bed for stormwater management and reduction of urban heat island effect in Singapore. *British Journal of Environment and Climate Change* **2** 410–420
[8] Carson T B, Marasco D E, Culligan P J and McGillis W R 2013 Hydrological performance of extensive green roofs in New York City: observations and multi-year modeling of three full-scale systems *Environmental Research Letter* **8** 1-13
[9] Md Zahir M H, Raman S N, Mohamed M F, Jamil M and Nopiah Z M 2014 The perception of Malaysian architects towards the implementation of green roofs: A review of practices, methodological and future research *EDP Sciences, E3s Web of Conferences* **3** 1-8
[10] KieranTimberlake 2018 *Green roof vegetation study: How can we predict the future species composition of a green roof?* Retrieved on November 12, 2018 from https://kierantimberlake.com/pages/view/31/green-roof-vegetation-study/parent:4
[11] Thuring C E and Dunnet N 2014 Vegetation composition of old extensive green roofs (from 1980s Germany). *Ecological Processes: A SpringerOpen Journal* **3** 1-11
[12] De-Ville S Menon M Jia X Reed G Stovin V 2017 The impact of green roof ageing on substrate characteristics and hydrological performance *Journal of Hydrology* **547** 332-344

Acknowledgment

The authors would like to acknowledge the financial support provided by Universiti Tun Hussein Onn under the Postgraduate Incentive Research Grant (GIPS - 1351) and the Exploratory Research Grant Scheme (ERGS – E052) which made this study possible. The authors would also like to thank the Research Centre for Soft Soil (RECESS), UTHM for giving the opportunity to use the laboratory sites for the green roof test beds.

Water Quality Classification Using an Artificial Neural Network (ANN)

Khadijah Sulaiman[1*], Lokman Hakim Ismail[1], Mohd Adib Mohammad Razi[1] Mohd Shalahuddin Adnan[1] and Rozaida Ghazali[2]

[1] Faculty of Civil and Environmental Engineering, University Tun Hussein Onn Malaysia
[2] Faculty of Computer Sciences and Information Technology, University Tun Hussein Onn Malaysia
*Corresponding author: khadijahhumaira94@gmail.com

Abstract. Malaysia is currently a rapidly developing country to achieve a 2020 vision. However the development that has been carried out contributed to a negative impact on the environment especially on water quality. Due to the deterioration of water quality, serious management efforts on water quality has been taken. Thus, the aim of this study is to investigate a technique that can automatically classify the water quality. The technique is based on the concept of Artificial Neural Network (ANN). Since the greater part of their methodologies depend on the idea of `pattern recognition'. Thus, it is convenient to inspect its ability in classify water quality. There are six environmental data were used in this study such as pH, total suspended solids (TSS), dissolved oxygen (DO), chemical oxygen demand (COD), biological oxygen demand (BOD), and ammonia. The data was obtained by in-site measurement and laboratory analysis. Then, the data was used as the feeder of input variables in the ANN database system. After training and testing the network of ANN, the result showed that 80.0% of accuracy classification with 0.468 of root mean square error (RMSE). This showed the encouraging results for classification.

1. Introduction

Poor administration of water quality can be disastrous to human civilisation as appeared by occurrences of the disease all over the world [1]. Because of the effects on human wellbeing, most nations have set-up the frameworks for water quality administration to guarantee the quality of water. The extension of water quality administration began in 1985, with the development of water quality criteria and benchmarks based on the water quality index [2].

The evaluation of water quality based on the index system has been used for over the past decades. However, there is still a limitation on the index system because there are many used data cannot correlate with an index [2]. In addition, the model requires a series of calculation before a final index value and class grade value can be obtained.

In response to these limitations, this research aims to investigate a technique that can efficiently and easily to classify the water quality. One of the approaches that conceivable in the evaluation of water quality is the Artificial Neural Networks (ANN), where there is a growing tendency of soft computing techniques and primarily data-driven approaches. Its function is to complement or even replace the knowledge-based model, which may deliver assessment systems that are operationally straightforward, adaptable and comprehensive, where the outcomes are effortlessly justifiable [3]. Hence, many types of researches have been studied by including the modelling of water quality [4-7].

2. Materials and Methods

2.1. Study Area and Water Quality Data

The Straits Malacca is subjected to a great variety of pollutants due to its strategic location as a major international shipping lane and the concentration of agriculture, industry and urbanisation which predominate on the west coast of Peninsular Malaysia. According to [8], Strait Malacca have a serious environmental problems because of many river discharge and due to agricultural and industry along the Strait Malacca.

The data set used in this study was generated through a data collection at site and laboratory analysis. Three locations of this study were chosen which are Pontian River, Batu Pahat River and Muar as shown in figure 1.

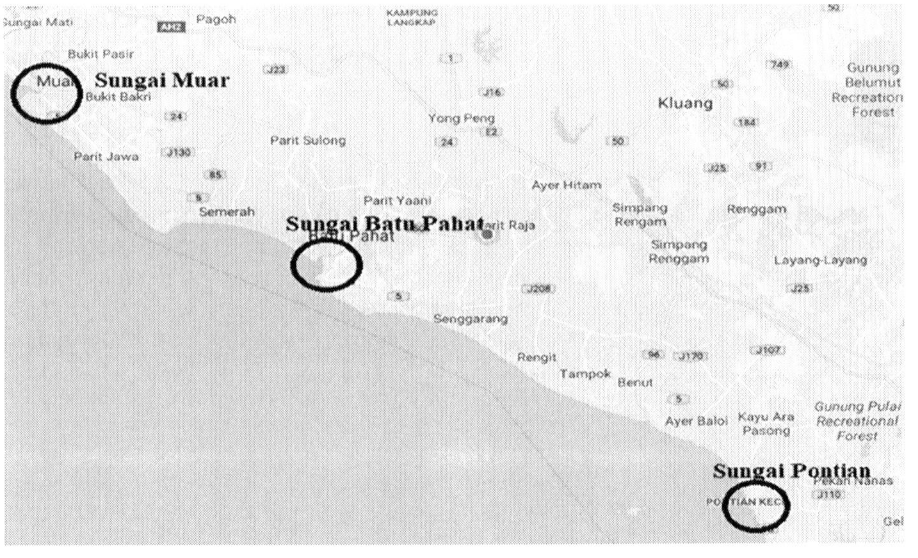

Figure 1. The location of sampling site.

The water quality parameters were collected from river and estuaries. The sampling site spread along the river including the river mouth and downstream of the river. Grab water samples were collected for 30 checkpoints for Pontian River and 35 checkpoints for each Batu Pahat River and Muar River. The two in-situ parameters, pH and DO were collected using HORIBA water quality monitor, meanwhile water samples were collected for lab analysis using HACH DR 6000. Parameters of lab analysis were BOD, COD, ammonia nitrogen and TSS.

2.2 Water Quality Standard

In Malaysia, the existing methodology for river water quality classification and monitoring is quite extensive. At the moment, Malaysia has over 1000 manual and automatic river water quality monitoring stations in 146 basins maintained by the Department of Environmental (DOE) alone [9]. These exclude other agencies such as Department of Irrigation and Drainage (DID) as well as the respective state level agencies.

Data on water quality is used to determine the status by examining whether it is clean, slightly polluted or polluted. The quality of river water has been monitored by the Department of Environment (DOE) since the late seventies. The primary aims were to establish and detect any changes in the water quality status and to identify the pollution sources. The classification has been continued on an annual basis by putting rivers in Class I, II, III, IV, or V based on the Interim National Water Standards for Malaysia (INWQS) and based on the guidelines for water use as shown in table 1.

Table 1. Guidelines for Water Use [10]

Class	Definitions
I	• Conservation of the natural environment. • Water supply I - Practically no treatment necessary (except by disinfection or boiling only). • Fishery I - Very sensitive aquatic species.
IIA	• Water supply II – Conventional treatment required. • Fishery II - Sensitive aquatic species.
IIB	• Recreational use with body contact.
III	• Water supply III – Extensive treatment required. • Fishery III - Common of economic value, and tolerant species; livestock drinking.
IV	Irrigation.
V	None in above

2.3 Artificial Neural Network

The terminology of ANN is developed from the biological model of the brain. A neural network consists of a set of connected cells, which is called the neurons. The neurons receive impulses from either input cells or other neurons and perform some kind of transformation of the input and transmit the outcome to the neurons or to output cells. The neural networks are built from layers of neurons connected so that one layer receives input from the preceding layer of neurons and passes the output on the subsequent layer. According to [11], the data fed through the input layer is scaled by an initial weighting through the connecting between neurons. The relationship between inputs (x), weights (w) and outputs (y) as shown in figure 2.

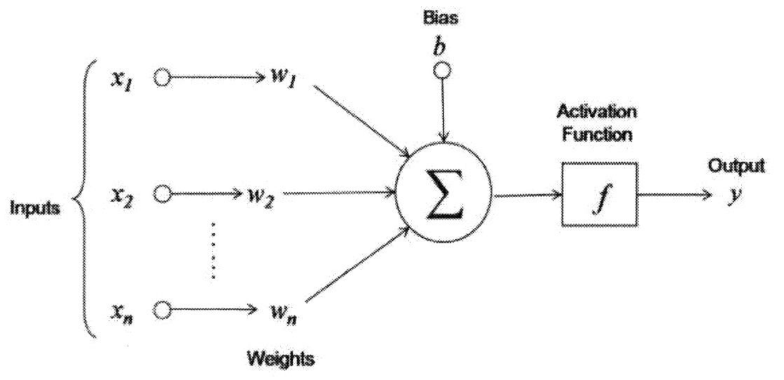

Figure 2. Computation Process of Neuron in ANN

3. Experiments and Results

3.1 Input Selection

The total of 100 data set from excel file was imported from excel file to develop the ANN model. Six parameters that had been collected were used in this input selection and five categories classes based on water quality standard were set as a target variables to produce an output.

3.2 Data Pre-processing

Normalization procedure before presenting the input data to the network is required since mixing variables with large magnitudes and small magnitudes will confuse the learning algorithm on the importance of each variable and may force it to finally reject the variable with the smaller magnitude. Normalization will scale the minimum value to 0 and maximum value to 1.

3.3 Training and testing network

In order to train and test the network, experimental data was categorized into two sets which were for training and testing sets. The training set was used to identify the biases and weight of the network. Meanwhile the testing set was used for calibration to prevent the overtraining of the network and could be used to measure the performance of the network. According to the [12], generally testing set should consist of 10-40% from the data set.

3.4 Artificial Neural Network model

The ratio of 80:20 was used in this experimental which means from 100 data sets, 80 data set was used for training and 20 data was used for testing. The algorithm of the neural network was used to train and test the data set. All the testing was recorded into the network to make it learn the potential relationship between water quality parameter and their corresponding categories. Six parameters represented the input layer nodes of the ANN model, while the five output layer nodes of the ANN model were represented the five different class categories. Percentages of classification accuracy and root mean square error (RMSE) was used to represent the performance of ANN as shown in Equation (1). The lower the RMSE, the more accurate is the estimation. In order to speed up the training process and maintain the error reduction, the hidden layer, momentum and learning rate were adjusted. From the trial and error of experimental, there were 1 hidden layer, 0.2 momentums, 0.3 learning rate of the network produced 0.468 of RMSE after 1000 iterations (cycle).

$$RMSE = \sqrt{\frac{1}{N}(X_{observed} - X_{predicted})^2} \tag{1}$$

Where, N is total number of observation in the data set.

3.5 Results

Based on table 2, it illustrated the confusion matric of the actual class (true class) and output class (predict class). From the illustration of confusion matric its shows 20 data of testing set. Vertically reading, there are 7 data in class II but only 6 data were correctly classified with 85.71% accuracy and 12 data in class III but only 9 data were classified correctly with 75.00% accuracy. Meanwhile 1 data in class IV were classified correctly with 100% accuracy. There no record matches the class I and V. The overall data correctly classified 16 out 20 data set with 80.0% of classification accuracy as shown in table 3.

Table 2. Confusion matric for classification

	True I	True II	True III	True IV	True V
Predict I	0	0	0	0	0
Predict II	0	6	3	0	0
Predict III	0	1	9	0	0
Predict IV	0	0	0	1	0
Predict V	0	0	0	0	0
Class accuracy	0 %	85.71%	75.00 %	100 %	0 %

Table 3. Overall results of the testing set

Number of record	Number of correctly classified	Accuracy percentage
20	16	80.0%

4. Conclusion

In this study, it was identified that water quality classification using ANN produced encouraging results of classification. As a conclusion, it can be concluded that, ANN is a model that can easily classify the water quality with the justifiable output. This provided performance results can be used as a reference to managing a water quality at study area.

5. References

[1] Fredrick T, Ponnaiah M, Murhekar M V, Jayaraman Y, David J K, Vadivoo S and Joshua V 2015 Cholera outbreak linked with lack of safe water supply following a tropical cyclone in Pondicherry, India *Journal of Health, Population and Nutrition* **33** 31-38

[2] Tyagi S, Sharma B, Singh P and Dobhal R 2013 Water Quality Assessment in Terms of Water Quality Index, *American Journal of Water Resources* **1** 34-38

[3] Ibrahim D, 2016 An Overview of Soft Computing, *12th International Conference on Application of Fuzzy Systems and Soft Computing (Vienna, Austria)* vol 102 (Vienna, Austria : Elsevier) pp 34-38

[4] Wechmongkhonkon S, Poomtong N and Areerachakul S 2012 Application of Artificial Neural Network to Classification Surface Water Quality *World Academy of Science, Engineering and Technology* **6** 199-203

[5] Gazzaz N M, Mohd Kamil Y, Ahmad Z A , Hafizan J and Mohammad F R 2012 Artificial neural network Modelling of the water quality index for Kinta River (Malaysia) using water quality variables as predictors *Marine Pollution Bulletin* **64** 2409-2420

[6] Rao V 2013 A Classical Way of Finding Water Pollution by Using Artificial Neural Network, *School of Computing Science and Engineering* **3** 449-457

[7] Gour S and Gour M 2014 Neural Network Approach in Water Quality Data Analysis for the River Narmada *Binary Journal of Data Mining & Networking* **4** 49-53

[8] Alzahrany A A, Elias S and Siong W B 2007 Determination of Heavy Metal Pollution In Depth Profile Of Marine Sediment Samples From The Strait Of Malacca *Fresenius Environmental Buletin* **16** 1279-1287

[9] Department of Environment Malaysia (DOE) 2006 *Malaysia Environmental Quality Report* In: Chapter 3: River Water Quality (Malaysia: Sasyaz Holdings Sdn Bhd,) pp 24

[10] Ismail Z, Sulaiman R and Karim R 2014 Evaluating Trends of Water Quality Index of Selected Kelang River Tributaries *Environmental Engineering and Management Journal* **13** 61-72

[11] Mansourkhaki A, Berangi M and Haghiri M 2018 Comparative Application of Radial Basis Function and Multilayer Perceptron Neural Networks to Predict Traffic Noise Pollution in Tehran Roads, *Journal of Ecological Engineering* **19** 113-121

[12] Ali N, Ahmed E, Othman A K and Othman J 2009 Prediction of Johor River Water Quality Parameter Using Artificial Neural Networks *European Journal of Scientific Research* **28** 422-435

Acknowledgement

The authors would like to thank the Ministry of Education Malaysia (MOE) and University Tun Hussein Onn (UTHM) for financial and technical support. This study has been supported by the Fundamental Research Grant Scheme (FRGS), VOT 1524.

The Advanced Analysis of Hydrodynamic and Sediment Transport Modelling in Agathis Lake

Rian Mantasa Salve Prastica[1*], Herr Soeryanton[1] and Dwinanti Rika Marthanty[1]

1 Civil Engineering Department, Faculty of Engineering, Universitas Indonesia *Corresponding author: rianmantasasp@gmail.com

Abstract. Resource Modelling Associates (RMA) was used to estimate the hydrodynamic process in Agathis Lake. The lake is situated in Universitas Indonesia, Depok, West Java, Indonesia. The model detects the velocity and total suspended solid distribution trends in the lake. RMA program generated the result of the trends. The program needs a calibration analysis to make the program output valid. Calibration analysis is conducted through sampling activities of field data. This research aims to do calibration analysis between the program output and the field data. The calibration analysis concludes that the present sampling locations are not sufficient to complete the calibration analysis. The research suggests adding sampling locations both from upstream to downstream and from surface water to bed surface of the lake.

1. Introduction

Suspended sediments have a primary role as pollutant source not only in urban areas but also coastal regions. The sediments get the role to determine the quality of coastal water [1, 2]. Moreover, sediments could role as flooding disaster root. The sediments attach the water physical body and fulfill the water body capacity.

Sophisticated technology today enforce scientist to develop tools to make users easier to understand environmental phenomena. Natural disasters like flood, earthquake, and sediment erosion become urban areas threat due to lack of environmental-friendly policy. The government, academia, and related stakeholders need to unite to achieve the policy. Hydrologic models are widely used by engineers, scientists, and hydrologists to understand and manage both natural and human activities that affect the watershed [3]. The hydrodynamic of hydraulic and total suspended solids are affected by several parameters. They are turbulent exchange coefficient both X and Y directions, turbulent diffusion both X and Y directions, turbulence factor, diffusion both X and Y directions and settling rate. All parameters are difficult to adjust to make the output program could represent the actual data. The successful output of the model depends on the calibration analysis [4].

The program model is based on the finite element method [5]. Most numerical models to simulate sediment transport use hydrodynamic models. The models could be classified according to their dimensionality, i.e. 1-Dimensional, 2-Dimensional, and 3-Dimensional [6,7]. Neary et al. [8] used 1-D models to study cohesive sediment while Zheng et al. [11] and Rao et al. [9] applied 2-D dimensional models. 3-D models are also widely used by several scientists to understand sediments phenomena like Gleizon et al. [10] and Zheng et al. [11]. The research uses the 3-Dimensional model to study the hydrodynamic process in the study area.

2. Research Method

Calibration analysis is conducted by analyzing field data and the output program. The velocity and total suspended solid distribution trends are observed 24-hour from the model. According to this limitation, sampling activities to obtain field data are conducted in 24 hours. The sampling locations are spread out from the upstream to the downstream part of the lake. The expected location and time are studied in the previous research by Prastica [6]. There are three main locations of sampling activities. The field data would be sampled in the inlet of the lake, water body, and the outlet. The sampling times range from 08.30 on March, 13th 2018 to 08.00 on March, 14th 2018. There are two main data that would be sampled: velocity and total suspended solids (TSS).

The sampling data would be analyzed in the laboratory to obtain the total suspended solids concentration. The RMA program output data would be calibrated by using the final field data from sampling activities.

3. Result and Discussion

3.1. Study area

The research studies Agathis Lake that is situated in Universitas Indonesia, West Java, Indonesia. The lake is located in the upstream part of the cascade-pond system in Universitas Indonesia. The lake is directly connected the urban settlements. Figure 1 depicts the lake and sampling locations for calibration analysis [6].

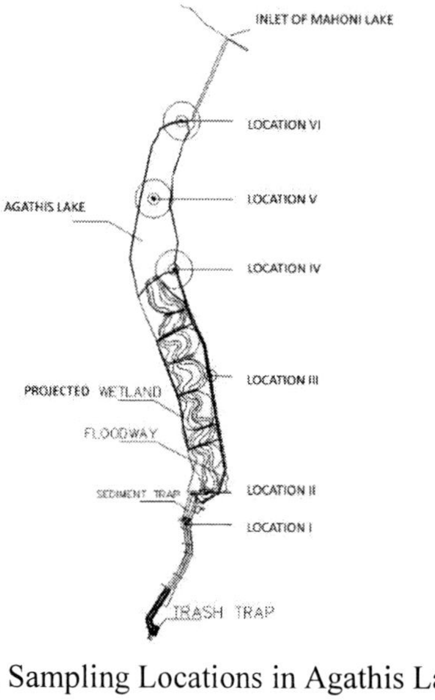

Figure 1. Sampling Locations in Agathis Lake [6].

Agathis lake has a catchment area approximately 871,188 m^2. The lake only 0.87% of the total catchment area meanwhile the percentage of impervious cover reaches 70.21%. Agathis Lake is directly connected to the urban settlements, so most of the pollutants that enter the lake's body come from the urban activities.

3.2. Resource Modelling Associates Program

There are four programs of Resource Modelling Associates (RMA): RMA-Gen, RMA-10, RMA-11, and RMA-PLT [6]. RMA-Gen transforms geometry to finite element method, RMA-10 simulates the 3-Dimensional model of velocity distribution, RMA-11 simulates the 3-Dimensional model of total suspended solid trends, while RMA-PLT visualises the result of the program running.

3.3 Sampling Activities

Sampling activities are studied from Prastica [6] and calibrated with the RMA output data in this research. Table 1 describes the location and the time of sampling activities from Prastica [6]. Table 2 tells the exact location and time of sampling activities. The research calls several codes to understand the model easier. Figure 2 depicts the codes number to recall the sampling locations.

Table 1. Field Sampling Location and Time [6]

Day	**Time**	**Location I Sediment Trap Inlet**	**Location II Sediment Trap Outlet**	**Location III Floodway Path**	**Location IV Inlet of Agathis Lake**	**Location V Waterbody of Agathis Lake**	**Location VI Outlet of Agathis Lake**
March, 13th 2018	08.30-09.00						
	10.00-11.00						
	12.00-12.30						
	12.30-14.30						
	15.30-16.00						
	16.30-17.00						
	19.00-20.00						
March, 14th 2018	07.00-08.00						

The location to take samples in the water depth is determined by dividing its width (W) and depth (D) into approximately half of the total each parameter. Figure 3 depicts the lake's water column location to take water samples.

3.4. Calibration Analysis

The calibration analysis is conducted by analyzing the sum of squared error (SSE) percentage among the data. This analysis aims to make the model could represent the actual data. There are three kinds of approach to doing calibration: expert intuition, field sampling with actual system, and theoretical analysis. This research uses field sampling study [6]. RMA-10 model calibration uses velocity prediction data and actual velocity data and RMA-11 uses TSS prediction data and their actual data from Agathis Lake system. There are 9 nodes of RMA program that represent the field sampling location. Inlet of sediment trap, outlet of sediment trap, inlet of floodway, floodway channel, inlet of Agathis lake, water body of lake I, water body of lake II, water body of lake III, and outlet of Agathis lake are represented by nodes: 4422, 1344, 4188, 3823, 5, 504, 2015, 912, and 580, respectively. Table 2 figures the analysis of SSE percentage for calibration of velocity and TSS data.

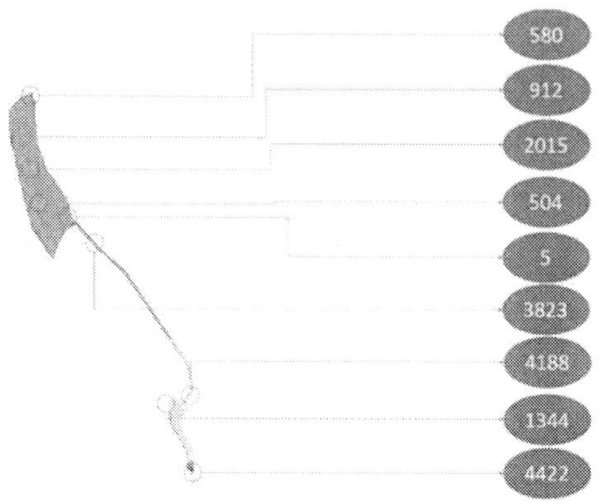

Figure 2. Nodes Location for Model

This analysis needs further analysis to obtain the least percentage of SSE. In this research, the unknown variables that are used for RMA-10 are 0.1 m^2/s for turbulent coefficient of x-direction, 0.1 m^2/s for turbulent coefficient of y-direction, 0.1 m^2/s for diffusion coefficient of x-direction, 0.1 m^2/s for diffusion coefficient of y-direction, and 0.02 for turbulence factor, whereas the unknown variables that are used for RMA-11 are 0.002 m/s for settling velocity, 0.001 m^2/s for diffusion coefficient of x-direction, and 1000 m^2/s for diffusion coefficient of y-direction. It is predicted that the changes of variables could give the different percentage of SSE that are needed to be analyzed in the further analysis.

Table 2. RMA-10 and RMA-11 Calibration Analysis

Time	**Node**	**V Field Data (m/s)**	**V Model (m/s)**	**SSE (%)**	**C field data (mg/l)**	**C model (mg/l)**	**SSE (%)**
8:00:00 AM	4422	0.470	0.508	0.3171	30	29.3738	2.08
9:00:00 AM	4422	0.450	0.449	0.0001	26	25.5236	1.83
10:00:00 AM	1344	0.320	0.322	0.0012	12	11.4253	4.78
11:00:00 AM	1344	0.300	0.322	0.1615	8	8.6780	8.47
12:00:00 PM	4188	0.400	0.424	0.1559	6	5.6594	5.67
1:00:00 PM	3823	0.150	0.211	2.4847	8	8.7928	9.91
2:00:00 PM	3823	0.180	0.211	0.5355	8	8.3827	4.78
3:00:00 PM	5	0.003	0.006	0.4583	1	0.9265	7.35
4:00:00 PM	5	0.003	0.006	0.4687	1	0.9140	8.60
5:00:00 PM	504	0.003	0.005	0.2908	2	1.9757	1.21
6:00:00 PM	2015	0.003	0.006	0.3958	3	2.5871	13.76
7:00:00 PM	2015	0.003	0.006	0.3975	3	2.6023	13.25
8:00:00 PM	912	0.050	0.041	0.1425	1	1.0193	1.93
9:00:00 PM	580	0.040	0.041	0.0060	1	1.0193	1.93
7:00:00 AM	580	0.040	0.041	0.0060	2	2.1930	9.65
8:00:00 AM	580	0.040	0.041	0.0060	2	2.1942	9.71

According to the calibration analysis, the highest percentage of SSE for the velocity parameter is 2.49%. This result indicates that the RMA output data are well-represented with the actual condition. The highest SSE of velocity parameter is located in the channel that connected the inlet of sediment trap and inlet of Agathis Lake. The condition could happen due to the dynamic of a narrow channel. The least percentage of SSE for TSS distribution is located in the water body of the lake meanwhile the highest

IOP Conf. Series: Materials Science and Engineering **601** (2019) 012006 doi:10.1088/1757-899X/601/1/012006

percentage could be found in the water body of the lake near the outlet. The range of SSE for TSS distribution is 1.21% - 13.76%. The analysis concludes that the RMA data could represent the actual data but still need more elaboration to be valid.

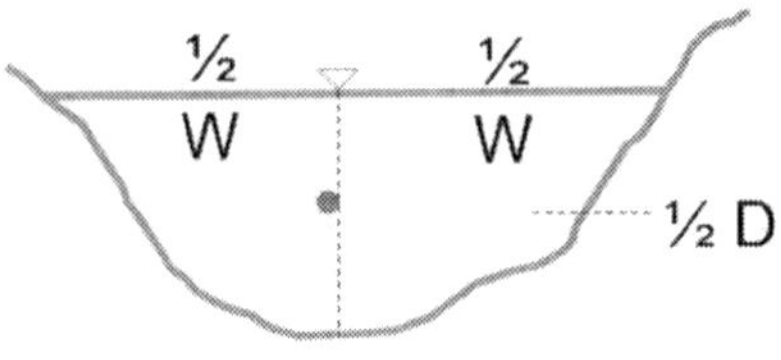

Figure 3. Water Sampling Location

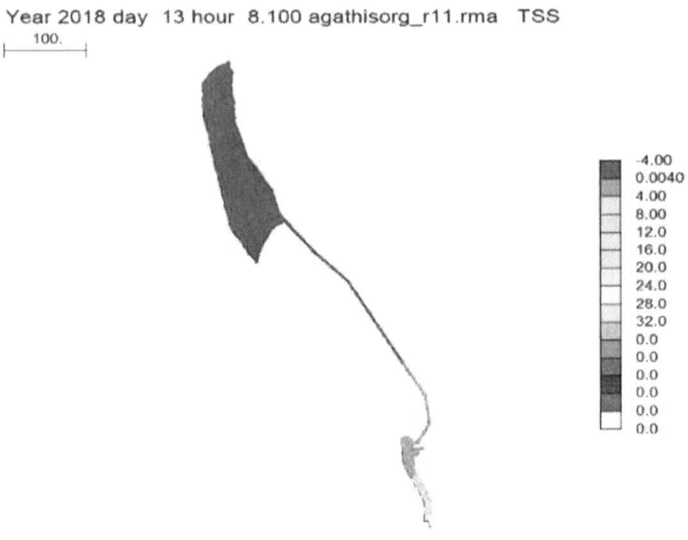

Figure 4. RMA-11 Program Running Result at 8.00 a.m.

Samples are taken in the surface water. The 3-Dimensional model should represent X, Y, and Z directions of the water body. It means that the depth parameter is important to notice. The representative locations from water depth should be analyzed to capture the best representative samples locations.

There are two suggestions for future calibration analysis. Firstly, the default parameters that affect the RMA program result data should be analyzed and calibrated. Secondly, samples should be taken in various depth locations to obtain the best representative samples. To conclude, the advance calibration should notice both parameters and samples data from various water depth.

3.5. RMA Program Data Output

The program output data are visualized by RMA-PLT. Figure 4 – 6 depict the RMA-11 program running at 8.00 a. m., 9.00 a. m., and 12.00 p. m., respectively.

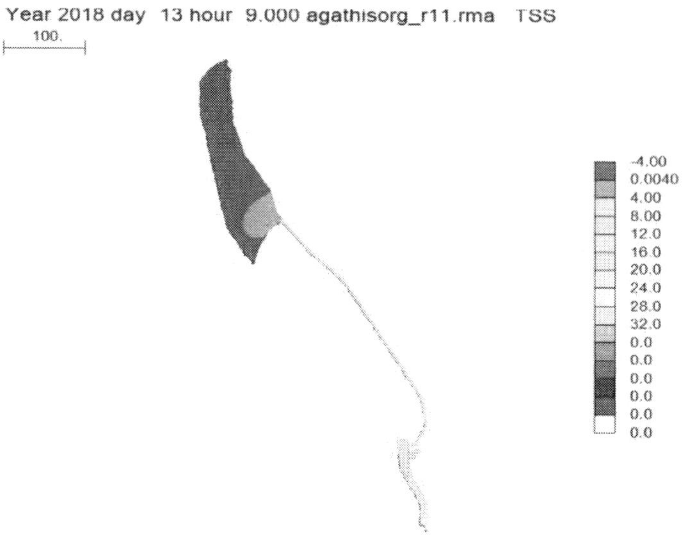

Figure 5. RMA-11 Program Running Result at 9.00 a.m.

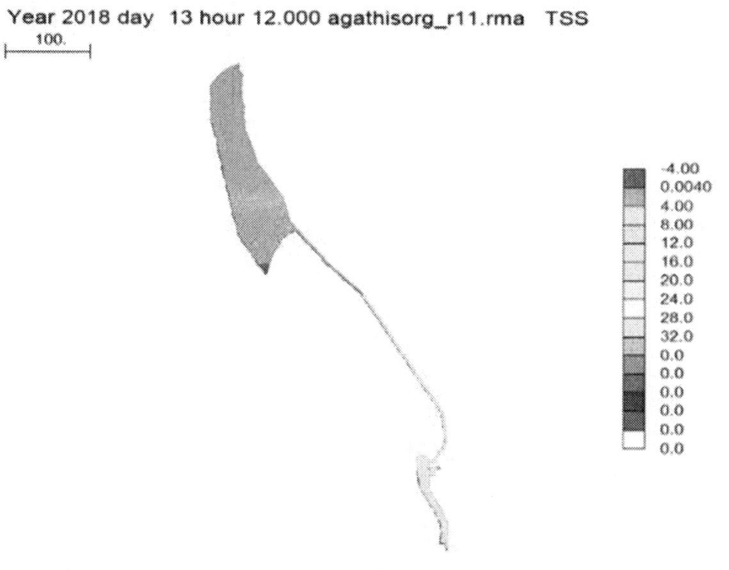

Figure 6. RMA-11 Program Running Result at 12.00 a.m

4. Conclusion

This research concludes that the model prediction of velocity and TSS distribution could represent the actual data according to calibration analysis. The research suggests more iterations of parameters to obtain the least percentage of SSE for RMA-10 and RMA-11 calibrations. In addition, the three-dimensional model could be performed with various depth layers of the lake to observe the TSS distribution in the z-direction. The observation of various depth layers could lead to the finding of the best representative sample location of the lake depth for the hydrodynamic model.

5. References

[1] Hani Muftah 2010 Faulty *Design and Construction Factors In Building Maintenance* Master Thesis (Johor : Universiti Tun Hussein Onn Malaysia)

[2] Eisma D., 1990 Transport and deposition of suspended matter in the north-sea and the relation to coastal siltation, pollution, and bottom fauna distribution, *Reviews in Aquatic Sciences.* ***3*** 181-216
[3] Tett P B, Joint I R, Purdie D A, Baars M, Oosterhuis S, Daneri G, Walne A W 1993 Biological consequences of tidal stirring gradients in the North Sea, *Phil. Trans. R. Soc. Lond. A.* ***343*** 493-508
[4] Premanand B D, Satishkumar U, Babu B M, Parasappa S K, Dandu M M, Kaleel I,Rajesh N L, Biradar S A 2018 Morphometric Analysis of Patapur Micro-watershed in North-Eastern Dry Zone of Karnataka Using Geographical Information System: A Case Study, *Int. J. Curr. Microbiol. App. Sci.* **7** 853-866
[5] Duan Q Y, Vijai K, Gupta Soroosh Sorooshian 1993 Shuffled complex evolution approach for effective and efficient global minimization, *Journal of optimization theory and applications.* **76** 501-521
[6] Prastica R M S 2018, September. Employing Galerkin method and field study for model calibration to perform sediment transport modelling in Agathis Lake, Universitas Indonesia, *AIP Conference Proceedings* vol 2014 (AIP Publishing) pp 020070
[7] Chapra S C 1997 *Surface Water Quality Modelling*, (New York: McGraw-Hill International Editions, New York)
[8] Neary V S, Wright S A Bereciartua P 2001 Case study: Sediment transport in proposed geomorphic channel for Napa River, *Journal of Hydraulic Engineering.* **127** 901-910.
[9] Rao A D, Dash S, Babu S V Jain I 2007 Numerical Modeling of Cyclone's Impact on the Ocean— A Case Study of the Orissa Super Cyclone. *Journal of Coastal Research*, pp.1245-1250.
[10] Gleizon P, Punt A G, Lyons M G 2003 Modelling hydrodynamics and sediment flux within a macrotidal estuary: problems and solutions, *Science of the total environment.* **314** 589-597
[11] Zheng L, Chen C, Alber M, Liu H 2003 A modeling study of the Satilla River Estuary, Georgia. AI: Suspended sediment, *Estuaries*, **26** 670-679

Acknowledgements
The authors wish to thank PITTA research grant No. 2511/UN2.R3.1/HKP.05.00/2018 as financial support to this research. The present work was performed as a part of the main research entitled Developing Sediment Transport Modelling on Agathis Lake, Universitas Indonesia. The authors wish to thank the Hydrology, Hydraulic, and River Laboratory of Engineering Faculty, Universitas Indonesia and Indonesian Institute of Science of Limnology (LIPI Limnology). The first author wishes to thank Indonesia Endowment Fund for Education scholarship (LPDP), which legislated by the Indonesian Finance Ministry, as financial assistance to pursue a master program in Universitas Indonesia.

IOP Conf. Series: Materials Science and Engineering **601** (2019) 012007 doi:10.1088/1757-899X/601/1/012007

Potential Use of Locally Available Filter Media in a UAFB Reactor Coupled with Natural Treatment in Soybean Industry Wastewater Treatment

Satoto E. Nayono[1*] and Didik Purwantoro[1]

[1]Department of Civil Engineering and Planning, Faculty of Engineering, Yogyakarta State University, Jalan Colombo 1, Yogyakarta 55281 INDONESIA
*Corresponding author: satoto.nayono@uny.ac.id

Abstract. Many food industries in developing countries cannot afford proper wastewater treatment and directly discharge their wastewater into surface water bodies causing several problems such as aesthetical, health and environmental problems. A new approach for low-cost wastewater treatment which is reliable, allows nutrients and energy recovery should be promoted. One of the alternative wastewater treatment technologies is anaerobic wastewater treatment which offers a number of advantages such as low energy requirement, low production of sludge and the possibility of energy recovery. As one of the disadvantages of using anaerobic method is that the effluent usually does not meet the authority's regulation to be discharged directly into water bodies, to further treat the effluent in order to meet the standard, the up flow anaerobic fixed-bed reactor will be coupled with a 'natural treatment'. This paper presents preliminary results of a research which has the main goal to examine the of several locally available materials as packing media in UAFB reactor and the appropriate vegetations for constructed wetland and/or macrophyte pond. Additionally, potential energy recovery from soybean wastewater is also calculated.

1. Introduction

Many food industries in developing countries fail to provide an adequate wastewater treatment installation and directly discharge their wastewater into surface water bodies such as creeks, rivers or lakes. This practice causes several problems such as aesthetical problems (odour nuisance, turbid water, *etc.*), health problems (skin infection, diarrhoea, *etc.*) and environmental problems (damage to surface or ground water, eutrophication, *etc.*). A new approach for wastewater treatment which is reliable and sustainable but requires low-cost and allows recovery of valuable nutrients and energy should be promoted. Selection of wastewater management systems depends on local environmental conditions, socio-economic situation, perceptions and cultural community as well as wastewater treatment technologies available.

In Indonesia, soybean products play a very important role in the dietary of the people. Since the protein of soybean is equal to the quality of protein found in meat and milk products, soybean products such as tofu (*tahu*), *tempe* (fermented soybean), or soya milk can substitute the need of protein requirement [1]. As a consequence of increasing number of soybean industries, wastewater volume resulting from soybean industry activities is also increasing. Similar with other wastewater generated by food processing plants, wastewater from soybean industries generally contains organic materials such as starch (carbohydrate), fat (lipids) and proteins. This type of wastewater, which mostly is a by-product of the soybean steaming process, has a typical COD content of 2,000 to 12,000 mg/l and a total nitrogen content of up to 100 mg/l [2, 3]. Therefore, this type of wastewater is considered as high-strength wastewater that has to be treated before it is released to the environment.

Wastewater treatment plants are designed to create optimum conditions for the enhancement of the rate of the natural processes. According to Van der Steen *et al.* [4], the main feature of sustainable wastewater treatment systems is aimed at recycling of nutrients, water and energy. Since there is no mechanical oxygen input is required, anaerobic microbiological processes for organic removal are preferred in sustainable wastewater treatment systems. In aerobic microbiological processes, the oxygen for process is supplied by photosynthesis (algae, plants) or natural re-aeration. Although the application of anaerobic processes for industrial and domestic wastewater treatment has been intensively investigated since more than 40 years ago [5], only very few studies reported the application of anaerobic process for wastewater treatment in soybean industries [6]. Compared to the aerobic wastewater treatment process, anaerobic wastewater treatment offers a number of advantages. Among them are low energy requirement, low nutrient requirement, production of only little excess microbial biomass and the possibility of energy recovery in the form of methane. However, beside these advantages, the anaerobic process still has some limitations (such as its high cost) to be considered.

One of the most widely used anaerobic reactor is upflow anaerobic fixed bed (UAFB) reactor. It is a bioreactor in which the anaerobic organisms grow as a biofilm on the surface of non-moving supporting material. The population of microorganisms in this reactor take the substrates and nutrients from and releases the product to the medium that flows over the film. The reactor is classified as an open system since the biomass is retained on the surface of supporting material until it reaches its maximum thickness. When this condition is achieved, the biofilm will loose its adhesiveness and it will be washed out through the effluent, therefore, this system does not require a biomass recirculation [6, 7].
One of the most important part in a UAFB reactor is its packing media or supporting materials. The high cost of a UAFB reactor is the contribution of media packing price. The media packing materials that have been examined in several researches include: plastic media packing [8], blast furnace cinders, polyurethane foam and crushed stone [9], charcoal [10], polystyrene and expanded clay [11], and clay ceramic particles [12].

The UAFB reactor has also been applied to treat several types of wastewater either in full-scale reactor or in laboratory-scale reactor. These wastewater include: whey [7], pharmaceutical-containing wastewater [8], coffee bean industry [9], winery [13], landfill leachate [14], slaughterhouse wastewater [15] and phenolic wastewater [16].

Considering the aforemention background problems, the main goals of this study was to examine the suitability of several local materials as packing (filter bed) media, either by physical/biological properties and economic considerations. In this study, several different packing materials have been evaluated and compared. The most suitable packing media will be used in the laboratory-scale upflow anaerobic fixed-bed reactors treating wastewater from soybean industry. Furthermore, several vegetations were also examined in order to find the most suitable method to enhance the quality of the UAFB effluent using "natural treatment" in the form of constructed wetland or macrophyte pond.

2. Materials and methods

The first step of this research is to examine the suitability of several locally available materials as packing material in an UAFB reactor. In order to determine which materials should be further examined, literature review and survey were conducted. The same procedure was also applied to find the suitable plants for constructed wetlands/macrophyte pond. The schematic diagram of laboratory-scale UAFB reactor coupled with "natural treatment" is shown in figure 1.

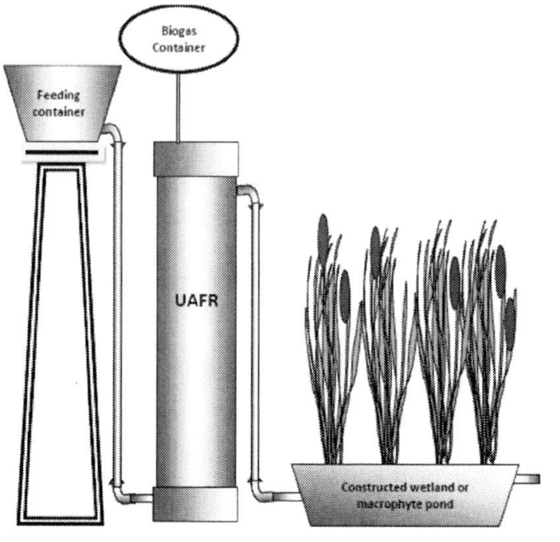

Figure 1. Schematic diagram of the upflow anaerobic fixed-bed reactor used coupled with constructed wetland/macrophyte pond.

2.1. Selection of locally available filter media

Before conducting material selection to find the suitable materials, this study set-up a criteria matrix in order to evaluate the materials. The material selection method is a way of ranking so that the top ranking of the six materials will be used as media packing in the laboratory scale reactors. The matrix was based on literature study and previous experiences which resulted in several requirements that become the focus of the suitable packing media material determination. These requirements include:

1. It should be locally available and also must be easily found in abundant reserve,
2. The price of the packing media material should be cheap for cheaper construction of UAFB,
3. The weight of the material should be light in order to have lighter construction which will lead to a cheaper construction since it does not need to use very high-quality materials to build the reactor,
4. The surface area of the material must be large enough to provide a place for the living microorganisms (biofilm), which will be in charge to reduce the pollutants,
5. The material is non-toxic and can provide a good environment for the growth of microorganisms that will form a biofilm on the surface of the filter media,
6. If possible, a recycled material so as to reduce the burden on the environment and prevents unused material to be discharged into the environment.

From several pre-surveys and observations that have been made, there are several potential materials to be used as packing media. These materials are (see also figure 2): *bantak* of Merapi (Merapi lava stone), woven bamboo ball that contains smaller diameter of *bantak,* bamboo ring, clay roof-tile fragments, clay ring, and irregular clay ball

2.2. Selection of vegetation for constructed wetland/macrophyte pond

Similar to the selection of packing materials, before conducting surveys to find the suitable vegetation, this study set-up a criteria matrix in order to evaluate the potential vegetation for constructed wetland/macrophyte pond. The requirements to determine appropriate vegetation include:

1. Should be locally available with abundant reserve on the location of the reactor,
2. Should have cheap price for procurement,
3. Does not require complex and expensive treatment,
4. Should have a function to supply oxygen to the microorganisms that reduce the organic pollutant,

IOP Conf. Series: Materials Science and Engineering **601** (2019) 012007 doi:10.1088/1757-899X/601/1/012007

5. When the vegetation is harvested, it should have an economic value (could as animal feed or handicraft material

The vegetation to be assessed are: bulrush/cattail (*ekor kucing*) and water bamboo for constructed wetland, duckweed (*lemna*) and *eichhornia* (*eceng gondok*) for macrophyte pond (see figure 3).

Figure 2. Selected materials to be observed as potential material for packing media in a UAFB. From top-left, clockwise: woven bamboo balls, bamboo rings, rooftile fragments, and *bantak*.

Figure 3. Selected plants to be observed as potential vegetation to be used for constructed wetland or macrophyte pond. Top, left-right: cat-tail and water bamboo, both for constructed wetland. Bottom, left-right: duckweed (*lemna*) and eichhornia (*eceng gondok*), both for macrophyte pond.

3. Results

3.1. Assessment results of filter media selection

Adapted from the requirements criteria, a weighing matrix is developed to search rankings of the materials. A value of 1 was given to the material with least compliance with the criteria and value 4 to the material with most suited with the criteria. The summary of the assessment and ranking of the potetial packing materials is presented in Table 1.

Table 1. Assessment and Ranking of the Potetial Packing Materials

Criteria	Type of material					
	1	**2**	**3**	**4**	**5**	**6**
Availability in nearby local/volume	4	3	3	2	2	2
Material price	4	3	3	3	2	2
Weight (specific gravity)	1	4	4	2	2	2
Surface area	3	4	4	3	3	4
Non-toxicity	4	4	4	4	4	4
Recycled material	3	3	3	4	3	3
Total score	19	21	21	18	16	17

Description of the type of material:

1. *Bantak* of Merapi (Merapi lava stone)
2. Woven bamboo ball that contains small diameter of *bantak*
3. Bamboo ring
4. Roof-tile fragments
5. Clay ring
6. Irregular clay ball

Based from the above assessment (on the best ranking basis), the materials will be used as packing media in the laboratory scale reactors are: woven bamboo balls containing small diameter *bantak* and bamboo rings. The price difference of among chosen materials with the one of more sophisticated product from western industries is very large. The most sophisticated media packing from industry will cost about USD 1,000/m^3 media while bamboo rings and woven bamboo balls cost about USD 150/m^3 and USD 250/m^3 respectively. Therefore, the use of this locally available materials as filter bed in the reactor will reduce the construction cost and at the same time give economic improvement opportunity to the community by enabling new workplace.

According to Joo-Hwa *et al.* [17], media roughness and pore size were more important over surface area on the reactor performance. Within the same media porosity, pore size plays a more significant role than surface area in term of organic pollutant removal efficiency. The study indicates that the ability of a support medium to retain biomass either as suspended growth entrapped in the void space, or by attachment to the media surface is a significant consideration in an UAFB reactor.

3.2. Assessment results of vegetation selection

Aquatic plant systems (macrophyte pond) are shallow ponds with floating or submerged aquatic plants. These systems include two types based on the dominant plant types. The first type uses floating plants and is distinguished by the ability of these plants to derive their carbon-dioxide and oxygen needs from the atmosphere directly. The plants receive their mineral nutrients from the wastewater. The second type of system consists of submerged plants and is distinguished by the ability of these plants to absorb oxygen, carbon-dioxide, and minerals from the water column. Submerged plants are easily inhibited by high turbidity in the water because their photosynthetic parts are below the water.

With the same method applied for packing material selection, bulrush/cattail (*ekor kucing*) and water hyacinth were selected as vegetation for constructed wetland and macrophyte pond (see also Table 2).

Table 2. Assessment and Ranking of the Potential Vegetation

Criteria	Type of vegetation			
	1	**2**	**3**	**4**
Availability in nearby local	3	2	4	4
Vegetation price	2	2	4	4
Treatment complexity	4	3	4	3
Environment disturbance potential	4	4	3	2
Potential oxygen supply	4	4	4	3
Economic value	4	4	4	2
Total score	21	19	23	18

Description of the type of material:

1. Bulrush (cattail)
2. Water bamboo
3. Water hyacinth (*Eichhornia*)
4. Duckweed (*lemna*)

Cattail, easily recognizable by their tall leaves and brown pistillate spike, is a common wetland plant. Cattails provide shelter for birds and mammals, shade water to cool fish, and their energy-rich rhizomes and shoots are a nutritious food source for many animals including humans [18]. With its high growing rate [19], *Eichhornia* is an excellent source of biomass which can be used for animal feed, energy sources and fertilizer/soil conditioner [20]. Several researchers also examined the potential use of *Eichhornia* as herbicide [21] and its use for medicinal use, *e.g.* its activity against human cervical cancer [22].

3.3 Methane production and potential energy recovery

The potential of a substrate to be converted to biogas during anaerobic treatment is essential to determine the suitability of a substrate as a feeding in anaerobic digester. The higher methane potential owned by a substrate the more attractive the substrate as a feeding, since it will give the higher energy recovery rate in the form of biogas. The biogas production potential of biodegradable solid wastes depends on the content of digestible carbohydrates, lipids and proteins, as well as on the content of more resistant cellulose, hemicellulose and lignin. According several researchers, COD elimination efficiency of UAFB reactors reached a range of approximately 60 - 90% with a range organic loading rate (OLR) of 1,5 – 4,6 kg COD/$m^3.d^{-1}$ [8, 23, 24]. This data will be used to predict the potential biogas production of soybean industry wastewater and also to calculate the energy balance of the treatment.

Assuming that the soybean wastewater has a COD content of 8,000 mg/l and applying an average hydraulic retention time (HRT) of 5 days, a reactor volume of 250 m^3 should be provided to treat wastewater from the semi-centralized (clustered) soybean industry. In this paper, it is assumed that COD removal is about 60%. With this assumption, 240 kg COD can be removed daily from wastewater by UAFB reactors. Using assumption that each kilogram of COD removed can produce 0.4 m^3 methane, the reactors can produce 96 m^3 biogas daily. It is equal to 57.6 m^3 methane daily as it is assumed that 60 % of methane is pure methane. If the energy content of methane is assumed 50.1 kJ/g CH_4 and methane density at 35°C is 0.6346 g/l [25], total energy recovered from wastewater would be 503.8 kWh/day. In Indonesia, electricity is sold for 0.065 USD/kWh in 2014. If the electricity produced by UAFB reactor is sold to the National Electricity Company (the PLN), it means that energy recovery activity from soybean industry wastewater can generate money of 11,951 USD per annum. See also Table 3 for the calculation of reactor dimension and potential energy recovery.

Table 3. Reactor dimension and potential energy recovery

Parameter	Unit	Value	Remarks
Wastewater production/household - industry	m^3/d	5	
Number of household - industry/cluster	HH	10	
Wastewater production	m^3/d	50	
Designed HRT	d	5	COD: 8.0 kg/m^3 OLR: 1.6 $kg/m^3/d$
Active reactor volume	m^3	250	
Daily biogas production	m^3	96	Removal efficiency: 60%, each kg COD removed produce 0.4 m^3 biogas
Daily methane production	m^3	57.6	Methane content: 60%
Energy recovered	MJ	1,812.1	Each m^3 methane: 31.46 MJ
Energy recovered	kWh/day	503.8	1 MJ: 0.278 kWh
Potential benefit	USD/year	11,951.8	1 kWh: USD 0.065

4. Conclusion

UAFB reactor can be constructed from locally available materials. The use of this kind of materials will reduce the cost of construction and made this technology affordable for developing countries. Moreover, from the above calculation simulation and prediction of energy recovery, it can be concluded that the application of UAFB reactor with locally available materials to treat wastewater from soybean industries has a very good prospect. This application can reduce the environmental risk of wastewater discharge and in the same time give the opportunity to recover energy from wastewater.

5. References

[1] Roughead Z K, Hunt J R, Lykken G I, and Johnson L K 2003 *Effects of Daily Substitution of Soy Protein for Meat Protein on Calcium Retention and Biomarkers of Bone and Cardiovascular Health in Postmenopausal Women* (Orlando: The 5th International Symposium on The Role of Soy in Preventing and Treating Chronic Disease) pp: n.a.

[2] Direstiyani L C and Bagastyo A Y 2016 *Tempe Industrial Wastewater Treatment by Using Combined Anaerobic Baffled Reactor and Biofilter Processes* (Surabaya: The 3rd International Postgraduate Conference on Biotechnology) pp: n.a.

[3] Japanese Advanced Environment Equipment 2004 *Membrane Separation Type Waste-water Processing Equipment for Fermented Soybean Production* Retrieved on November 8, 2018 from http://nett21.gec.jp/JSIM_DATA/WATER

[4] Van der Steen N P, Polprasert C, Veenstra S, and Gijzen H J 2001 *Wastewater Treatment: Natural System* (Delft: IHE Delft)

[5] Hickey R F, Wu W M, Veiga M C, and Jones R 1991 Start-up, Operation, Monitoring and Control of High-rate Anaerobic Treatment Systems, *Wat. Sci. and Tech.* **24** 207-255.

[6] Böhnke B, Bischofsberger W and Seyfried C S (eds.) 1993 *Anaerobtechnik-Handbuch der anaeroben Behandlung von Abwasser und Schlamm* (Berlin: Springer Verlag)

[7] Handajani M 2004 *Degradation of Whey in an Anaerobic Fixed Bed (AnFB) Reactor* PhD Thesis (Karlsruhe: Universitaet Karlsruhe)

[8] Chelliapan S, Yuzir A, Md Din M F, and Sallis J 2011 Anaerobic Pre-treatment of Pharmaceutical Wastewater using Packed Bed Reactor, *Int'l Journal of Chem. Eng. and Appl.* **2** 32 - 37

[9] Fia F R L, Matos A T, Borges A C, Fia R, and Cecon P R 2012 Treatment of Wastewater from Coffee Bean Processing in Anaerobic Fixed Bed Reactors with Different Support Materials: Performance and Kinetic Modeling, *Journal of Env. Man.* **108** 14–21

[10] Acharya B K, Hilor P, Sarayu M, Shouche Y, Singh V, and Datta M 2011 Kinetic Modelling and Microbial Community Assessment of Anaerobic Biphasic Fixed Film Bioreactor Treating Distillery Spent Wash, *Wat. Res.* **45** 4248–4259

[11] Barros A R, de Amorim E L C, Reis C M, Shida G M, and Silva E L 2010 Biohydrogen Production in Anaerobic Fluidized Bed Reactors: Effect of Support Material and Hydraulic Retention Time, *Int'l Journal of Hyd. Energy* **35** 3379 – 3388

[12] Han W, Yue Q, Wu S, Zhao Y, Gao B, Li Q, and Wang Y 2013 Application and Advantages of Novel Clay Ceramic Particles (CCPs) in an Up-flow Anaerobic Bio-filter (UAF) for Wastewater Treatment, *Biores. Tech.* **137** 171-178

[13] Ganesh R, Rajinikanth R, Thanikal J V, Ramanujam RA, and Torrijos M 2010 Anaerobic Treatment of Winery Wastewater in Fixed Bed Reactors, *Bioprocess and Biosystems Eng.* **33(5)** 619-628

[14] Thabet O B D, Bouallagui H, Cayol J, Ollivier B, Fardeau M-L, and Hamdi M 2009 Anaerobic Degradation of Landfill Leachate Using an Upflow Anaerobic Fixed-Bed Reactor with Microbial Sulfate Reduction, *Journal of Hazardous Materials* **167** 1133–1140

[15] Méndez-Romero D C, López-López A, Vallejo-Rodríguez R, León-Becerril E 2011 Hydrodynamic and Kinetic Assessment of an Anaerobic Fixed-Bed Reactor for Slaughterhouse Wastewater Treatment, *Chem. Eng. and Proces.* **50** 273–280

[16] Bakhshi Z, Najafpour G, Neya B N, Kariminezhad E, Pishgar R, Moosavi N 2011 Recovery of Upflow Anaerobic Packed Bed Reactor from High Organic Load during Startup for Phenolic Wastewater Treatment, *Chem. Industry & Chem. Eng. Quart.* **17** 517–524

[17] Joo-Hwa T, Kuan-Yeow S, and Jeyaseelan S 1997 Media Factors Affecting the Performance of Upflow Anaerobic Packed-Bed Reactors, *Env'tal Monit. and Assess.* **44** 249–261

[18] Northwest Research and Outreach Center 2016 *Cattail Management in the Northern Great Plains: Implications for Wetland Wildlife and Bioenergy Harvest* (Minnesota: University of Minnesota)

[19] National Research Council 1976 *Making Aquatic Weeds Useful: Some Perspectives for Developing Countries* (Washington, DC: The National Academies Press, doi:10.17226/19948)

[20] Su W, Sun Q, Xia M, Wen Z and Yao Z 2018 The Resource Utilization of Water Hyacinth (*Eichhornia crassipes* [Mart.] Solms) and Its Challenges, *Resources* **7** 48

[21] Chai T-T, Ngoi J-C and Wong F-C 2013 Herbicidal Potential of *Eichhornia crassipes* Leaf Extract Against *Mimosa pigra* and *Vignaradiata. Int'l Journal of Agri. and Bio.* **15** 835–842

[22] Lenora L M, Senthil Kumara J, Murugesan S and Senthilkumar N 2015 Anticancer Activity of Water Hyacinth [*Eichhornia Crassipes* (Mart) Solms] on Human Cervical Cancer Cell Line, *Octa Jour. of Env. Res.* **3** 327-331

[23] Nursetyowati P, Soewondo P, and Handajani M 2015 Performance of Upflow Anaerobic Fixed Bed Reactor with Pumice Stone Support in Organics Removal and Biogas Formation from Liquid Phase Biowaste (only in Indonesian), *Jurnal Rek. Kim. dan Ling.* **10** 148 - 156

[24] Yousefzadeh S, Ahmadi E, Gholami M, Ghaffari H R, Azari A, Ansari M, Miri M, Sharafi K, and Rezaei S 2017 Comparative Study of Anaerobic Fixed Film Baffled Reactor and Up-Fow Anaerobic Fixed Film Fixed Bed Reactor for Biological Removal of Diethyl Phthalate from Wastewater: A Performance, Kinetic, Biogas, and Metabolic Pathway Study. *Biotech. for Biofuels* **10**:139

[25] Metcalf & Eddy, Inc. 2003 *Wastewater Engineering: Treatment and Reuse 4th edition* (New York: McGraw-Hill)

IOP Conf. Series: Materials Science and Engineering **601** (2019) 012008 doi:10.1088/1757-899X/601/1/012008

Impact of Surrounding Soils on Surface Water Quality in Sembrong River

Syarifah Nur Fahirah Tuan Besar[1], Syarifuddin Sofik[1] andAzra Munirah Mat Daud[1*]

[1] Department of Water and Environmental Engineering, Faculty of Civil and Environmental Engineering, Universiti Tun Hussein Onn Malaysia, 86400 Parit Raja, Batu Pahat, Johor, Malaysia
*Corresponding author: azra@uthm.edu.my

Abstract. Peatlands enclosed around 2.7 million ha in Peninsular Malaysia, 1.66 million ha in Sarawak, 0.8 million ha in Sabah, and are commonly found in Johor. Previous studies showed that waters in the peat soils area are acidic (pH 5.7 and pH 6.6). Anthropogenic activities such as mining and agriculture, will in turn impact water quality. Agriculture activities, through the usage of fertilizers, produces cadmium and zinc. Furthermore, soils in acidic conditions could liberate aluminum into the river. This study aims to characterize acidified peat soils based on geochemical properties (pH and heavy metals concentrations), identify the impact of the geochemical variability on surface water quality, and determine the effects of hydrologic cycle on the geochemical variability. Sembrong River was chosen as the study site; soils and water samples were collected nearby and from the river. Testing methods for soils and water covered both in-situ measurements and laboratory analyses. All parameters were taken thrice in each of the sampling point (labelled as S1, S2 and S3). The data of precipitation a week before sampling was recorded for early prediction of the water level. Real water level (depth of the river) was measured on the day of sampling. Based on the relation between month and geochemical properties according to water depth, results showed that the properties were affected by the rainfall intensities. Results also revealed the increased the water depth, the decreased the pH value for soil and water. The most acidic for pH value is peat soils with pH 2.52. Besides, the concentration of heavy metals in the surface water for Sembrong River followed the sequence: aluminum > iron > zinc. Overall results of this study have indicated that properties of peat soils, anthropogenic activities and hydrologic cycles have a major impacts on surface water quality of Sembrong River.

1. Introduction

Surface water is the vital source in the world and could be known as the clear liquid which had the appearance colorless or pale. The function of waters cannot be denied for the consumers and high demand for clean water is very critical nowadays. Besides, 2.5% of Earth's water contains freshwater and very peril by the days because of climate change and human actions (economic) [1]. Nowadays, the issue of the water pollution is very highlighted because of due the contamination factor.

The main source of water contamination comes from anthropogenic activities; agricultural, industrial and municipal. According to United States Environmental Protection Agency, the human activities comprise of municipal, industrial, agricultural, residential and commercial, and these activities would attack the quality of both surface and ground water [2]. Besides, the industrial consumption of water can produce an effluent which can lead to poor water quality which carries chemicals and non-chemicals and also suspended materials [3]. In addition, when all of these substances are going through into the aquatic surrounding, it will form intricate mixtures that might contribute to the illnesses to human being and also the organisms that consume the polluted waters [4]. Previous study also has shown, acidity of upland freshwater can be influenced by the decay of blanket peat process [5]. The impact from decay of blanket peat process will influence the water quality, hence, will contribute acidity to water body as compared to the previous natural conditions. For instance, acid sulfate soils can produce harmful effects by liberating the toxic mixture (between acidity and heavy metals) to the river and release amount of Al, Cd, Co, Ni, and Zn [6]. Acidic soils could liberate Al to the soil and outward water [7]. High concentration of Al will lead to the diseases and a study by Muthuraman, has shown that aluminum sulfate will create the illness such as Alzheimer's disease in human [8].

Fertilizer and pesticides are the main factor blooming of crop yields. The micronutrients and macronutrients can get from the fertilizers or pesticides. The macronutrients comprise of calcium and magnesium while micronutrients comprise of iron, copper, cobalt, zinc and also manganese [9]. All of these nutrients will exist the toxicity in the soil that will lead to destroying the environment and also will produce retard crops. All of these factors that occur is important to check the water quality because the elements that contain in the soil could influence the surface water.

Water quality monitoring is needed in understanding the palatable conditions of a water body. Study by Mhlongo, has mentioned that highest quantity of chemicals is used in human routine life which could influence the quality of surface water, and raised challenges in maintaining the surface water quality [10]. Hence, the monitoring and management are crucial in ensuring the sustainability of water quality is preserved [11]. Thus, the measurement of geochemical properties of surface water will provide a better understanding of a surface water conditions. This will lead to sustainable environmental friendly to both human beings and other aquatic life.

This present study was proposed to identify the surface water conditions through the monitoring of geochemical properties and hydrological impacts on the conditions of the acidic peat soils surrounding the river as well as the geochemical properties. The objectives of this study comprise of three: 1) characterize acidified peat soils based on geochemical properties (heavy metals and pH), 2) identify the impact of the geochemical variability on surface water quality and 3) determine the effects of hydrologic cycle on the geochemical variability (determined by rainfall intensities). This will lead to the understanding on the surface water quality and soil management based on the elements and factors that contribute to the soil and surface water acidity.

2. Material and Methods

2.1 Study area

The study was conducted at Sembrong River, Batu Pahat, Johor. Sembrong River has been taken as natural water supply for the residence in Parit Raja and Sri Gading. Besides, Sembrong River also collects natural water for Sembrong Dam. According to Botero, the length of Sembrong River is 22 km while the river container wrapping about 200 km^2 [12]. The surrounding of study zone (partially of Sembrong River) was covered with anthropogenic activities which comprise agricultural such as palm oil plantations and paddy cultivation, and also covers residential areas.

In this study, three stations were selected along the study river, namely S1, S2 and S3 as shown in figure 1 for studying the geochemical properties condition of water quality and acidified peat soils. The latitude and longitude of the station: S1 (N 1° 55' 17.47'', E 103° 8' 48.50''), S2 (N 1° 53' 12.41'', E 103° 7' 29.04'') and S3 (N 1° 52' 57.99'', E 103° 5' 13.44''). The station points of this sampling were chosen because appropriate for sample to be taken.

*yellow arrow = Streamline of Sembrong River.

Figure. 1 Sampling station S1, S2 and S3 at Sembrong River.

2.2 Methods

2.2.1 Sample collection and preservation

Soil and water samples were collected from three sampling stations (S1, S2 and S3) within three different sampling events, while the soil samples were collected at the river; nearby the water sampling points. Water levels (depth) were also measured in every sampling. Water samples were collected at approximately one third of the water depth using cylindrical water bailer [13], and soil samples were taken at approximately 0.5 to 1.0 m depth using hand auger [14]. Water samples were kept in a 1-L polypropylene or polyethylene bottles [15]. The sampling bottles were acid washed and rinsed with deionized water prior to sampling session to elude contaminations [16]. Soil samples were kept in a 500 g radiation-sterilized polyethylene zipper bag to ensure the moisture content of the soils is well preserved [17]. All soil and water samples were then kept in ice during transportation and kept in dark cold room at 4°C in the laboratory [18], prior to analyses.

2.2.2 Testing methods

In this study, a multi-parameter meter; HANNA Instruments (Model H19828) was used for in-situ measurement. The in-situ parameters include pH, temperature, total dissolved solids (TDS), electrical conductivity (EC), and oxidation reduction potential (ORP) and dissolved oxygen (DO) [19, 20]. Echo sounder was utilized to measure the water depth [21]. Table 1 shows the summary of methods for in-situ measurement and laboratory analyses.

Maintenance is defined as "the combination of all technical and administrative actions, including supervision actions, intended to retain an item, or restore it to, a state in which it can perform a required action [4]. Another definition of maintenance also being offered by the Chartered Institute of Building, which is "work undertaken in order to keep, restore or improve every facility, every part of the building, its services and surrounds, to an agreed standard, determined by the balance between need and available resources". Meanwhile The Committee of Building Maintenance recommended the adoption of the maintenance definition as follows, "Work undertaken in order to keep, restore or improve every facility, every part of the building, its services and surrounds to a currently acceptable standard and to sustain the utility and value of the facility" [5].

Table 1. Parameters for in –situ measurements and laboratory analysis on this study.

Type	Data
Water Parameter	In- situ measurements : Temperature, pH, DO, ORP, EC, Depth Laboratory analysis : Al, Fe and Zn
Soils Parameter	Laboratory analysis : pH, Al, Fe and Zn,
Methods and References	EPA Method 9045D (7.2.1): Determination for soil pH EPA Method 300.0. : Extraction for soil and water sample

All laboratory analyses were conducted in FKAAS Laboratories. The laboratory analyses comprise of determination of Fe, Al and Zn concentrations (for both water and soil samples) and pH (for soils only). The determination of pH for soils, the slurry method need to be done with ratio water and soils (1:1) with followed the EPA Method 9045D (7.2.1). For the determination for heavy metals comprise of Fe, Zn and Al, an Inductively Couple Plasma-Mass Spectrometer, (ICP-MS ELAN 9000) were utilized [12].

The method of extraction for water and soil sample were based on EPA Method 300.0. The solid materials such as soils, extraction need to be done with ratio soils and distilled water (1: 10). After that, the mixture of water and soils were stirred using magnetic stirring equipment for 10 minutes and filter the occur slurry. Then, after the filter was done, used the syringe and attached the 0.45 μ membrane type filter at the nozzle of the syringe [22].

After all of the extraction were completed, the dilution method need to be done before using ICP-MS. The dilution factor calculation is shown below:

$$Df = Vt / Vs \quad (1)$$

where, Vt is volume total; Vs is volume sample; and Df is dilution factor.

All data were recorded and examined with using the software which is Microsoft Excel 2016. This application will produce the accurate and fast results for all of the information and data that occur in the all of the experiments that will be conducted in future research. The appropriate results will ease the future study to accomplish the aim of the future research and all of the results will be as the guideline to maintain and manage the acidified peat soils that impact to natural water quality in Batu Pahat.

3. Results and Discussion

3.1 Water and soils characteristics analysis

The geochemical parameters of this study comprises of pH and heavy metals (Al, Zn, Fe). All of the parameters were taken in each of the sampling site and labelled as 3 station point (S1, S2 and S3). All of the results were taken thrice for each station point for more precise data. All data that acquired from all the three stations to recognize the features of water and soils that concerning the geochemical properties.

The data of precipitation a week before sampling was recorded in Table 2 to estimate the accuracy of water depth (measured using depth sounder) which could produce the grow of surface water that would touch the near soil surroundings. Based on the Table 2 the precipitation data for the first sampling shows the sample of each site was taken a day before with the overall average were rainfall for 30 minutes while for the second sampling, the sample was taken 3 days before which occurred two heavy rainfall events with the overall average of rainfall session was 4 hours. Lastly for the third sampling, the sample was taken 3 days before with the overall average were rainfall for 1 hour.

Table 2. Precipitation data a week before sampling.

	Rain Data	
Date	Type of rain	Periods
5/3/2018	Heavy rain	6:00-7:00p.m
6/3/2018	Light rain	9:00-9:30p.m
7/3/2018	Rain	12:00-4:00p.m
8/3/2018	-	-
9/3/2018	-	-
10/3/2018	Rain	10:00-10:30a.m
29/3/2018	-	-
30/3/2018	-	-
31/3/2018	-	-
1/4/2018	Heavy rain	4:00-7:00p.m
2/4/2018	-	-
3/4/2018	Heavy rain	2:00-3:00p.m
26/7/2018	-	-
27/7/2018	-	-
28/7/2018	-	-
29/7/2018	Rain	8.00– 9.00a.m.
30/7/2018	-	-
31/7/2018	-	-

3.2 Analysis the effects of soils surrounding features on surface water

The analysis effects of soils surrounding features on surface water comprise of effects month on geochemical properties according to water depth. All of the results could be guide as monitoring for the future presents. The data collection from Sembrong River which contains three dissimilar sampling point were evaluated.

3.2.1 Effects of month on geochemical parameters

Effects of month on geochemical parameters comprise of pH, Fe, Al and Zn, Figure 2 and Figure 3 shows the connection between month and pH for sampling station 1 (S1), sampling station 2 (S2) and sampling station 3 (S3). pH is one of the crucial parameters in water quality while the different water depth is occurred depending on the precipitation prior to the sampling duration.

Based on the Figure 2 and Figure 3, S1, S2 and S3 showed the higher the water depth, the decreased pH value for water and soil. Results can be proven in april month by Figure 2 with highest water depth is 2.2 m at S2 with pH water 2.55 meanwhile for Figure 3 highest water depth is 2.7 m at S3 with pH soils of 2.52. Besides, compared to precipitation data in april month, 2 heavy rains was occurred before the water and soils sampling were started. Overall, rain is influenced the pH of water and soils based on the increased of water depth. The factors that impact the pH water and soil more decreased because of the elements in the soil and also the precipitation. A study by Wei, stated that, the outcome of acid rain influenced the ecological comprise of ground and water to be acidic [23].

Figure 4 and Figure 5 show the connection between month and heavy metals for Fe. Figure 4 and Figure 5 showed the more lower the water depth value, the more reduced the Fe concentration. In august month for water and soils results, all of the results tend to increased compared to other both month in march and april meanwhile for the Fe soils concentration, the results for S3 is the highest compared to other sampling site. Overall results, not spotted rain influenced the Fe concentrations to be more increased. The elements in the soil (such as: pyrite (FeS_2)) tend Fe concentrations were rise. Based on the previous study, study had reported that, pyrite was spotted in the compound of stone (metamorphic) and sedimentary [24]. In conclusion, the Fe concentrations could rise even no precipitation because of due the elements in the soil.

Figure 6 and Figure 7 show the connection between month and Al. The outcomes revealed that, the more lower the water depth value, the more reduced the Al concentration for water and soil and can be proven by Al concentration value decreased in water and soil (0.08 mg/l and 0.05 mg/l) at lowest water depth (1.5 m) for S1.

Furthermore, Figure 7 showed in april month, all of the sampling site results tends to increased compared to al water concentration. Moreover, two heavy rains in april month can affect the increased of the al soils concentration. A study by Wei, revealed that, the accretion of acid rain may increase the poisonous of aluminum [23]. Furthermore, the lessen of pH values could contains the rise leaching waste of Al^{3+} [25].

Figure 8 and Figure 9 shows the connection between month and heavy metals for Zn. Based on the graph and results in Figure 8 and Figure 9, what had been observed that the decreased the water depth value, the more decreased the Zn concentration and be proven by S1 (1.5 m), Zn concentration value reduced for water (0.02 mg/l) meanwhile for S2 with water depth (2 m), the results low in Zn concentration value for soil (0.13 mg/l).In august month, all of the sampling points (S1, S2 and S3) results were higher than other months for water and soils. Generally, zinc concentrations were grow with not spotted rain in august month. Besides, the comparison results between pH (water and soils) and Zn concentrations in march month, the more higher the pH, the lessen the zinc concentrations. According to Martínez, claim that, the zinc poisonous reduced when pH ascend [26].

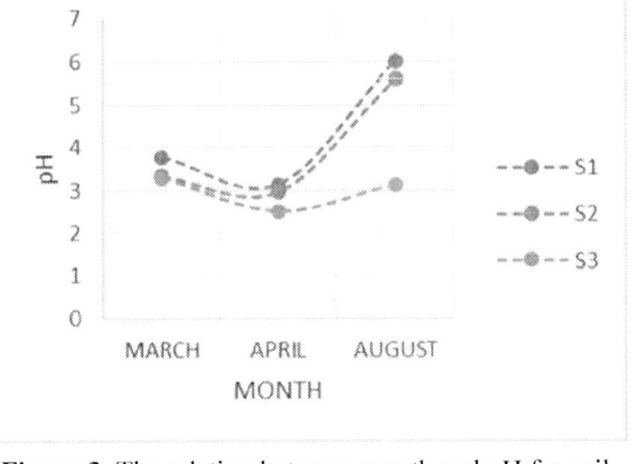

Figure 2. The relation between month and pH for water. **Figure 3.** The relation between month and pH for soil.

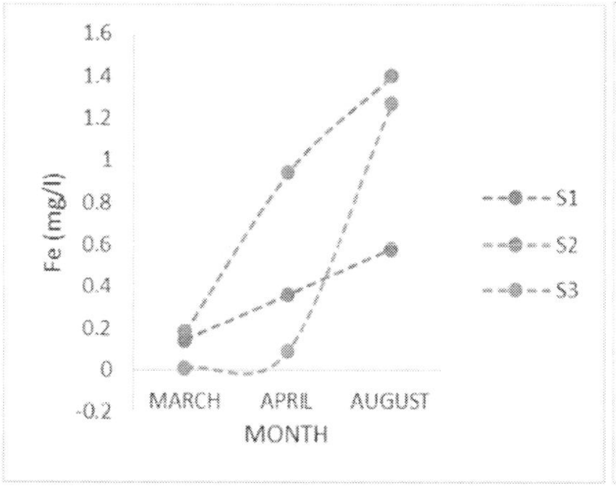

Figure 4. The relation between month and Fe for water

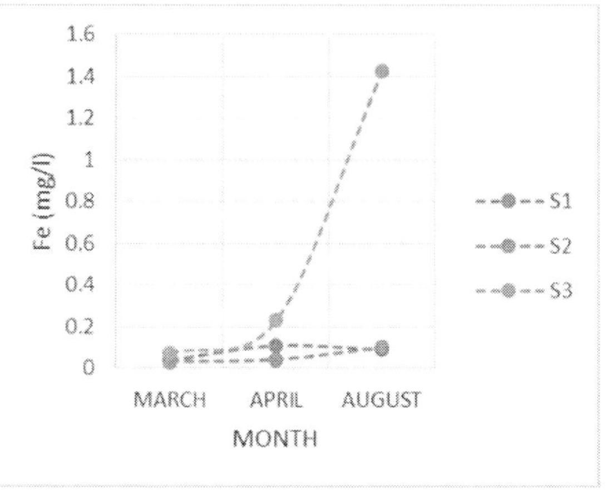

Figure 5. The relation between month and Fe for soil.

Figure 6. The relation between month and Al for water.

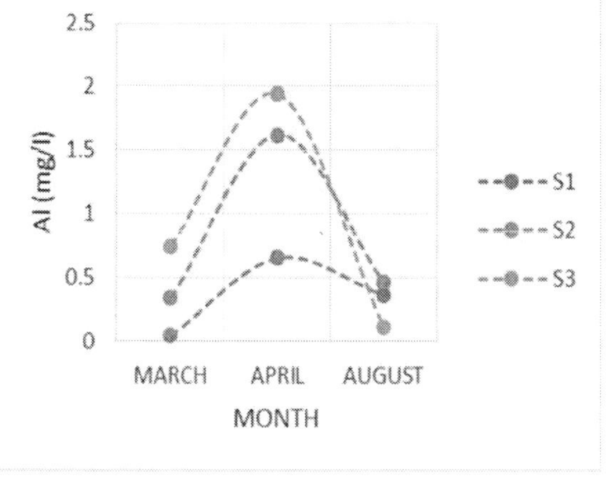

Figure 7. The relation between month and Al for soil.

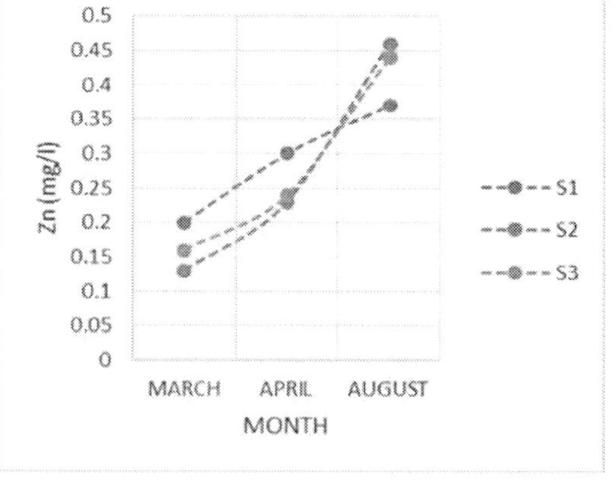

Figure 8. The relation between month and Zn for water.

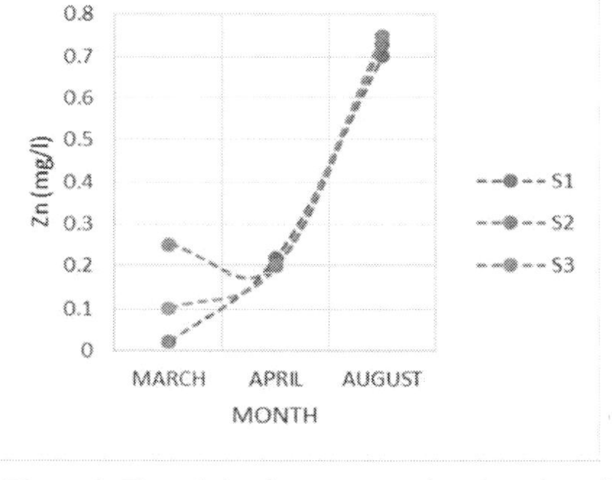

Figure 9. The relation between month and Zn for soil.

4. Summary

To sum up, the characteristics of acidified peat soils based on the geochemical properties (pH, Al, Fe and Zn concentrations), identify the impact of the geochemical variability on surface water and determine the effect of hydrolysis cycle on the geochemical variability (determine the rainfall intensities). Overall of the study based on the results that had achieved for the relation between month and geochemical properties according to water depth, the raining was influenced the results by the increased the number of water depth, the more decreased the pH value for soil and water that tends to pH 2. Besides, the most acidic for pH value is peat soils with pH 2.52 compared the other results for peat soils. Besides, the concentration of heavy metals in the surface water for Sembrong River followed the sequence: Al > Fe> Zn. The results showed that, aluminium is the most contaminated compare Fe and Zn and this tends the Sembrong River to be polluted. Moreover, results had shown the acidified peat soils have a major impact on surface water quality for Sembrong River.

5. References

[1] Distefano T and Kelly S 2017 Are we in deep water? Water scarcity and its limits to economic growth, Ecol. Econ **142** 130–147
[2] United States Environmental Protection Agency (US EPA) 2002 *Ground Water Contamination*, Get. Up to Speed, pp 10
[3] Garg K K and Prasad B 2017 Treatment of toxic pollutants of purified terephthalic acid waste water: A review *Environ. Technol. Innov.* **8** 191–217
[4] Batista N J C, Carvalho Melo Cavalcante A A, Oliveira M G, Medeiros E C N, Machado J L, Evangelista S R, Dias J F, Santos C E I, Duarte A, Silva F R and Silva J 2016 Genotoxic and mutagenic evaluation of water samples from a river under the influence of different anthropogenic activities *Chemosphere* **164** 134–141
[5] Martin-Ortega J, Allott T E H, Glenk K and Schaafsma M 2014 Valuing water quality improvements from peatland restoration: Evidence and challenges *Ecosyst. Serv* **9** 34–37
[6] Beucher A, Fröjdö S, Österholm P, Martinkauppi A and Edén P 2014 Fuzzy logic for acid sulfate soil mapping: Application to the southern part of the Finnish coastal areas," *Geoderma*, 226–227 pp. 21–30
[7] Drábek O, Kipkoech K I, Komárek M, Tejnecký V and Borůvka L 2015 Study of interactions between relevant organic acids and aluminium in model solutions using HPLC and IC *Soil Water Res* **10** 172–180
[8] Muthuraman G and Sasikala S 2014 Removal of turbidity from drinking water using natural coagulants *J. Ind. Eng. Chem* **20** 1727–1731
[9] Botero W G, Oliveira L C de, Rocha J C, Rosa A H and dos Santos A 2010 Peat humic substances enriched with nutrients for agricultural applications: Competition between nutrients and non-essential metals present in tropical soils *J. Hazard. Mater* **177** 307–311
[10] Mhlongo S, Mativenga P T and Marnewick A 2018 Water quality in a mining and water-stressed region *J. Clean. Prod* **171** 446–456
[11] Misaghi F, Delgosha F, Razzaghmanesh M and Myers B 2017 Introducing a water quality index for assessing water for irrigation purposes: A case study of the Ghezel Ozan River *Sci. Total Environ.*, **589** 107–116
[12] Latiff* A A A, Karim A T A, Muhamad A and Hashim N H 2009 Study of metal pollution in Sembrong River, Johor, Malaysia *Int. J. Environ. Eng* **1** 384–404
[13] Matamoros V 2012 *Equipment for water sampling including sensors* **1** Elsevier
[14] Shien P T, Salma D and Ismail A 2011 A Study on Factors Influencing the Determination of Moisture Content of Fibrous Peat *UNIMAS E-J. Civ. Eng* **2** 39–47
[15] Myers, Mark D 2006 Handbooks for Water-Resources Investigations: National field Manual for the Collection of Water-Quality Data U.S. Geol. Surv 231
[16] Chan-Keb C A, Agraz-Hernández C M, Perez-Balan R A, Gómez-Solano M I, T D N J Maldonado-Montiel, Ake-Canche B and Gutiérrez-Alcántara E J 2010 Acute toxicity of water and aqueous extract of soils from Champotón river in Lactuca sativa *L Toxicol. Reports* **5**, pp. 593–597
[17] Phang I R K, Chan Y S, Wong K S and Lau S Y 2018 Isolation and characterization of urease-producing bacteria from tropical peat *Biocatal. Agric. Biotechnol* **13** 168–175

[18] Awang H, Daud Z and Hatta M Z M 2015 Hydrology Properties and Water Quality Assessment of the Sembrong Dam, Johor, Malaysia," *Procedia - Soc. Behav. Sci.* **195** 2868–2873
[19] Sallam G A H and Elsayed E A 2018 Estimating relations between temperature, relative humidity as independed variables and selected water quality parameters in Lake Manzala, Egypt *Ain Shams Eng. J* **9** 1–14
[20] Al-Badaii F, Othman M S, Gasim M B 2013 Water Quality Assessment of the Semenyih River *J. Chem* **2013** 31–34
[21] Baharim N B, Yusop Z, Askari M, Yusoff I, Wan Muhd Tahir W Z, Othman Z, Zainal Abidin M R 2012 Preliminary Results of Stratification Study in Sembrong *The 2nd International Conference On Water Resources (ICWR2012)* pp. 1–7
[22] Pfaff J D 1993 Determination of inorganica anions by ion chromatography. Method 300.0 EPA Methds pp. 1–28
[23] Wei H, Liu W, Zhang J, Qin Z 2017 Effects of simulated acid rain on soil fauna community composition and their ecological niches *Environ. Pollut* **220** 460–468
[24] Oliveira C M, Machado CM, Duarte G W, Peterson M 2016 Beneficiation of pyrite from coal mining *J. Clean. Prod* **139** 821–827
[25] Jiang J, Wang Y P, Yu M, Cao N, Yan J 2018 Soil organic matter is important for acid buffering and reducing aluminum leaching from acidic forest soils *Chem. Geol.* **501** 86-94
[26] Martínez-Cortijo J, Ruiz-Canales A 2018 Effect of heavy metals on rice irrigated fields with waste water in high pH Mediterranean soils: The particular case of the Valencia area in Spain *Agric. Water Manag* **210** 108–123

A Case Study: Water Quality Content between Water Supply and Water Vending Machine at Parit Raja, Johor

Suhaila Sahat[1*], Nadia Kariya[2], Masiri Kaamin[1], Norbaizura Hamid[1], and Nor Farah Atiqah Ahmad[1]

[1] Centre of Diploma Studies, Universiti Tun Hussein Onn Malaysia, Pagoh, Johor, Malaysia
[2]Faculty of Civil and Environmental Engineering, Universiti Tun Hussein Onn Malaysia, 86400 Parit Raja, Batu Pahat, Johor, Malaysia
*Corresponding author: suhailasa@uthm.edu.my

Abstract. Water quality study was conducted on water vendor machines with different types of systems that are available around Parit Raja, Batu Pahat such as reverse osmosis systems, systems nanotechnology nano-silver system and water supply pipes. The aim of this study is to distinguish the water quality from the vending machine that follow the standard quality whether authorized or not. Standard water quality parameters such pH, iron concentration experiments, experimental manganese, aluminum and experimental testing ammoniacal nitrogen were computed as comparison. Nanotechnology system water produces better quality when comparing to the pipe water as its pH value shows 6.62. From the experiments the concentration of iron, manganese, aluminium and ammoniacal nitrogen shows 0.08-, 0.07-, 0.05-, and 0.01 mg/L respectively which all the values are in compliance with the National Drinking Water Quality Standards (NDWQS). As for water from reverse osmosis system and pipe water supply both its pH is 6.2 and 6.4 correspondingly which both are not concurrence with National Drinking Water Quality Standards. However, iron content for water from reverse osmosis (0.30 mg/L) is higher than water supply pipe (0.12 mg/L) and both are in the standard range. Iron content of the water supply pipe which is not allowed by 0.14 mg/L. Compared with water from a reverse osmosis system in which the iron content is at a maximum of 0.10mg / L. The parameters of aluminium and ammoniacal nitrogen, the two systems are at acceptable levels. For water from nano-silver system, all parameters are in compliance with the National Drinking Water Quality Standards.

1. Introduction

Water is the most important nutrient in life after air. Water is also one of the natural resources, which play an important role in human life for daily activities directly or indirectly. Now days, all around the world are facing with water crisis due to population growth, global economic and industrial growth rates. Biswas [1] found that population growth of middle class is estimated to have 4.9 billion by 2030 and it expand from 3.2 billion in 2020 and 1.8 billion in 2009. World water demand keep on increasing each year [2-3] and in Malaysia, the demand for agricultural, industrial and domestic purpose has steadily increased started in 1980 until 2000 from 8.9 billion m^3 to 15.5 billion m^3 [4]. This kind of statistic has emerge an awareness among citizen about water quality for their daily used especially drinking water.

According to the Malaysian Standards, MS 1228 [5] average rate of water use requirements is 225 L/person/day. On average, the volume of water in the human body is 65% of total body weight and volume are highly variable in every person. There are few human organs that contain a lot of water, which is crucial for the water to be clean and quality water. This is because the water is clean and of high quality needed for the body, but it also plays an important role in the growth of the community every day there are about 25,000 people will die from the use of untreated water in everyday life [6].

A research among urban resident in Negeri Sembilan, Malaysia on water issues found that colour, odour and taste were the main problem with their tap water. Due to this matters, 85% respondents has purchased domestic water filters, 41% boiled water and 17% bought bottled water and almost quarter that had water filter clarify it is more convenience for them[7]. The awareness begins when people concern about health, poor water quality and increasing water pollution in the country. The diseases cause by polluted water that attack humans are transmissible and spread through water. Such conditions can cause epidemics of course everywhere.

Based on these concerns, water filter companies have supply many type of water vending machine that easily found in public area. Vending machine is a self-service machine produce purified water dispensing into customer supplied container such as a water bottle [8]. Malaysian citizen perception on vending machine is the water produced by it safe to drink because they on rely on the advertisement by the company provider not from the public health point of view. A good service of water vending machine is when connected to an approved, potable water supply and when they are properly designed, manufactured, installed and maintenance [9].

In Malaysia, several studies have been recorded on the presence of ammonia and manganese in drinking water treatment [10] and minerals content in drinking water [11]. Therefore, through this study, there are two objectives to achieve. The objectives are, (i) identify the contents of water quality based on tests conducted in the laboratory and (ii) determine whether water quality from vending machine follows the standards required.

2. Material and Method

In this study, there are four types of water taken from different water filtration systems. These systems comprises of nano technology, nano silver and reverse osmosis. Water supply pipes, which produced from a water treatment system were also used in the comparison between the other water filtration. Selection of samples according to several parameters involving pH, concentration of iron, manganese, ammoniacal nitrogen and aluminum. Water quality of the content will be either comply with National Drinking Water Quality Standards (NDWQS) or not and this standard issued by the Ministry of Health Malaysia. Table 1 shows the National Drinking Water Quality Standards was used as a standard for this study.

Five experiments were conducted including pH test [13], experiment concentrations of iron [12], manganese experiment [13], experiment aluminum and experiment ammoniacal nitrogen [13] based on severel studies have done [10-11]. In carrying these experiments, equipment and materials used should be determined properly in order to reduce the error.

Table 1. National Drinking Water Quality Standard (NDWQS) [12]

Parameter	Group	Raw Water		Treated Water	
		Minimum	Maximum	Minimum	Maximum
Total coliform(TC)	1	0.00	5000.00	0.00	0.00
e-coli (E-Coli)	1	0.00	5000.00	0.00	0.00
Turbidity (NTU)	1	0.00	1000.00	0.00	5.00
Color	1	0.00	300.90	0.00	15.00
pH	1		9.00	6.50	9.00
Chlorine (Cl)	1	-	-	0.20	5.00
Ammoniacal nitrogen(NH_3-N)	2	0.00	1.50	0.00	1.50
Fluoride (Fl)	2	0.00	1.50	0.40	0.60
Ferum (Fe)	2	0.00	1.00	0.00	0.30
Aluminium(Al)	2	-	-	0.00	0.20
Mangan (Mn)	2	0.00	0.20	0.00	0.10
Chemical Oxygen Demand (COD)	2	0.00	10.00	-	-
Biochemical Oxygen Demand (BOD)	2	0.00	6.00	0.00	0.00

3. Results and Discussion

Figure 1 to Figure 5 shows the average value for each water filtration system for pH, iron, manganese, aluminium and ammonical nitrogen concentration, respectively. According to National Drinking Water Quality Standards, the value allowable pH range is 6.5 to 9.0. The data shows in Figure 1 indicate that, ph value for water produced from reverse osmosis system and pipe supply is less than 6.5. Figure 2 shows the data for iron concentration. Reverse osmosis system has recorded the highest value of 0.30mg / L of iron concentration value. Meanwhile, pipe water supply has the second highest value of 0.12mg / L and nano technology system with the lowest value, 0.08mg/L. According to the National Water Quality Standards (KKM, 2009), the permissible value of iron for treated water is between 0.00mg / L to 0.30mg / L. From Figure 3, shows the supply pipe for the highest value for manganese with 0.14mg/L. The allowable value is between 0.00mg/L to 0.10mg/L. It is may be caused the pipes were not well maintained. Nano technology has the lowest value. Figure 4 shows the analysis of aluminium. According To National Drinking Water Quality Standards[12], the value allowable aluminium range is 0.00mg/L to 0.20 mg/L. In the study, all of the water has allowed value. Figure 5 shows the analysis of ammoniacal nitrogen. According To National Drinking Water Quality Standards (Table 1), the value allowable ammoniacal nitrogen ranged from 0.00 mg/L to 0.15 mg/L. In the study, all of the water has allowed value.

Based on the laboratory results, the values for all four water supply system are acceptable as according to the National Drinking Water Quality Standards. There are few types of metal that are allow in water due to some of it are naturally exists while others are from pollution. Natural metal like aluminum exists in tap water as a result of its use as a coagulant in water treatment such as excessive aluminum sulfate. Ammoniacal this nitrogen results from microbiological activity and usually exists in surface water and underground water.

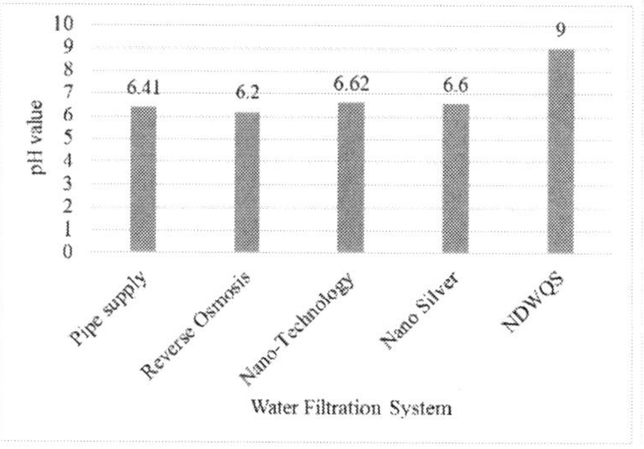

Figure 1. Analysis of pH

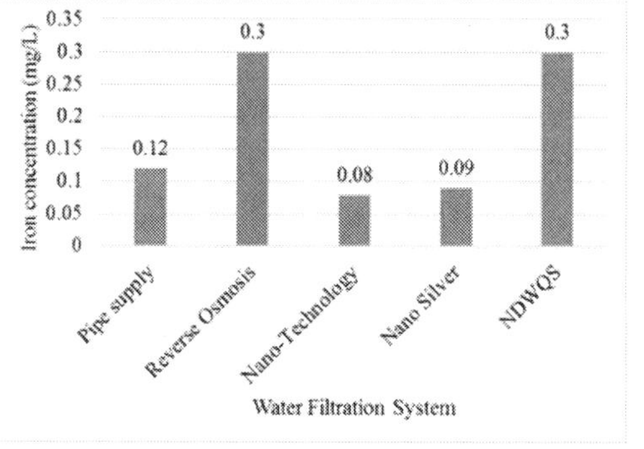

Figure 2. Analysis Concentration of Iron

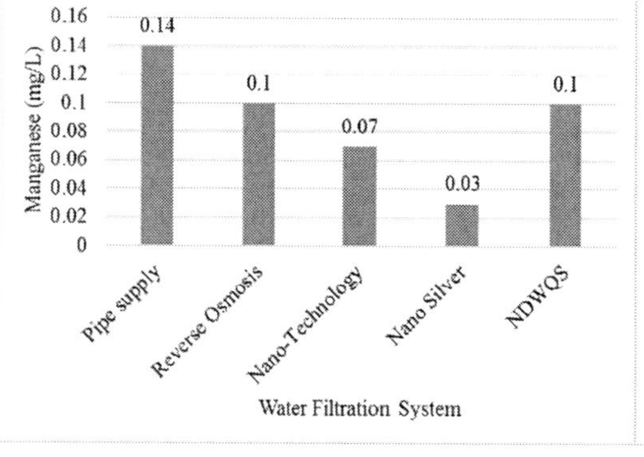

Figure 3. Analysis of Manganese

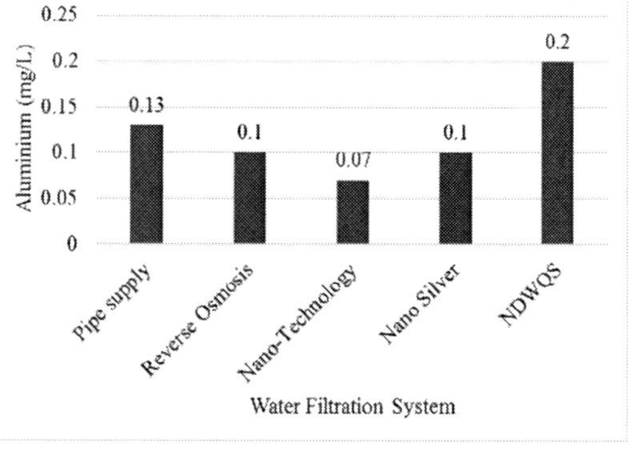

Figure 4. Analysis of Aluminium

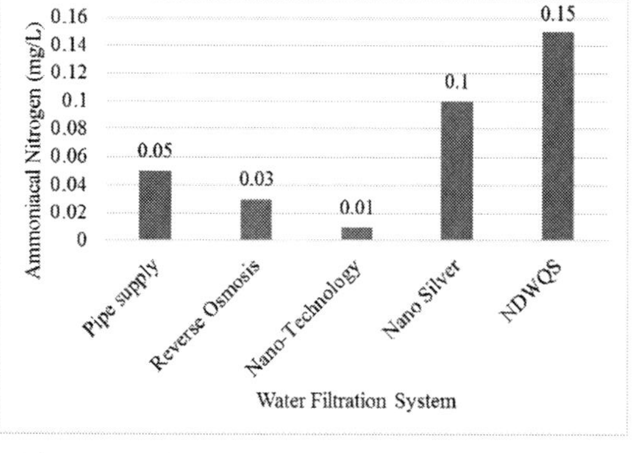

Figure 5. Analysis of Ammoniacal Nitrogen

4. Conclusion

As a conclusion, based on analysis result obtained, it shows that nano- technology system produced better water quality compared to the others different system. This can be seen when all the selected parameters, the water system using nano-technology are comply with the values in the National Drinking Water Quality Standards. Meanwhile, reverse osmosis system is categories as the worst water filtration system compared to the other three system in water vendor machine. Among all the parameters that has been tested it shows that pH parameter has the unfavourable value. Besides that, there are two parameters that have the maximum permitted concentrations of iron and manganese. These two parameters are sourced from the pipes that are not maintained due to the presence of iron and manganese can be detected if there are on the pipe. Other that pipe water supply, water filtration machine system or water vending machine also not properly maintained. Water sources from the pipe water supply has low water quality because there are some parameters that do not follow the National Drinking Water Quality Standards such as pH and manganese parameters. Water quality is also affected due to the pipes that are not properly maintained.

5. References

[1] Biswas A K 1999 Water Crisis: Current Perceptions and Future Realities *Water Int* **24** 363–367
[2] Seckler D, Amarasinghe U, De Fraiture C, Keller A. Molden D and Sakthivadivel R 2000 *World Water Supply and Demand: 1995 to 2025* Colombo Sri Lanka: IWMI
[3] Rosengrant M W, Cai X, and Cline S A 2002 *World water and food to 2025: dealing with scarcity* International Food Policy Research Institute Washington District of Columbia
[4] Ahmed F, Siwar C and Begum R A 2014 Water resources in Malaysia: Issues and challenges *J. Food, Agric. Environ* **12** 1100–1104
[5] Pepper I L, Gerba C P and Brusseau M L 2006 *Environmental and pollution science, 2nd ed* (San Diego, Calif: Elsevier Science)
[6] Aini M S, Fakhrul-Razi A, Mumtazah O and Chen J C M 2007 Malaysian households drinking water practices: A case study *Int. J. Sustain. Dev. World Ecol* **14** 503–510
[7] McSwane D Z, Oleckno W A and Eils L M 1994 Drinking water quality concerns and water vending machines *Journal of Environmental Health* **56** 7-12
[8] Hasan H A, Abdullah S R S,. Kamarudin S K and Kofli N T 2011 Problems of ammonia and manganese in Malaysian drinking water treatments *World Appl. Sci. J.* **12** 1890-1896
[9] Azlan A, Khoo H E, Idris M A, Ismail A and Razman M R 2012 Evaluation of minerals content of drinking water in Malaysia *The Sci. World J.* **2012** 1-10
[10] Jamie B and Richard B 1996 *A Practical Guide to the Design and Implementation of Water Quality Studies and Monitoring Programmes* (Chapman & Hall, London.)
[11] Flinn Scientific 1994 *Spectrophotometer Laboratory Manual* (Flinn Scientific: Batavia, IL) pp 55–60
[12] Frisbie S H, Mitchell E J, Dustin H, Maynard DM and Sarkar B 2012 World Health Organization discontinues its drinking-water guideline for manganese. *Environ Health Perspect.* **120** 775–778
[13] American Public Health Association, American Water Works Association, Water Environment Federation, 1998. *Standard methods for the examination of water and wastewater. 20th ed.,* (Washington, DC)

IOP Conf. Series: Materials Science and Engineering **601** (2019) 012010 doi:10.1088/1757-899X/601/1/012010

The Effectiveness of Raw and Dried Artocarpus Heterophyllus (Jackfruit) Seed as Natural Coagulant in Water Treatment

Mahmood N. A. C[1*] and Zaki Z. Z. M[1]

[1*] Faculty of Civil Engineering, Universiti Teknologi MARA, Shah Alam, 40450 Selangor, Malaysia
*Corresponding author: noratikah095@yahoo.com

Abstract. Coagulation is one of the process in water treatment that mainly reducing turbidity of water by adding the coagulant such chemical or natural coagulant. Chemical coagulant such as Alum is widely used as it is efficient in removing tubidity. However, there are some adversed impact of chemical as coagulant to human and environment specifically. In this study, the effectiveness of Artocarpus Heterophyllus seed as natural coagulant were measured. There are two types the natural coagulant from Artocarpus Heterophyllus were considered in this study such as raw and dried seed. Besides, two different extraction process of coagulant by using distilled water and Sodium Chloride (1M NaCl) were used to determine the efficiency. The result showed dried Artocarpus Heterophyllus seed (1M NaCl) at 20mg/L achieved the optimum dosage of coagulant compared to dried Artocarpus Heterophyllus seed (distilled water), raw Artocarpus Heterophyllus seed (distilled water) and raw Artocarpus Heterophyllus seed (1M NaCl) in removing turbidity. It reduced up to 50% turbidity, and 70 %suspended solid after the treatment process. It also showed that salt Sodium Chloride (NaCl) can be used as addictive in treating the water.

1. Introduction

Almost all types of productions in agriculture, industry, energy, transport need water as their source of life [1]. All organisms need water as essential element for survival in maintaining their life [2]. For millennia, it has been known that ecosystem conservation and human survival depend on the reliable availability of adequate quality of water.

Population growth and rapid industrial developments increase the issues that related to cleanliness and sanitized water globally. Many of fresh water supplies can be contaminated with inorganic, organic and biological matters that rise potential harm towards people [3].

Clean water is such an important thing in human life. It is because without clean water, people cannot continue with their daily life. Groundwater, surface water and rainwater are three sources of water. There are some processes on how people can obtain the clean and safe surface water to be drinking and use them by water treatment process that comprised of coagulation and flocculation, followed by settling, filtration and disinfection process [3].

According to [5], the objective of water treatment is to ensure effluent that contain clean water is discharged back to environment after removal of suspended solids to prevent the solid material uses up oxygen (O2) in the process of decay. Oxygen gas is needed for the aquatic life to survive.

The use of chemical coagulants can be reduced by the availability of the natural coagulants. Usually alum is widely used as chemical coagulant because it promised turbidity reduction but it also causes health problem to human and voluminous sludge production [3]. Water quality guidelines can be reaches by monitor of raw water. It can be done by measure colloidal and suspended matters which affects cloudiness of water that call as turbidity [3]. Level of sulphate is high in treated water when the alum is used as chemical coagulant. It is because alum is an inorganic coagulant which increases concentration of Total Dissolve Solid (TDS) in finished water [6].

Jar test is a process of mimics of full scale operation that give a great help to determine the relationship of coagulant and the water sample that being use in jar test. In addition, beakers or jar is use to hold water sample and then certain amount dosage of coagulant by varying the amount of dosage to obtain the optimum dosage amount of coagulant by added them into the beaker. After that, the water sample is stirred so that the floc formation, development, and settlement can be happen like in a real water full scale treatment plant. When the materials in sample of water react with coagulant, the floc will be form and clump together [7]

Clean water production cause many coagulants are widely used in conventional water treatment processes. These coagulants can be inorganic coagulants such as aluminum sulfate and polyaluminum chloride or synthetic organic polymers such as polyacrylamide de-rivatives and polyethylene imine. Lastly, natural coagulants such as chitosan and microbial coagulants are also being used in the water treatment [8]. Chitosan helps to remove turbidity in the range of 50% up to 87.5% which more than half of turbidity in water sample. The study shows E.Coli can be reduced by using chitosan as naural coagulant [9]. As for bioflocculant produced by Aspergillus niger microbe, the optimum pH required to result in lowest turbidity was in the range 4-5 [10]. Coagulant is an agent to remove colour and turbidity in coagulation process of water treatment.

A study by [11], shows that aluminium can be obtain easily as it is natural abundance, mammal species like human being can be harm by aluminum. People who diagnose with Alzheimer's disease show the presence of aluminium in their neuron. Alzheimer's dis-ease makes its patients to loss their memory because the brain cells are dead. It happens when human drink the treated water that con-tain residue of aluminium and the water reach the gastrointestinal system following oral exposure and lastly small amount of alumini-um penetrate into nervous system tissues.

Actually, majority of natural coagulant is coming from the plant based coagulants such as cactus, cereals, fungus, nuts, shrubs and spice. Currently, natural coagulant is made up of fruit waste that surprisingly become a great help in treating the water [4]. Furthermore, by using natural coagulants such as the Moringa Oleifera tree instead of aluminium salts might give a lot of advantages such as smaller costs of water production, less sludge production and can be obtain easily [5].

Nowadays, there a lot demand to change the chemical coagulant to the natural coagulant in the coagulation process. It is because natural coagulant does not give negative effect towards human health and natural coagulant also can remove turbidity of water successfully. Aluminum Sulfate (alum) and ferric chloride that has been used in water treatment process may have high potencies to remain some metal residual in treated water which is not good for people use.

This study related to Artocarpus Heterophyllus (Jackfruit) seed that act as natural coagulant to replace chemical coagulant such as alum in this study. The weight of Artocarpus Heterophyllus seed is 8%-15% of the total fruit weight. Its seeds comprised of 67% of carbo-hydrates and 20% of protein contents [12]. The coagulant is more effective when the concentration of protein in the coagulant is higher [13]. Besides, Artocarpus Heterophyllus seed also has jacalin and artocarpin which both of them are lectin. These lectins are antibacterial, antifungal and ant carcinogenic properties [14].

2. Materials and Methods

2.1. Sampling Procedure

The water samples were taken from Sungai Klang at Kampung Padang Jawa, Shah Alam. The colour of water sample is green-brownish and turbid. The plastic container must be clean from any impurities to ensure the water sample is not affected by the impurities as it will change the quality of water sample. The plastic containers were then be labelled to differentiate them with others container. This research involved laboratory work where the jar test experi-ment is carried out (ASTM D2035). Twenty litres (20) of water sample is taken to undergo the Jar Test process. In this study, 500 mL of water sample were filled in each beaker for six (6) numbers of beakers. The apparatus for this study are spec-trophotometer, turbidity meter, jar test apparatus, weighing balance, spatula, magnetic stirrer, magnetic stir bar, beaker of 500 mL and 1 L, pipets, oven and blender. Alum, Artocarpus Heterophyllus seed, distilled water and 1M NaCl are material that is needed in this la-boratory work. Figure 1 shows summary of methodology.

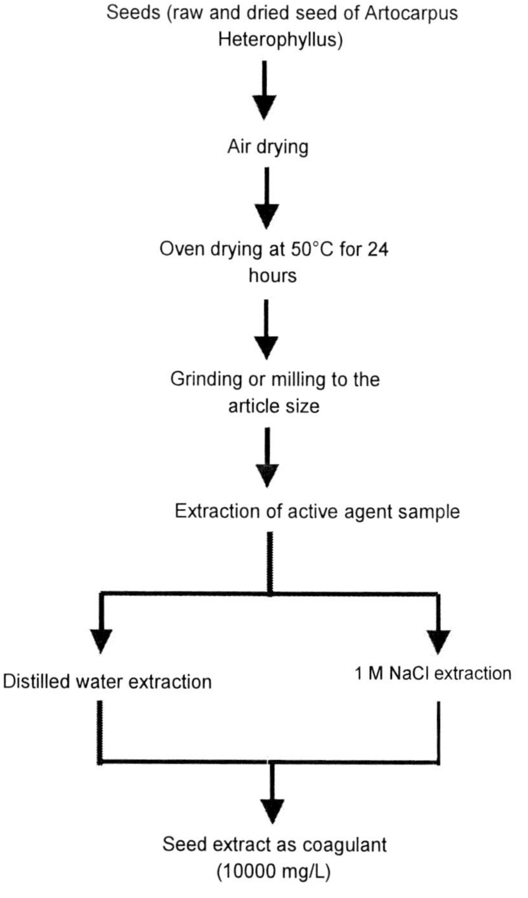

Figure 1. Summary of methodology [15]

2.2. Preparation of raw and dried Artocarpus Heterophyllus seed in powder form

The Artocarpus Heterophyllus was obtained from Masjid Tanah, Melaka. Figure 2 shows raw Artocarpus Heterophylllus seeds and dried Artocarpus Heterophyllus seeds. The seeds were dried in oven for 24 hours at 50°C. After that, the dried and raw Artocarpus Heterophyllus was blended by using domestic blender to obtain the powder form [15].

1. Artcarpus Heterophyllus seeds.

2. Dry the seeds in the oven at 50°C for 24 hours.

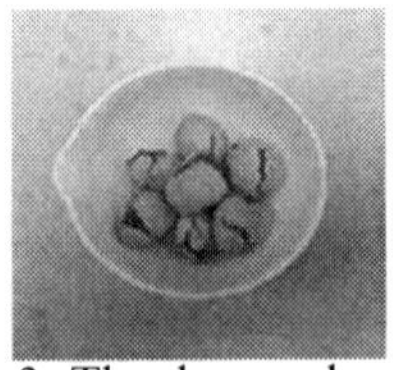

3. The dry seeds.

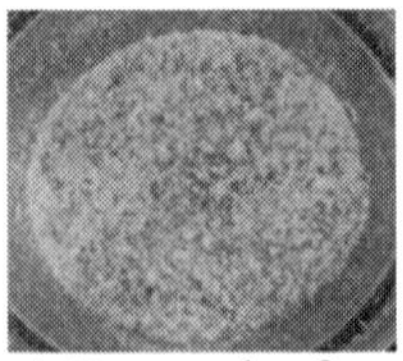

4. Dry seeds after blended.

5. Artocarpus Heterophyllus coagulant.

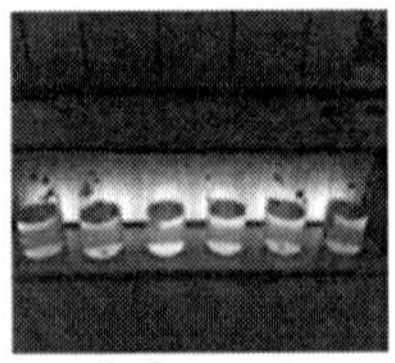

6. Jar test.

Figure 2. process of coagulant preparation [15]

2.3. Extraction method of Aluminum Sulphate (Alum)

Use 10000 mg of Alum powder to be dissolved in one (1) liter of distilled water by using a magnetic stirrer mixer to obtained 10000 mg/L concentration of Alum as chemical coagulant.

2.4. Extraction method of dried Artocarpus Heterophyllus

Dried Artocarpus Heterophllus coagulant was obtained by dis-solved 5000 mg of powder of dried Artocarpus Heterophyllus in 500 mL of distilled water or Sodium Chloride (1M NaCl) by using magnetic stirrer mixer for 30 minutes to get 10000 mg/L in concen-tration. Lastly, filter it with muslin cloth to remove larger particle of seeds [15]. 1 M NaCl is used in this study because of the previous research shows that the coagulant solutions extracted with 1 M salt solution is more effective in reducing colour and turbidity of water sample compared to the solution extracted with water [14]

2.5. Extraction method of raw Artocarpus Heterophyllus

The 10000 mg/L concentration of raw Artocarpus Heterophyllus coagulant was produced by mix 500 mL of distilled water or Sodi-um Chloride (1M NaCl) with 5000 mg of raw Artocarpus Hetero-phyllus powder by using magnetic stirrer mixer for 30 minutes as shown in. Lastly, remove large particle of seeds from the solution by using muslin cloth [15].

2.6. Jar test experiment

Coagulation, flocculation and sedimentation test (ASTM D2035) was conducted by using Jar Test apparatus. 500 mL of water sam-ple were poured into six (6) beakers. After that, 0 mg/L, 20 mg/L, 40 mg/L, 60 mg/L, 80 mg/L and 100 mg/L coagulants were added into each beaker. Start the jar test with rapid mixing process for one (1) minute at 80 rpm followed by slow mixing process for 30 minutes at 20 rpm. Lastly, allowed the water sample to undergo sedimentation process for 30 minutes at 0 rpm. The experiment was repeated for each type of coagulants (Alum, dried and raw Artocar-pus Heterophyllus).

HACHDR-2800 Spectrophotometer was required to determine colour (Method 8025), COD (Method 8000) and suspended solid (Method 8006) of water sample. The river water sample in each beaker need to be checked with colour, COD and suspended solid because there is different amount of them in each of beaker even though it was from the same water sample. Turbidimeter (HACH 2100P) was used to evaluate the turbidity in water sample in unit of NTU. Before evaluation of COD value by spectrophotometer, the water sample was filled into the vial that contained oxidizing agents that called as silver sulphate, mercury, potassium dichromate and sulphuric acid and need to be digested in DRB 200 COD Reactor.

3. Results

The purpose of this research is to identify the water quality parame-ter such as turbidity, suspended solid and colour and COD before and after the process, to determine the optimum dosage concentration of natural coagulant of raw and dried Artocarpus Heterophyl-lus seed to be compared with Alum in surface water treatment pro-cess, to determine the effectiveness of raw and dried Artocarpus Heterophyllus seed in improving water quality and lastly, to deter-mine the effect of salt Sodium Chloride, NaCl as addictive in treat-ing the water. Table 1, Table 2, Table 3, Table 4 and Table 5 show the quality of water sample and result of Alum and Artocarpus Heterophyllus seeds.

Table 1. The quality of water sample and result of Alum for distilled water as solvent in extraction process of coagulant.

Water sample	Turbidity (NTU)	Turbidity removal (%)	Suspended solid (mg/L)	Suspended solid removal (%)	Colour (PtCo)	Colour removal (%)	COD (mg/L)
Raw	26.70	-	37	-	238	-	62
0 mg/L	14.00	47.57	14	62.16	129	45.8	-
20 mg/L	8.25	69.10	7	81.08	83	65.13	-
40 mg/L	5.98	77.60	5	86.49	66	72.27	-
60 mg/L	3.99	85.06	3	91.89	45	81.09	-
80 mg/L	2.76	89.66	2	94.59	28	88.24	36
100 mg/L	3.03	88.65	3	91.89	30	87.39	-

Table 2. The quality of water sample and result of dried Artocrpus Heterohyllus seed for distilled water as solvent in extraction process of coagulant.

Water sample	Turbidity (NTU)	Turbidity removal (%)	Suspended solid (mg/L)	Suspended solid removal (%)	Colour (PtCo)	Colour removal (%)	COD (mg/L)
Raw	26.7	-	37	-	238	-	62
0 mg/L	21.60	19.10	30	26.83	171	29.92	-
20 mg/L	14.19	46.85	12	70.73	128	47.54	39
40 mg/L	15.13	43.33	14	65.85	137	43.85	-
60 mg/L	16.67	37.57	15	63.41	146	40.16	-
80 mg/L	17.19	35.62	17	58.54	147	39.75	-
100 mg/L	18.08	32.28	20	51.22	152	37.7	-

IOP Conf. Series: Materials Science and Engineering **601** (2019) 012010 doi:10.1088/1757-899X/601/1/012010

Table 3. The quality of water sample and result of raw Artocrpus Heterohyllus seed for distilled water as solvent in extraction process of coagulant.

Water sample	Turbidity (NTU)	Turbidity removal (%)	Suspended solid (mg/L)	Suspended solid removal (%)	Colour (PtCo)	Colour removal (%)	COD (mg/L)
Raw	47.8	-	47	-	308	-	62
0 mg/L	43.10	9.83	44.5	5.32	294	4.55	-
20 mg/L	27.70	42.05	24	48.94	203	34.09	-
40 mg/L	24.20	49.37	23	51.06	193	37.34	17
60 mg/L	25.10	47.49	23	51.06	196	36.36	-
80 mg/L	29.40	38.49	30	36.17	226	26.62	-
100 mg/L	25.80	46.03	27	42.55	194	37.01	-

Table 4: The quality of water sample and result of dried Artocrpus Heterohyllus seed for Sodium Chloride (1M NaCl) as solvent in extraction process of coagulant.

Water sample	Turbidity (NTU)	Turbidity removal (%)	Suspended solid (mg/L)	Suspended solid removal (%)	Colour (PtCo)	Colour removal (%)	COD (mg/L)
Raw	8.12	-	10	-	50	-	18
0 mg/L	6.82	16.01	9	10	49	2	-
20 mg/L	3.96	51.23	3	70	41	18	3
40 mg/L	4.25	47.66	5	50	43	14	-
60 mg/L	4.31	46.92	6	40	45	10	-
80 mg/L	4.34	46.55	6	40	46	8	-
100 mg/L	5.84	28.08	8	20	48	4	-

Table 5: The quality of water sample and result of raw Artocrpus Heterohyllus seed for Sodium Chloride (1M NaCl) as solvent in extraction process of coagulant.

Water sample	Turbidity (NTU)	Turbidity removal (%)	Suspended solid (mg/L)	Suspended solid removal (%)	Colour (PtCo)	Colour removal (%)	COD (mg/L)
Raw	13.21	-	17	-	104	-	18
0 mg/L	10.5	20.51	16	5.88	102	1.92	-
20 mg/L	8.88	32.78	12	29.41	89	14.42	-
40 mg/L	7.26	45.04	9	47.06	93	10.58	9
60 mg/L	10.3	22.03	14	17.65	96	7.69	-
80 mg/L	11.53	12.72	15	11.76	101	2.88	-
100 mg/L	12.3	6.89	16	5.88	102	1.92	-

4. Discussion

4.1. Colour

The water sample after treatment should be colourless. From the Table 1 until Table 5, Alum is the most efficient coagulant to remove the colour from water sample at 80mg/L of dosage coagulant while the raw Artocarpus Heterophyllus seed (1M NaCl) at 100mg/L is the less efficient to remove the water sample colour. The most efficient natural coagulant in reducing the colour of water sample is dried Artocarpus Heterophyllus seed (distilled water) at 20mg/L with the colour removal of 47.54% compared to other natural coagulant. The raw Artocarpus Heterophyllus seed (distilled water), dried Artocarpus Heterophyllus seed (1M NaCl) and raw Artocarpus Heterophyllus seed (1M NaCl) are efficient to remove the colour at 40mg/L, 20mg/L and 20mg/L with the colour removal of 37.34%, 18 and 14.42% respectively. Previous researh shows Moringa Oleifera seed removed 82% of colour by using 1 M KCl solution in the extraction of coagulant active component compared to water solution with 33% colour removal [14].

4.2. Suspended solid

Based on the results in tables provided, Alum, dried Artocarpus Heterophyllus seed (distilled water), dried Artocarpus Heterophyllus seed (1M NaCl) and raw Artocarpus Heterophyllus seed (1M NaCl) removed the suspended solid after the treatment at 80mg/L, 20mg/L, 20mg/L and 40mg/L. Raw Artocarpus Heterophyllus seed (distilled water) remove the most of suspended solid at 40mg/L and 60mg/L which is up to 70% removal. However, the percentage of reduction is slightly lower as compared to Moringa Oleifera seed after oil extraction (MOAE) with the removal of 95% suspended solids [16]

4.3. Turbidity

The result of turbidity removal is shown in Table 1 until Table 5 for five (5) sample of different coagulant that comprised of Alum, dried Artocarpus Heterophyllus seed (distilled water), raw Artocarpus Heterophyllus seed (distilled water), dried Artocarpus Heterophyllus seed (1M NaCl) and raw Artocarpus Heterophyllus seed (1M NaCl).

Water, organic compound, or salt solution can be used as different types of solvent for the seed extraction process but water is the most popular compared to the other because of its availability, good polarity and cheap in the cost. Extraction protein also widely used by using the salt solution which gives a great turbidity removal [17].

The optimum dosage for Alum, dried Artocarpus Heterophyllus seed (distilled water), raw Artocarpus Heterophyllus seed (distilled water), dried Artocarpus Heterophyllus seed (1M NaCl) and raw Artocarpus Heterophyllus seed (1M NaCl) are 80mg/L, 20mg/L, 40mg/L, 20mg/L and 40mg/L respectively.

When the coagulant reached the optimum dosage, the values of turbidity removal is slightly increase but still shows improvement compared to the raw water. The highest turbidity removal (50% removal) for natural coagulant is dried Artocarpus Heterophyllus seed (1M NaCl). However the result is not effective as rambutan seed at dosage of 50 mg/l gave a maximum turbidity removal of 92% for distilled water solvent while 99% turbidity removal for 1M NaCl solution. The use of salt solution to extract protein is also widely used. Previous study has also reported that better turbidity removal can be achieved by using salt solution to extract the coagulant agent (protein) from the seed [16]

4.4. COD

The reading of COD value is taken by using spectrophotometer before the water sample was treated by the coagulants. After finish each of the jar test, the water sample that has reached the optimum coagulant dosage for each test must be taken to test the COD. From the result in Table 1, Table 2, Table 3, Table 4 and Table 5 shows that Alum, dried Artocarpus Heterophyllus seed (distilled water), raw Artocarpus Heterophyllus seed (distilled water), dried Artocarpus Heterophyllus seed (1M NaCl) and raw Artocarpus Heterophyllus seed (1M NaCl) are very effective in reducing the COD value in water sample especially raw Artocarpus Heterophyllus seed (distilled water). Previous study shows that when cactus was used with Aluminum Chloride Hexahydrate (AlCl36H2O) synchronously to treat water sample, the removal efficiency of turbidity and COD were higher than that of cactus or AlCl36H2O was used solely [17].

5. Conclusion

Based on this study it can be concluded that raw Artocarpus Heterophyllus seed and dried Artocarpus Heterophyllus seeds have potential to be used as an alternatives natural coagulant for the water treatment purposes whether by using distilled water or Sodium Chloride (NaCl) as coagulant solution.

From Table 4, it shows that dried Artocarpus Heterophyllus seed (1M NaCl) at 20mg/L which removes turbidity up to 51.34% as the optimum dosage coagulant in this research is the most effective of natural coagulant compared to dried Artocarpus Heterophyllus seed (distilled water), raw Artocarpus Heterophyllus seed (distilled water) and raw Artocarpus Heterophyllus seed (1M NaCl) to remove turbidity. It reduced turbidity, suspended solid and colour and COD in water sample after the treatment process. It shows that salt Sodium Chloride (NaCl) can be used as addictive in treating the water. From the results, Artocarpus Heterophyllus seed can remove the turbidity in the low turbidity of water sample. As the results show, it can be said that the result of Alum is better than dried Artocarpus Heterophyllus seeds as it can remove large amount of colour, suspended solid and turbidity from water sample compared to Artocarpus Heterophyllus seed. Table 1 exhibits that Alum at optimum coagulant dosage 80mg/L is more effective in the water treatment compared to the natural coagulant but Alum is not suitable to be consumed by the human body while Artocarpus Heterophyllus seed as natural coagulant is more environmental friendly and more safe to be use by human.

This study investigated Artocarpus Heterophyllus seed and found its potency for water treatment comparable to Alum. The Artocarpus Heterophyllus seed demonstrated turbidity removal comparable to the conventional water treatment coagulant alum, making community water safe for use. Artocarpus Heterophyllus seed should be investigated further individually and in combination with other natural coagulants for community water treatment because of its being eco-friendly (Green) and having demonstrated great potentials as water treatment. Artocarpus Heterophyllus is easy to be finds in Malaysia at all season. By using the Artocarpus Heterophyllus seed as natural coagulant, it helps to reduce the waste generation as the seeds cannot be eat by people. Natural coagulants also prove that it gradually reducing the reliance on chemical coagulants and still can maintained the efficiency of water treatment.

6. References

[1] Grey, D., & Sadoff, C. W. 2007. Sink or Swim? Water security for growth and development. *Water Policy*. **9** 545–571
[2] A. Azlan, H.E. Khoo, M.A. Idris, I. A. and M. R. R. 2012 Consumption Patterns and Perception on Intake of Drinking Water in Klang Valley, Malaysia. *Pakistan Journal of Nutrition* **11** 584-590
[3] Choy, Y. S., Prasad, K., Wu, T. Y., & Raghunandan, M. (014. Utilization of Pant-based natural coagulants as future alternative towards sustainable water clarification. *Journal of Environmental Sciences* **26** 2179–2181.
[4] Arnoldsson, E., Bergman, M., Matsinhe, N., & Persson, K. M. 2008. Assessment of drinking water treatment using Moringa oleifera natural coagulant. *VATTEN* **64** 137–150.
[5] United States Geological Survey 2015. *Wastewater treatment: Water use. The USGS Water Science School* Retrieved from http://water.usgs.gov/edu/wuww.html
[6] Gebbie, P. (2006). An operator's guide to water treatment coagulants *Earth Tech Engineering 31st Annual Qld Water Industry Workshop – Operations Skills an operator's guide to water treatment* **14** pp 14–20.
[7] Satterfield, Z. 2005 *Jar Testing* Tech Brief (The Environmental Service Centre) **5** pp 1–4.
[8] Okuda, T., Baes, A. U., Nishijima, W., & Okada, M. 1999. Research Note Improvement of Extraction Method of Coagulation Active Components From Moringa Oleifera Seed. *Water Resources* **15** 3373–3378.
[9] Bina, B., Mehdinejad, M. H., Nikaeen, M., & Attar, H. M. 2009 Effectiveness of chitosan as natural coagulant aid in treating turbid waters *Iran. J. Environ. Health. Sci. Eng* **6** 247–252.
[10] Fatin Nabilah Murad, Md. Zahangir Alam, M. N. S., & Al-Mamun, A. A. 2017 Optimization of flocculation process by microbial coagulant for removal of turbidity in river water. *IIUM Engineering Journal* **18** 63–70.
[11] Perl, D. P. 1985 Relationship of aluminum to Alzheimer's disease. *Environmental Health* **63** 149– 153
[12] Resendiz-Vazquez, J. A., Ulloa, J. A., Urías-Silvas, J. E., Bautista-Rosales, P. U., Ramírez, J. C., Rosas-Ulloa, P., & González-Torres, L. 2017 Effect of high-intensity ultrasound on the technofunctional properties and structure of jackfruit (Artocarpus heterophyllus) seed protein isolate. *Ultrasonics Sonochemistry*, 37, 436–444.
[13] Madrona, G. S., & Serpelloni, G. B. 2010 Study of the effect of saline solution on the extraction of the moringa oleifera seed's active component for water treatment *Water Air and Soil Pollution* **211** 409–415.
[14] Bhatia, S., Othman, Z., & Ahmad, A. L. 2007 Pretreatment of palm oil mill effluent (POME) using Moringa oleifera seeds as natural coagulant *Journal of Hazardous Materials* **145** 120–126.
[15] Natumanya R., & Okot-okumu J. 2015 Evaluating coagulant activity of locally available Syzygium cumini , Artocarpus heterophyllus and Moringa oleifera for treatment of community drinking water *International Journal of Biological and Chemical Sciences* 2535–2554.
[16] Abidin, Z. Z., Mohamed, M. F., Ghani, A., & Abdullah, L. 2014 Preliminary Study of Rambutan (Nephelium Lappaceum) Seed as Potential Biocoagulant for Turbidity Removal *Advanced Materials Research* **917** 96–105
[17] Zhang, J., Zhang, F., Luo, Y., & Yang, H. 2006 A preliminary study on cactus as coagulant in watertreatment *Process Biochemistry* **41** 730–733

Acknowledgement
The authors would like to thank to Faculty of Civil Engineering, UiTM Shah Alam for the support and fund given for this study. This project would not be completed without the help and support from the management of faculties.

IOP Conf. Series: Materials Science and Engineering **601** (2019) 012011 doi:10.1088/1757-899X/601/1/012011

Home Water Purification System in Malaysia: Qualitative and Quantitative Study

Nur Shahirah Abd Rahim[1] and Norzila Othman[1*]

[1] Faculty of Civil and Environmental Engineering, Universiti Tun Hussein Onn Malaysia, 86400 Parit Raja, Batu Pahat, Johor, Malaysia
*Corresponding author:norzila@uthm.edu.my

Abstract. Concern over tap water quality has led to the expansion of water filtration system in Malaysia. This study is important to ensure that brands of water filters that have dominated the market in Malaysia do not exceed the quality of water released by safe water filters and no other pollutants leach from the machine. An exploratory study was undertaken to determine the pattern and selection factor of a home water filter system through a set of questionnaires. The survey results indicated that the most popular brands in Malaysia is Brand A from reverse osmosis filtration system. In the experimental study, five most frequently used brands of home water filters were evaluated and compared. Among the 27 parameters studied for each home water filter system, all parameter of physical, chemical and bacteriological quality is complied with the guideline values of the Ministry of Health Malaysia. Parameter of bacteriological quality shows absence bacteria parameter was detected in all samples tested. Meanwhile, physical quality parameter was given best performance removal from brand E and chemical water quality was efficient removal from brand C. In most cases user apply the home treatment of tap water to ensure that the quality of water physically look better and safe to be consumed.

1. Introduction

Water is the comprehensive dissolvable of our planet where both inorganic and natural issue used as a part of all procedures of living things is broken down or travels, dealing with the life of each and every living creature, including people [1]. Water question is unquestionably one of the primary factors that are related with the human improvement considering its effect on human lives.

The Malaysian government has determined that all water sources for drinking should comply with the National Drinking Water Quality Standards guidelines. Increasing awareness on the matter has caused many Malaysian consumers to have installed a water filtration system at home. Hence, the private sector has developed and brought water for household use filtration technology aimed at improving the quality of drinking water and reducing the threat of water borne diseases. This filter is used to hold the water impurities and produce clean water after filtering [2]. The effectiveness of the water filter system can be improved by attempting a mix of two or more treatment methods media for the complete and successful removal.

Various types of water filter systems are available in the Malaysian market using filtering media such as silica, sand, activated carbon, polypropylene, UF membrane, zeolite, fiber, fine woven fabrics [3]. Activated carbon filters is a material commonly used in water filtration systems as it is able to absorb various pollutants found in water such as pesticides, industrial chemicals, taste and odor [4].

Drinking water purposes by the used of solid adsorbents shows potential as one of the most efficient methods for the treatment and removal any contaminant in water [5].

Home water filtration system is needed for consumer to improve the public health and concern for water contamination [6].In most cases the filtration systems were tested at the origin place on the capability to purify the water. Therefore, extensive study should be conducted for ensuring the most preferable brand by the consumer resulted to produce high quality of water after through filtration process. The theories and technologies principle were studied behind a few popular types of home water filter system available in Malaysian market. This result was conducted to identify the dominant water filter type in Malaysia by conducting a laboratory experiments on water quality parameters include physical, chemical and bacteriological test before and after through the water filters system. Therefore, this study is important to ensure that brands of water filters that have dominated the market in Malaysia do not exceed the quality of water released by safe water filters and no other pollutants leach from the machine.

2. Methodology

2.1. Secondary data collection.

The set of question forms was provided in the form of google form as the study instrument. This questionnaire form was opened within a certain period within 6 months of collection period. 140 numbers of respondents throughout Malaysia who give feedback on the questionnaire. The data was analysed and the selected brand of water filter system were conducted in laboratory experiments to check the quality water filter. The form of questions contained in the set of questionnaires is as follows:

i. The location of residence
ii. Brand of home water filter machine used in home
iii. The type of water source used
iv. Why install the home water filter system
v. The period for using a water filter machine
vi. Are respondent satisfied with the quality of tap water in home
vii. Are respondent satisfied with the water filter quality after install home water filter system
viii. How many water filter machines are installed at home
ix. How long respondent use water filter machine
x. How often does a water filter machine be maintained
xi. Where respondent get information about water filter system products
xii. What is the primary factor that emphasized the consumer in the selection of water filters products

2.2. Analysis of water quality and filter media.

2.2.1. Sampling.

The water sample was collected in 1000 ml plastic bottles or glass bottles [7]. The water sample is taken directly from the pipe and after the water passed through the filter. All bottled sample was label with information such as type of samples, dates and times. All sampling bottles was acid washed prior to sampling. At the start of experimentation, water samples were collected after running the water for 5 minutes [8]. Sample bottles was fully fill with water samples to ensure the elimination of air bubbles and organic particulate matter entrapped in the bottles. The sample was stored in an icebox and transported back to the lab for analysis. Water samples was stored at 4°C before further analysis [7].

2.2.2. Method of test and analysis instruments.

Sampling, sample preservation and laboratory analysis was based on Standard Methods for the Examination of Water and Wastewater [7]. Table 1 shows all methods and equipment used in the testing of water quality.

Table 1. Equipments used in the experiment.

Equipment	Model	Analysis
pH meter	HACH Sension	pH measure
Turbidity meter	HACH 2100P	Turbidity measure
Atomic Absorption Spectrometer (AAS)	Perkin Elmer Analyst 800	Measure of magnesium and sodium
Inductively Coupled Mass Spectrometry (ICP-MS)	Perkin Elmer Elan 9000	Measure of aluminium, arsenic, cadmium, chromium, copper, iron, lead, manganese, selenium, silver and zinc
Ion Chromatography (IC)	Dionex Ics- 2000	Measure of chloride, fluoride, nitrate and sulphate
Mercury Analyzer	MA 3000 RD/SC	Measure of mercury
DR 6000	HACH UV-Vis Spectrophotometer	Measure of colour, ammonia, phenol, cyanide, hardness, residual chlorine and anionic surfactant
Chromocult agar	-	Total coliform and Escherichia Coli

3. Results and Discussions

3.1. Survey analysis: Statistical results

From the survey results, respondent's distribution covers the entire state in Malaysia. Majority of household consumers are those who live at Malacca. Follow by Johor and Selangor.

The most popular types of water filter system used by consumers is Brand A with 50 respondents. The second commonly used water filter is Brand B with 45 respondents, followed by Brand C, D, and E with 20, 8, 6 respondents respectively. Table 2 presents that the properties and mechanism in the filter media. According to Table 2, the common filter media was used in each types of water filter system are activated carbon filter, sediment filter as sand filter and natural filter for example zeolite filter.

There are several reasons that the consumers choose to install Brand A and B as the dominant type of water filter. The main reasons are might due to its cheaper price and easiest unit to install compare to other types of water filters. Other study also revealed that consumers are more affordable to install this kind of water filter [9]. The first popular brand A is applying reverse osmosis system to produce reverse osmosis water to consumers. Through the reverse osmosis filtration system, it will filter up to 99.99% of bacteria and also suspended matter in the water. The disadvantage from these systems is about this filter eliminates the good minerals and bacteria that the body needs [10]. Furthermore, the second main brand is Brand B from the nano positive technology filter media to produce alkaline drinking water. This system is a water-stripping technology that filters out almost all the hazardous matter and hazardous suspended matter in the water by leaving only the useful bacteria and minerals salts needed by the body [11].

The water sources used in the water filter system are derived from tap water sources of water treatment plant. From the data, only two respondents used water filter system sources from groundwater and the case are from Perak and Kelantan.

Table 2. The properties and mechanism in the filter media.

Name of water filters	Types of water filters	Types of filter media in the systems
A	Reverse Osmosis Filtration System	Stage 1: Plus-sediment filter Stage 2: Pre-carbon filter Stage 3: RO membrane filter Stage 4 and 5: Post carbon and fine filters Stage 6: Ceramic filter
B	Nano Positive Filter	Stage 1: Sediment filter Stage 2: Pre-carbon filter Stage 3 to 5: Natural filter Stage 6: Nano-positive filter
C	Reverse Osmosis Purification System	Stage 1: Pre-sediment filter Stage 2: Sediment filter Stage 3: Organic granular activated carbon pre-filter Stage 4: Reverse Osmosis membrane Stage 5: Organic granular activated carbon post-filter
D	Water Purification Innovative Technology	i. Ultraviolet technology ii. Carbon filter technology iii. In one cartridge with an ultraviolet (UV) lamp inside a carbon block filter
E	Healthy Ionized Water	Stage 1: PP Polypropylene Stage 2: UF Membrane Stage 3: NSF GAC Carbon Stage 4: Japan UNITIKA filter Stage 5: Far infrared Stage 6: Electrolysis, Japan Platinum Baked Titanium

There are several reasons that the consumers choose to install Brand A and B as the dominant type of water filter. The main reasons are might due to its cheaper price and easiest unit to install compare to other types of water filters. Other study also revealed that consumers are more affordable to install this kind of water filter [9]. The first popular brand A is applying reverse osmosis system to produce reverse osmosis water to consumers. Through the reverse osmosis filtration system, it will filter up to 99.99% of bacteria and also suspended matter in the water. The disadvantage from these systems is about this filter eliminates the good minerals and bacteria that the body needs [10]. Furthermore, the second main brand is Brand B from the nano positive technology filter media to produce alkaline drinking water. This system is a water-stripping technology that filters out almost all the hazardous matter and hazardous suspended matter in the water by leaving only the useful bacteria and minerals salts needed by the body [11].

The water sources used in the water filter system are derived from tap water sources of water treatment plant. From the data, only two respondents used water filter system sources from groundwater and the case are from Perak and Kelantan.

Around 93% reported that the reasons for purchasing water filter system to ensure health. Less respondents (6%) agree that news and complaint about water pollution followed by influenced by friends and relatives. From the data shows that the Malaysian people is concern about the water supply as drinking water in human lives. According to study from [12], also stated that installing water filter system seem to be health concerns, followed by perception of poor tap water quality and increasing contamination and pollution in the country.

States that 90% of the respondents are not satisfied with the supplied tap water and only 10% of the respondents are satisfied with the current supplied tap water. Quality of tap water is difference from one place to another place. Respondents indicated that yellowish color (74%), the second reasons is chlorine smell (14%) and followed by problem from suspended solid (5%) and skin problem (1%). Same data studies from [12], stated that respondents said that color, odor, and taste were the main problems with their tap water quality. Other than that, unsatisfied tap water might also due to the distribution system at home rather than efficiency of treatment process at the water treatment plants.

The most respondents (98%) reported that satisfied with the quality of drinking water after installing a water filter system. Only 2% of the respondents not satisfied. The reasons some consumers not satisfied with the water filter system have been install at home may due to several reasons such as the filtration system is not working well. Most of respondents are satisfied with water from filtration system is because the expected function of each water filter system by removal of odor, inorganic waste, chlorine, microbes, organic waste, pesticides and an added of mineral in drinking water [12].

The results 118 (84%) respondents of the total users have installed one unit of water filter, 22 (16%) respondents have installed two water filters and no respondents installed more than two units of water filter at home. From the study by [13], there are two types of water treatment systems that are currently available include Point-of-Use and the Point-of-Entry water filter. From this study, the result shows that most of water quality parameters at the house with POE and POU water filters shows a significant reduction as compared to other houses that has either a POU or POE water filter only [13]. This can conclude that the water quality parameters become more perform when two water filter is installing at home.

The most respondents used water filters within one to five years (51%). Other than that, the use of water filter system for over five years and less than a year is 44% and 5% respectively.

The results show that, periodic maintenance every two to five months (43%) and six to twelve months (39%) is the dominant maintenance. While compared to the respondents who stated periodic maintenance period per month was 16% and 1% who had never maintained a water filter machine since its inception. The maintenance for water filter system is important to produce the good water quality. Previous study by [14], stated that the disease such as water borne come from unclean of drinking water.

In general, the dissemination of the primary information sources in the selection of water filters system is through friends or neighbour (31%), followed by information through salesman product (24%) and their own search on internet about the brand of water filter (18%). This is because they might be realized by friends and neighbour about the presence of the water filter machine is from the internet and media social. Based on the previous study, it has been stated that media social is the vital sources for supply the data based on the case of environmental [12].

The selection factors of water filter system have shown that most respondents agree that the price (23%) offered by a water filter system is a major factor in the selection of the product. The services offered (20%), while the design and diversification of water filter system function (17%) are the second and third factors that are the respondent's choice. From the results, the price is the most demand part to select the water filter machine.

3.2. The analysis of water quality parameter.

3.2.1. Bacteriological quality: Escherichia coli and coliform.

The experimental result of microbiological test shows that none of the water filter brand samples had detectable *Escherichia coli* in the water filter before and after filtering. Thus, each of the water tested met the MOH standard safe drinking water criteria for absence of bacteria [14]. The results is in agreement with [14], no *Escherichia coli* were detected at post-treatment in any of 320 samples of drinking water sources collected in developing countries. According to [13], was also found out that the house installed with both Point-of-Use (POU) and Point-of-Entry (POE) filters tend to provide the greatest reduction in *Escherichia coli* and total coliform. From the study by [15], it is shown that the removal of *Escherichia coli* and total coliform was found significant, which was supported by the reduction in diarrhoea cases observed by ceramic silver-impregnated pot filter users in a recent field study. The sources of coliform bacteria are due to re-growth in distribution systems, breakdown of the treatment barriers, due to corrosion control and also the availability of organic and inorganic nutrient [16].

3.2.2. Physical quality: Color and turbidity

Table 3. Summary results of physical water quality.

Parameter of chemical quality	Best removal of brand	Percentage removal	p-value for each brand		References
			Before filtration	After filtration	
Color	E	73.4%	$p > 0.01$	$p < 0.10$	[17,18]
Turbidity	E	94%	$p < 0.01$	$p < 0.01$	

Note: $p < 0.01$-highly significant, $p < 0.05$-significant, $p > 0.10$-not significant [19].

Based on the results in this study and the previous study can be proved that the natural filter and membranes filter give a best performances percentage removal of physical contaminant in drinking water.

3.2.3. Chemical quality

Table 4. Summary results of chemical water quality.

Parameter of chemical quality	Best removal of brand	Percentage removal	p-value for each brand		References
			Before filtration	After filtration	
pH	C	The treatments of filtration can increase water to be more neutral after filter process	$p > 0.10$	$p < 0.05$	[20]
Hardness	E	67%	$p < 0.10$	$p < 0.10$	[21]
Ammonia	A and C	100%	$p > 0.10$	Same data	[22,23]
Manganese	B	60%	$p < 0.01$	$p < 0.01$	[22]
Cyanide	D	67%	$p < 0.01$	$p < 0.01$	[24]
Residual Chlorine	A and C	96% and 90%	$p < 0.10$	$p < 0.01$	[12, 25]
Nitrate	C	94%	$p > 0.10$	$p > 0.10$	[9]
Sulphate	C and A	99% and 98%	$p > 0.10$	$p > 0.10$	[20]
Aluminium	C and B	99% and 95%	$p < 0.05$	$p < 0.01$	[26]
Arsenic	C	93%	$p > 0.10$	$p > 0.10$	[14]
Cadmium	C and E	100% and 98%	$p < 0.05$	$p > 0.10$	[27]
Chloride	C	94%	$p > 0.10$	$p > 0.10$	-
Chromium	C	52%	$p < 0.05$	$p < 0.01$	[26]
Copper	A	97%	$p > 0.10$	$p < 0.01$	[28]

Table 4. Summary results of chemical water quality (cont.)

Parameter of chemical quality	Best removal of brand	Percentage removal	p-value for each brand		References
			Before filtration	After filtration	
Magnesium	C and A	95% and 93%	$p < 0.01$	$p < 0.01$	[13]
Fluoride	C	84%	$p > 0.10$	$p > 0.10$	[6,8]
Iron	C and A	100% and 91%	$p < 0.05$	$p > 0.10$	[13,29]
Lead	A, C and E	100%, 100% and 97%	$p < 0.05$	$p > 0.10$	[25]
Mercury	B and C	100%	$p > 0.10$	$p > 0.10$	-
Selenium	C and A	99% and 98%	$p > 0.10$	$p > 0.10$	-
Silver	A, C and D	98%, 94% and 90%	$p < 0.01$	$p > 0.10$	-
Sodium	C	85%	$p < 0.01$	$p < 0.01$	-
Zinc	A and C	98% and 94%	$p > 0.10$	$p < 0.01$	[13]
Phenol	D	25%	$p > 0.10$	$p > 0.10$	[24]

Note: $p < 0.01$-highly significant, $p < 0.05$-significant, $p > 0.10$-not significant [19].

4. Conclusion

It is hoped that the output of the research will produce significance finding to be adopted by Ministry of Health Malaysia in ensuring home filter system will be closely monitored by the government for safety and health. The main factor in the selection of water filter systems is the cost offered by the supplier and main data about the water filter system is acquired through influence by advice of friends and neighbors respond. Ensure the drinking water quality through filtration process can be ensured, assessed and approved by the Ministry of Health and other authorities if there are guideline that must be followed by all manufacturer of home water filtration systems. Results of experimental study for water quality parameters testing before and after passing through a water filter system showed most of the parameters meets drinking water quality by MOH except for some parameters do not show a better removal. Among five brands that were investigated in this study, brand C is the best performance removal of reverse osmosis membrane purification system with five stage filter media in the systems. The quality and effectiveness of each home water filter system can be identified from the usage pattern by the filter media in each filter system.

5. References

[1] Ribeiro M A S, Murgu M, Silva V M, Sawaya A C H F, Ribeiro L F, Justi A and Meurer E C 2017 The screening of organic matter in mineral and tap water by UHPLC-HRMS *Talanta*, **174**(April) 581–586

[2] Jenkins M W, Tiwari S K and Darby J 2011 Bacterial, viral and turbidity removal by intermittent slow sand filtration for household use in developing countries: Experimental investigation and modeling *Water Res.* **45**(18) 6227–6239

[3] Ab Razak N H, Praveena S M, Aris A Z, and Hashim Z 2015 Drinking water studies: A review on heavy metal, application of biomarker and health risk assessment (a special focus in Malaysia), *J. Epidemiol. Glob. Health* **5**(4) 297–310

[4] Moyo S, Wright J, Ndamba J, and Gundry S 2004 Realising the maximum health benefits from water quality improvements in the home: A case from Zaka district, Zimbabwe *Phys. Chem. Earth* **29**(15–18) SPEC.ISS. 1295–1299

[5] Haarhoff N A J, Geldenhuys S J and Sibiya M 2008 *Assessments and Improvement of Filter Media Cleanliness in Rapid Gravity Sand Filters* **1525** 1-53

[6] Eftekhar B, Skini M, Shamohammadi M, Ghaffaripour J, Nilchian F and Author C 2015 The Effectiveness of Home Water Purification Systems on the Amount of Fluoride in Drinking Water, *J Dent Shiraz Univ Med Sci. J Dent Shiraz Univ Med Sci* 16(163) 278–281

[7] American Public Health Association 2012 Standard Methods for the Examination of Water and Wastewater pp 1–19

[8] Loh K H, Habibah Y, Noridah A, Sabarina O, and Mahrusah J 2010 A Study of the Effect of Home Water Filtration Systems on Fluoride Content of Drinking Water in Johor, *A Compend. Abstr. Res. Proj. Publ. Oral Heal. Pers. Minist. Heal.* **33**(2) 8-13

[9] Shaharudin N, Suradi N, Amani N and Mohd F 2017 Measurement of Water Quality Parameters for Before and After Maintenance Service in Water Filter System, vol. 06006 pp 1-7

[10] Greenlee L F, Lawler D F, Freeman B D, Marrot B and Moulin P 2009 Reverse osmosis desalination: Water sources, technology, and today's challenges, *Water Res.*43(9)2317–2348

[11] Mpenyana-Monyatsi L, Mthombeni N H, Onyango M S and Momba M N B 2012 Cost-effective filter materials coated with silver nanoparticles for the removal of pathogenic bacteria in groundwater," *Int. J. Environ. Res. Public Health* **9**(1) 244–271

[12] Aini M S, Fakhrul-Razi A, Mumtazah O and Chen J C M 2007 Malaysian households' drinking water practices: A case study, *Int. J. Sustain. Dev. World Ecol.* **14**(5) 503–510

[13] Annuar N A, Yaziz M I, Latif P A and Saad Z M 2010 A Study of the Performance of Selected Pous and Poes Water Filters on the Maintenance of Potable Water Quality, *Proc. Postgrad. Qolloquium 2009/2010*, pp 222–228

[14] Souter P F Souter P F, Cruickshank G D, Tankerville M Z, Keswick B H, Ellis B D, Langworthy D E, Metz K A, Appleby M R, Hamilton N, Jones A L and Perry J D. 2003 Evaluation of a new water treatment for point-of-use household applications to remove microorganisms and arsenic from drinking water, *J. Water Health* **1**(2) 73–84

[15] Halem D, Laan H, Heijman S G J, Dijk J C and Amy G L 2009 Assessing the sustainability of the silver-impregnated ceramic pot filter for low-cost household drinking water treatment *Phys. Chem. Earth* **34**(1–2) 36–42

[16] Chu C, Lu C, and Lee C 2005 Effects of inorganic nutrients on the regrowth of heterotrophic bacteria in drinking water distribution systems *J. Environ. Manage.* **74**(3) 255–263

[17] Budari N M, Ali M F, Kassim J and Omar S M 2013 Burnt oil palm shell filter media for bacterial removing by single-and dual-media filtration, *BEIAC 2013 - 2013 IEEE Bus. Eng. Ind. Appl. Colloq.* pp 465–470

[18] Kasim N, Mohammad A W and Abdullah S R S 2016 Characterization of hydrophilic nanofiltration and ultrafiltration membranes for groundwater treatment as potable water resources, *Desalination and Water Treatment* **57**(17) 7711–7720

[19] Senn S N G, Rothman K J, Carlin J B, Poole C and Altman D G 2016 Statistical tests, P values, confidence intervals, and power: a guide to misinterpretations, *Eur. J. Epidemiol.* **31**(4) 337–350

[20] Musa S, Ariff N A, Kadir M N A and Denan F 2016 Simplified Method for Groundwater Treatment Using Dilution and Ceramic Filter, *IOP Conf. Ser. Mater. Sci. Eng.* **136**(1) 1-9

[21] Su F, Luo M, Zhang F, Li P, Lou K and Xing X 2009 Performance of microbiological control by a point-of-use filter system for drinking water purification, *J. Environ. Sci.* **21**(9) 1237–1246

[22] Abu Hasan H, Sheikh Abdullah S R, Kamarudin S K and Tan Kofli N 2013 On-off control of aeration time in the simultaneous removal of ammonia and manganese using a biological aerated filter system, *Process Saf. Environ. Prot.* **91**(5) 415–422

[23] Roccaro P, Barone C, Mancini G and Vagliasindi F G A 2007 Removal of manganese from water supplies intended for human consumption: A case study, *Desalination* **210**(1–3) 205–214

[24] Reni Desmiarti A H 2016 Removal of Phenol from Water by Advance Oxidation Process Using Plasma System, vol. 11, no. 23, pp 14004–14007

[25] Smith E and El Komos S 2009 Tap water quality and performance of point-of-use treatment devices in Cairo, Egypt," *Water Environ. J.* **23**(2) 119–127
[26] Law B B 2005 *The Usage of Domestic Water Filtration Systems in Malaysia* Bachelor Civil Engineering Thesis (University of Southern Queensland Faculty of Engineering and Surveying) pp 1-116
[27] Ong C, Ibrahim S and Gupta B S 2007 A survey of tap water quality in Kuala Lumpur, *Urban Water J.* **4**(1) 29–41
[28] Moreno-Piraján J C, Garcia-Cuello V S and Giraldo L 2011 The removal and kinetic study of Mn, Fe, Ni and Cu ions from wastewater onto activated carbon from coconut shells, *Adsorption* **17**(3) 505–514
[29] Mohanad E 2010 Design of a Portable Dual Purposes Water Filter System, *J. Eng. Sci. Technol. Sch. Eng. Taylor's Univ. Coll.* **5**(2) 165–175

Acknowledgement
Authors would like to thank for the provision of funds that supplied and the cooperation provided by the Food Safety and Quality Division, Ministry of Health Malaysia to carry out this study. Besides that, this paper was partly sponsored by the Centre for Graduate Studies UTHM, Post Graduate Research Grants (GPPS) and Fundamental Research Grant Scheme (FRGS 1575) supported by Ministry of Higher Education (MOHE).

Kinetics Modelling of Batch Anaerobic Co-digestion of Domestic Primary Sewage Sludge and Food Waste in a Stirred Reactor

Siti Mariam Sulaiman[1] and Roslinda Seswoya[1]*

[1]Micro Pollutant Research Centre, Department of Water and Environmental, Faculty of Civil and Environmental Engineering, Universiti Tun Hussein Onn Malaysia, 86400 Parit Raja, Batu Pahat, Johor, Malaysia
*Corresponding author: roslinda@uthm.edu.my

Abstract. Anaerobic digestion was conventionally applied for treating sewage sludge. However, the accumulation of solid waste particularly food waste has reach the critical levels worldwide. In practice, the food waste was dumped into the landfill for ultimate disposal. However, the greenhouse gases produced in the landfill makes this is no longer a preferable option. Anaerobic digestion was seen as an alternative for managing the food waste in a sustainable way. Methane, a renewable energy is potentially in replacing fossil fuel. However, the methane yield from the digestion of food waste inefficient. Therefore, a study of the co-digestion of sewage sludge and food waste was conducted to investigate the improvement of the methane yield. This study was conducted by using a mixture of domestic primary sewage sludge and food waste as a co-substrate for the anaerobic digester. The kinetics modified Gompertz modelling was applied to describe the anaerobic digestion process. A series of batch biochemical methane potential (BMP) assay was prepared using Automatic Methane Potential Test System (AMPTS II) to investigate the anaerobic digestibility of the mixture of domestic primary sewage sludge and food waste. The BMP assay showed that the co-digestion improved the ultimate methane yield by 32.6% higher than domestic primary sewage sludge alone, indicated that the co-substrate characteristics influencing the methane yield. Besides that, the greater VS/TS ratio of the substrate also resulted in the greater methane yield. The kinetics parameter from the modelling analysis were slightly lower as compared to the laboratory data.

1. Introduction

Rapid development has leads to a significant environmental problem with the increasing of waste [1]. In Malaysia, Solid Waste Corporation of Malaysia (SWCorp) established under the Solid Waste Management and Public Cleansing Corporation Act 2007 (Act 673) to ensure solid waste and public management become more effective. Malaysia authority face great challenges in handling waste and treatment due to lack of expertise and fund [2]. Generally, organic waste is the highest amount of waste generated, with 50% by weight dominated by food waste (FW). Numerous of waste including food waste and sewage sludge were burned or disposed into landfill [3]. Once disposed into landfill, the waste will naturally biodegrade and release environmental harmful element such as leachate and gases [4].

Anaerobic digestion mainly applied for sewage sludge stabilization. Nevertheless, this treatment also actively developed for other waste such as FW [5]. According to [6] FW is suitable used as substrate for anaerobic digestion since its characteristics of high organic content and easily biodegradable. However, anaerobic digestion of FW as sole substrate is inefficient [7]. Anaerobic digestion of FW conducted by

[8] obtained ultimate methane yield of 201.5±11 mL CH_4/g VS. As alternative, co-digestion of sewage sludge and FW was conducted. Previous study conducted by [9] shows a significant improvement of methane yield of 393 mL CH_4/g VS with addition of FW as compared to sewage sludge alone (310 mL CH_4/g VS). On other study, co-digestion of sewage sludge and organic fraction municipal solid waste (OFMSW) consisted of restaurant and canteen waste obtained higher ultimate methane yield of 365.49 mL CH_4/g VS compared to sewage sludge as mono-digestion [10]. This difference possible due to the different mixture of co-substrate used and inoculum to substrate (I/S) ratio. Two mixing ratio of sewage sludge and food waste was used to investigate the effect of co-substrate mixture of sewage sludge and food waste on the methane yield conducted by [11]. The observation of 4:1 mixing ratio of sewage sludge and FW produced higher methane yield 355 mL CH_4 g/ VS compared to 2.4:1 (239 mL CH_4 g/ VS). Other study with similar mixing ratio of 4:1 of sewage sludge and FW conducted by [12] produced methane yield of 215 mL CH_4 g/ VS which almost 85.3% higher than mono-digestion.

Biochemical methane potential test (BMP) could be conducted in continuous, semi-continuous or in batch study to evaluate the methane production from anaerobic digestion [8], [13]. Recently, studies on batch BMP assay using Automatic Methane Potential Test System (AMPTS II) was used as its precise predictability [5]. As the BMP assay took long duration, the kinetics modelling applied to describe the anaerobic digestion process. Modified Gompertz applied to verified the validity of model by comparing the laboratory and predicting data [8]. Batch BMP assay on co-digestion of sewage sludge and OFMSW (consist of food waste) conducted by [8] showed that the modified Gompertz estimated the ultimate methane yield of 211.76 mL CH_4/g VS which is well fitted to the methane yield from the laboratory assay.

Due to the fact of less study of co-digestion of domestic sewage sludge and FW in Malaysia, this study was carried out to investigate the anaerobic digestibility of this mixture and the methane production kinetics.

2. Material and methods

2.1. Substrate and inoculum

This study used primary domestic sewage sludge (PSS) for mono-digestion. Meanwhile, 4:1 mixing ratio with wet mass ratio of PSS and FW was prepared adapting the ratio describe by [11]. PSS was collected from sewage treatment plant in Universiti Tun Hussein Onn Malaysia (UTHM). While the FW consist of leftover consists of rice, vegetables and meats collected from UTHM cafeteria. The FW was manually screened for non-biodegradable contaminants before roughly chopped and blended around two to three minutes using household blender into a thick consistency. Then, the substrate was frozen until use in the digestion experiment [14]. Inoculum is the anaerobically digested sludge was collected from existing full-scale anaerobic digester treating palm oil mill effluent (POME). The samples and inoculum was stored at 4°C until the anaerobic digestion assay [15].

2.2. BMP assay

AMPTS II was used for the batch BMP assay. Duran bottles of 500mL (reactor) were used and filled with substrate and inoculum [5] and illustrated in Figure 1. The BMP assay conducted with inoculum to substrate (I/S) ratio of 2:1 based on VS basis [9]. Duplicate blank and triplicate sample reactor were prepared each for mono-digestion (PSS) and co-digestion (PSS and FW). The pure nitrogen gas (N_2) was flushed for 2 minutes to create anaerobic environment in the headspace [9]. To avoid early reaction, the reactor was prepared one after another. The reactor was incubated in water bath at mesophilic temperature (37°C) and mixing speed at 100rpm [6]. The mass of inoculum and substrate filled into the reactor were calculated based on the manual of AMPTS II [16].

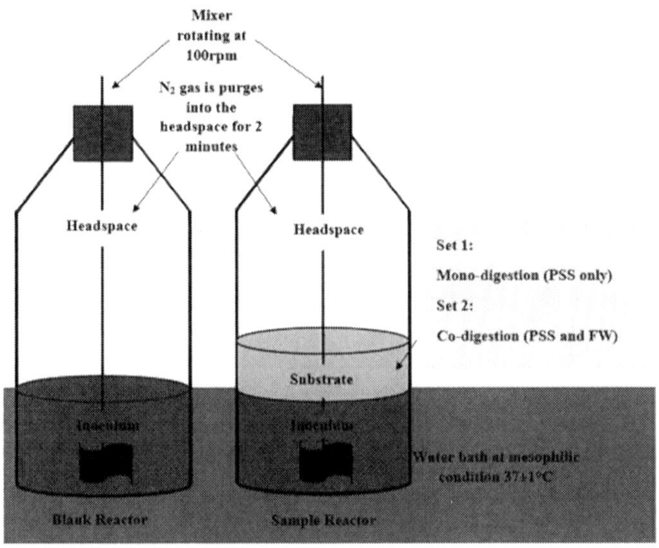

Figure 1. The reactor used in this study

2.3. Analytical method

Table 1 summarized the analytical method for determining the characteristics of PSS and mixture of PSS and FW.

Table 1. Analytical method used in this study

Parameter	Analytical method	Reference
COD (mg/L)	HACH 8000 method	[17]
TS (mg/L)	APHA Standard method (2540B)	[18]
VS (mg/L)	APHA Standard method (2540E)	[18]
Protein (mg/L)	Lowry method	[19]
Carbohydrate (mg/L)	Dubois method	[10]
Alkalinity (g/L as $CaCO_3$)	APHA Standard method: method 2320B	[10]

2.4. Kinetic modelling

Modified Gompertz modelling is a common equation used to describe the kinetic methane production for co-digestion of sewage sludge and FW [5], [8]. Modified Gompertz are well described as Equation 1.

$$M = M_0.\exp\left\{-\exp\left[\frac{R_m.e}{M_0}(\lambda - t) + 1\right]\right\} \quad (1)$$

Where y(t) is the specific methane yield (mL/g Vs) at anaerobic digestion time t (d), f_d is the maximum methane potential (mL/g VS), R_m is the maximum methane production rate (mL/g VS day), e is Euler's constant (2.7183), and λ is the lag phase time (days).

3. Result and discussion

3.1. Substrate characteristics

Table 2 summarized the characteristics of substrates in this study. Due to the mixing of PSS and FW, the TS, VS, COD, protein and carbohydrate were increased. This is possible to the additional of organic from the FW. From observation, the VS/TS ratio increase to 0.92 from 0.55 with additional of FW. [20] also observed VS/TS ratio of 0.82 from the mixture of sewage sludge and FW. High VS/TS indicated high organic transformation potential of substrate in anaerobic digestion process [10]. This study also showed increasing of COD. The similar result of COD also observed by [8]. Improvement of COD for co-digestion indicates adequate organic substrate to produce methane [21]. Additional of FW into the mixture relatively increase of protein and carbohydrates. From observation, the mixture was dominated by carbohydrate similar to mixture of FW and sewage sludge observed by [10].

Table 2: Characteristics analysed in this study (N=3)

Parameter	PSS	PSS+FW
Total solid, TS (mg/l)	7.33	43.33
Volatile solid, VS (mg/l)	4.00	39.67
VS/TS	0.55	0.92
COD (mg/l)	111.50±0.63	2246.50±0.91
Protein (mg/l)	72.75±6.02	711.43±11.93
Carbohydrate (mg/l)	79.83±8.57	1561.10±60.35

3.2. Methane accumulation

Figure 2 and Figure 3 showed the methane accumulation from the BMP assay for mono-digestion and co-digestion which ends on 25th days and 12th days respectively. The blank data was the average of duplicate blank reactor while sample reactor was the average of triplicate reactors. The net methane accumulation was obtained after subtracting the blank from each substrate. From the BMP assay, the net accumulated methane was 620.3mL and 183.5mL for co-digestion and mono-digestion respectively. As expected, higher net methane accumulation of co-digestion was observed compared to mono-digestion. The other study of co-digestion of sewage sludge and FW, the maximum methane production was observed at 8th day of BMP assay [22]. [23] focused on co-digestion study of FW and waste activated sludge, observed that the biogas production range from 170 to 443mL/g TS.

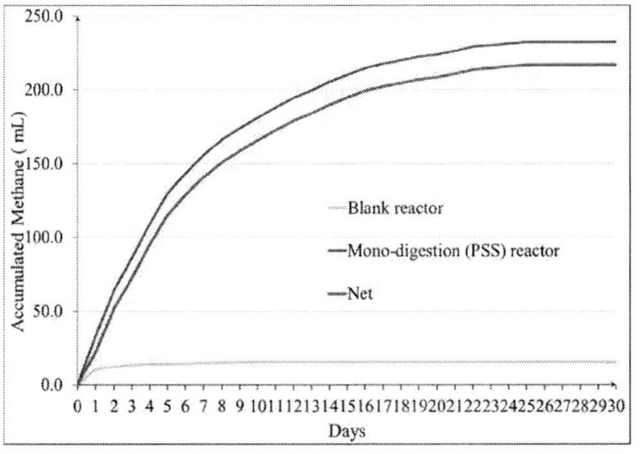

Figure 2: The accumulated methane for mono-digestion

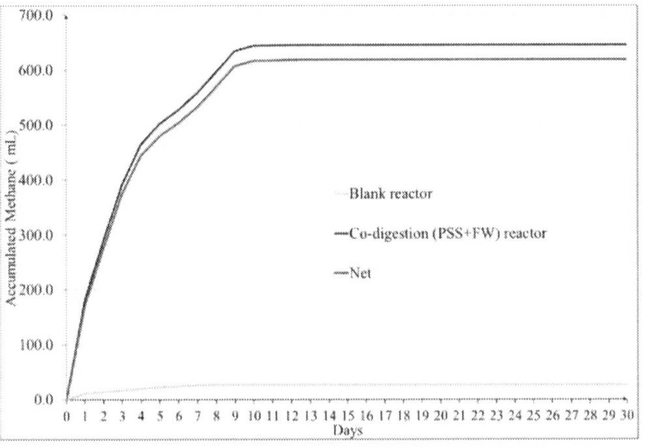

Figure 3. The accumulated methane for co-digestion

3.3. Methane potential

The ultimate methane yields were presented in Figure 4. Mono-digestion yields at 833.21 mL CH_4/g VS and the ultimate methane yield of the co-digestion was 1236.43 mL CH_4/g VS. The ultimate methane yield obtained slightly higher compared to previous studies. [10] observed ultimate methane yield of 365.49 mL CH_4/g VS of co-digestion of OFMSW consisting canteen and restaurant waste with sewage sludge. Similar co-digestion conducted by [8] observed the ultimate methane yield of 200.2 mL CH_4/g VS which higher than was observed from mono-digestion of secondary sludge (164.5 mL CH_4/g VS). This different most probably due to the fact of difference in the organic content, chemical composition or inoculum to substrate (I/S) ratio between each of BMP assay conducted [21].

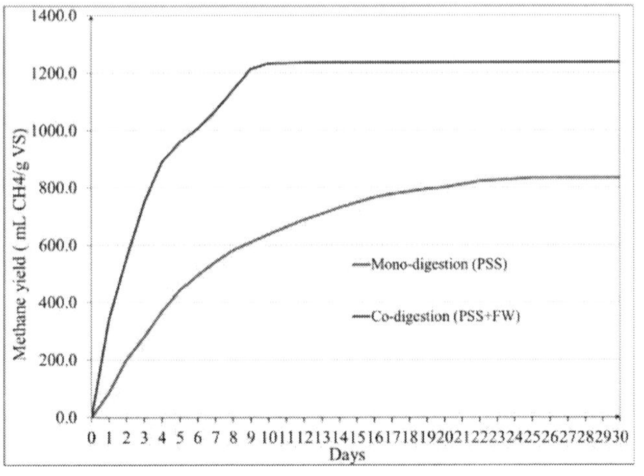

Figure 4. Methane yield for mono-digestion and co-digestion

According to [24], methane production rate can be determined using the steepest slope from the methane yield. In this study, the methane production rate of co-digestion is higher as compared to mono-digestion, at 342.27 mL CH_4 VS/day and 114.87 mL CH_4 VS/day respectively. Methane production rate was important as it is used to reflects the biodegradability and amount of degradable matter in assay [11]. [25] concluded that higher methane production rate could be achieved with higher carbohydrate content. In this study, the co-digestion mixture have high carbohydrate concentration (Table 2). Other study on co-digestion of FW and sewage sludge, [22] observed almost 37% of degradation rate and with minimum of 18% increase in methane yield production rate.

3.4. Kinetic modelling

Methane production for all parameter were modelled based on Equation 1. The kinetic constants of M_o, R_m, and λ of the model were determined using the least squares fitting method (non-linear regression approach) with the aid of solver function in MS Excel ToolPak [23]. Table 4 summarized the results of fitting the kinetic modelling to the BMP assay. By plotting laboratory and analysis of modified Gompertz data, Figure 5 and Figure 6 were prepared. Each figure was showing the methane yield curve from mono-digestion and co-digestion respectively.

Table 3 showed the ultimate methane yield, maximum methane production rate and lag phase from both laboratory and kinetics data. The modelling analysis showing slightly lower kinetics as compared to the laboratory observation. Study of mono-digestion and co-digestion by [8] observed the similar trends of lower kinetic modelling compared to laboratory data in which the ultimate methane yield observed from the laboratory data were twofold from the Gompertz modelling for both mono-digestion and co-digestion.

Lag phase of 0.04 day of mono-digestion and co-digestion was observed from the assay. However, the lag phase was neglected as it is low. The kinetic modified Gompertz observed no lag phase. Budiyono et al. [26] stated that short lag phase indicated no acclimation period and short time to produce methane. If the substrate used is not lack of nutrient, the degradation activity and methane production will generated immediately. Seswoya et al. [5] also observed the lag phase in the assay but no lag phase from analysis of kinetic modelling. The lag phase could be minimize by co-digestion with higher carbohydrates and lower protein [27].

Table 3. Summary of laboratory data with kinetic modelling

Parameter	Mono-digestion		Co-digestion	
	(L)	(MG)	(L)	(MG)
Ultimate methane yield, f_d (mL CH_4/g VS)	833.21	795.04	1236.43	1218.48
Maximum methane production rate, Rm (mL CH_4/g VS d)	114.87	88.12	342.27	267.35
Lag phase, λ (d)	0.04	0.00	0.04	0.00

*L- Laboratory data, MG- Modified Gompertz

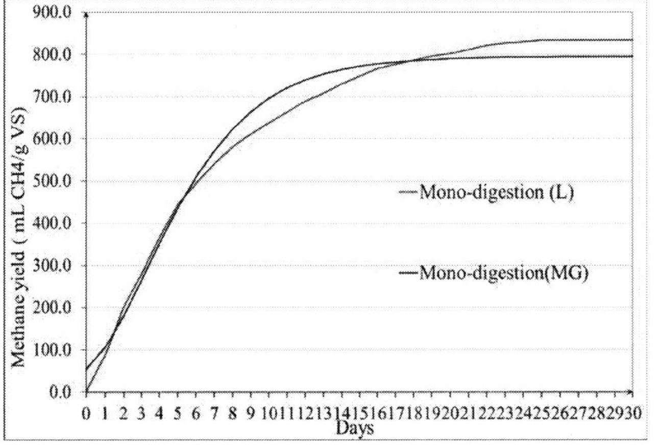

Figure 5. The laboratory data and modified Gompertz for mono-digestion

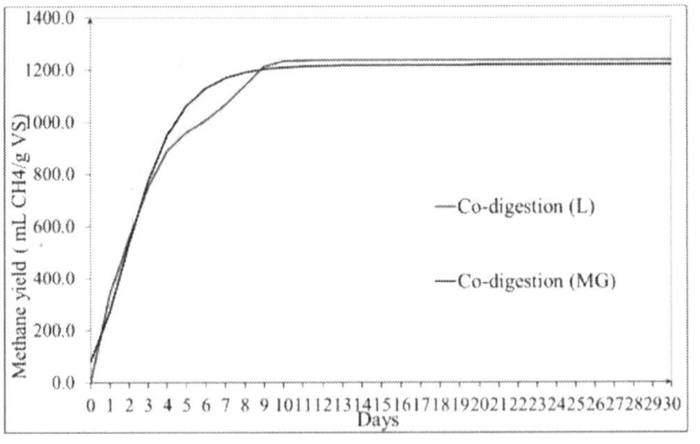

Figure 6. The laboratory data and modified Gompertz for co-digestion

4. Conclusion

Additional of FW as co-substrate not only enhance to the improvement of ultimate methane yield but also improve the characteristics of the co-substrate. With FW in anaerobic co-digestion, significant difference of ultimate methane yield was observed. This study showed the ultimate methane yield increased 32.6% higher than PSS alone. Although the kinetic Gompertz modelling used was slightly lower than laboratory data, about 90% of ultimate methane yield estimated within seven days of the BMP assay. Also, no lag phase was observed in the kinetic modelling. In future, various series of mixture of co-substrate and I/S ratio will be applied to estimate the methane yield. Besides, more kinetics modelling will be used to describe the anaerobic co-digestion of PSS and FW.

5. References

[1] Braguglia C M, Gallipoli A, Gianico A and Pagliaccia P 2018 Anaerobic bioconversion of food waste into energy: A critical review *Bioresour. Technol.* **248** 37–56

[2] Johari A, Ahmed S I, Hashim H, Alkali H and Ramli M 2012 Economic and environmental benefits of landfill gas from municipal solid waste in Malaysia *Renew. Sustain. Energy Rev.* **16** 2907–2912

[3] Grosser A, Neczaj E, Singh B R, R. Almas A, Brattebo H and Kacprzak M 2017 Anaerobic digestion of sewage sludge with grease trap sludge and municipal solid waste as co-substrates *Environ. Res.* **155** 249–260

[4] Fisgativa H, Tremier A and Dabert P 2016 Characterizing the variability of food waste quality: A need for efficient valorisation through anaerobic digestion *Waste Manag.* **50** 264–274

[5] Seswoya R A, Abdul Karim A T, Darnak N A and Abdul Rahman M F 2018 Methane Potential from the Digestion of Food Waste in a Batch Reactor *Int. J. of Eng and Techol.* **7** 36–39

[6] Ma Y, Yin Y and Liu Y 2017 New insights into co-digestion of activated sludge and food waste : Biogas versus biofertilizer *Bioresour. Technol.* **241** 448–453

[7] Liao X, Zhu S, Zhong D, Zhu J and Liao L 2014 Anaerobic co-digestion of food waste and landfill leachate in single-phase batch reactors *Waste Manag.* **34** 2278–2284

[8] Nielfa A and Cano R 2015 Theoretical methane production generated by the co-digestion of organic fraction municipal solid waste and biological sludge *Biotechnol. Reports.* **5** 14–21

[9] Koch K, Helmreich B and Drewes J E 2015 Co-digestion of food waste in municipal wastewater treatment plants : Effect of different mixtures on methane yield and hydrolysis rate constant *Appl. Energy.* **137** 250–255

[10] Cabbai V, Ballico M, Aneggi E and Goi D 2013 BMP tests of source selected OFMSW to evaluate anaerobic codigestion with sewage sludge *Waste Manag.* **33** 1626–1632

[11] Dai X, Duan N, Dong B and Dai L 2013 High-solids anaerobic co-digestion of sewage sludge and food waste in comparison with mono digestions: Stability and performance *Waste Manag.* **33** 308–316
[12] Wong J W C, Kaur G, Mehariya S, Parthiba O and Chen G 2018 Food waste treatment by anaerobic co-digestion with saline sludge and its implications for energy recovery in Hong Kong *Bioresour. Technol.* **268** 824-828
[13] Zahedi S, Rivero M, Solera R and Perez M 2018 Mesophilic anaerobic co-digestion of sewage sludge with glycerine : Effect of solids retention time *Fuel.* **215** 285–289
[14] Browne J D, Allen E and Murphy J D 2014 Assessing the variability in biomethane production from the organic fraction of municipal solid waste in batch and continuous operation *Appl. Energy.* **128** 307–314
[15] Seswoya R and Abdul Karim A T 2017 Sample Biochemical Methane Potential from the Digestion of Domestic Mixed Sewage Sludge in Batch Tests *Int. J. Adv. Sci. Eng. Inf. Technol.* **7** 431-437
[16] Bioprocess Control 2016 *AMPTS II & AMPTS II Light Automatic Methane Potential Test System Operation and Maintenance Manual* (Sweden: Bioprocess Control AB) pp 55-59
[17] Ergu T H, Tezel U, Gu E and Demirer G N 2001 Anaerobic biotransformation and methane generation potential of cheese whey in batch and UASB reactors *Waste Manag.* **21** 643–650
[18] Park N D, Thring R W and Helle S S 2012 Comparison of methane production by co- digesting fruit and vegetable waste with first stage and second stage anaerobic digester *Water Science & Tech.* **65** 1252-1257
[19] Wang H, Zhang Y and Angelidaki I 2016 Ammonia inhibition on hydrogen enriched anaerobic digestion of manure under mesophilic and thermophilic conditions *Water Res.* **105** 314–319
[20] Kim S, Choi K, Kim J O and Chung J 2013 Biological hydrogen production by anaerobic digestion of food waste and sewage sludge treated using various pretreatment technologies *Biodegradation.* **24** 753–764
[21] Li Z, Chen Z, Hong Y, Yuanpeng W, Weiang L Jo-Shu C, Qingbiao L and Ning H 2018 Anaerobic co-digestion of sewage sludge and food waste for hydrogen and VFA production with microbial community analysis *Waste Manag.* **78** 789–799
[22] Xie S, Wickham R and Nghiem L D 2017 Synergistic effect from anaerobic co-digestion of sewage sludge and organic wastes *Int. Biodeterior. Biodegradation.* **116** 191–197
[23] Dinh P V, Fujiwara T, Pham P and Hoang M G 2018 Kinetic of Biogas Production in Co-Digestion of Vegetable Waste, Horse Dung, and Sludge by Batch Reactors Kinetic of Biogas Production in Co-Digestion of Vegetable Waste, Horse Dung, and Sludge by Batch Reactors. *Proc. Int. Conf. on Environ and Renew. Energy (Vietnam)* vol 159 (IOP Earth and Environmental Science) pp 1-8
[24] Seswoya R and Abdul Karim A T 2017 Organics Characteristics of Sludge from a Full-Scale Anaerobic Digester Treating Domestic Mixed Sewage Sludge *Proc. Int. Symposium on Civil and Env. Eng (Melaka)* vol 103 (Malaysia: MATEC Web Conference) pp 1-8
[25] Li Y, Jin Y, Borrion A, Li H and Li J 2017 Effects of organic composition on mesophilic anaerobic digestion of food waste *Bioresour. Technol.* **244** 213–224
[26] Budiyono I, Syaichurrozi and Sumardiono S 2013 Biogas Production Kinetic from Vinasse Waste in Batch Mode Anaerobic Digestion *Malaysian J. Sci.* **32** 2–14
[27] Deepanraj B, Sivasubramanian V and Jayaraj S 2015 Experimental and kinetic study on anaerobic digestion of food waste: The effect of total solids and pH *J. Renew. Sustain. Energy.* **7** 1-12

Acknowledgement

This research was supported by Universiti Tun Hussein Onn Malaysia, Tier 1 Grant-Vot U844 and Postgraduate Research Grant (GPPS)-Vot U945.

Preliminary Assessment of Teknologi Lake Quality Status at Universiti Tun Hussein Onn Malaysia (UTHM) Campus in Parit Raja, Johor, Malaysia

Muhammad Safwan Miswan[1], Radin Maya Saphira Radin Mohamed[1*], Adel Ali Saeed Al-Gheethi[1] and Amir Hashim Mohd Kassim[1]

[1]Micro Pollutant Research Centre (MPRC), Department of Water and Environmental Engineering, Faculty of Civil and Environmental Engineering, Universiti Tun Hussein Onn Malaysia, 86400 Parit Raja, Batu Pahat, Johor, Malaysia
*Corresponding author:maya@uthm.edu.my

Abstract. A study was carried out to determine the status of the water quality condition in UTHM campus area located in Teknologi Lake, Batu Pahat, Johor, Malaysia. Sampling was conducted in two different times; the first sampling was in 22 March 2018 and the second sampling in 13 September 2018. A total of ten water quality parameters was selected, analyzed in-situ and exsitu, and classified based on NWQS Classifications. The status of lakes were determined by using Water Quality Index (WQI). Results for in-situ water quality parameters are: pH contents ranged from 6.42 to 6.61; DO from 9.0 to 9.8 mg/L; temperature from 28.7 to 28.9 ^{o}C. Ex-situ water quality parameters are; BOD was ranged from 105.2 to 106.0 mg/L, COD was ranged from 1 to 19 mg/L, TSS was ranged from 19 to 20 mg/L, turbidity was ranged from 12 to 13 NTU, ammoniacal nitrogen from was ranged 3.6 to 4.3 mg/L, nitrite was ranged from 0.13 to 0.11 mg/L and nitrate was ranged from 1.4 to 1.5 mg/L. The observed values for seven parameters in both sampling period of Teknologi Lake were classified as IIA/B. The results of the lake can be used for recreational purposes and suitable for sensitive aquatic species. However, based on WQI calculated value, the status of the lake during the study period was indicated as slightly polluted for Teknologi Lake. Overall, Teknologi Lake was classified as class IV due to agricultural activites and the water suitable for irrigation uses only.

1. Introduction

A lake is one of important water sources in Malaysia. A lake is enclosed body of water (usually freshwater) with considerable size, surrounded by land with no direct access to sea except with a river or stream that feeds or drains the lake [1]. Lake may occur anywhere within a river basin as natural or man-made lakes. Natural lakes consist of wetlands, marshes, estuarine lakes, or naturally occurring lakes whereas man-made lakes can be referred as reservoirs, retention pond, ex-mining pond, recreational lakes [2] or as urban landscape elements. In Malaysia, water quality of a lake can be influenced by external inputs entering the lake from the watershed as well as the inlake ecosystem, nutrients cycling and internal loading. External inputs can be organic and inorganic pollutants as well as nutrients which cause deterioration of water quality from point and non-point sources. Nutrient input from point sources is relatively easy to determine and well characterized because it is concentrated and sampling is easy [3]. Meanwhile, input from nonpoint sources is more difficult, but broad ranges of nutrient loss rates from different types of land uses are known [4]. Point sources include discharge from domestic and municipal wastewater, agricultural effluents including applied of fertilizers and industrial wastewaters whereas non-point sources include urban/ storm water runoff, agricultural runoff, septic tank overflow and construction sites runoff [5].

Excessive input of nutrients such as phosphorus and nitrogen into the lakes will lead to eutrophication. Agricultural and urban uses lead to the greatest degree of runoff, with human population density in a watershed demonstrating a significant positive correlation to nitrogen and phosphorus runoff [6]. Eutrophication is the process by which a water body becomes uncontrollably rich and abundance in aquatic plants such as algae and aquatic macrophytes (water weeds) due to plant nutrients enrichment, especially phosphorus and nitrogen as dissolved solutes and as compounds bound to organic and inorganic matters from natural and anthropogenic sources [1]. Huang et al., [1] stated that 56 lakes or 62% of the 90 lakes evaluated were eutrophic while 34 lakes or 38% were mesotrophic. These lakes were assessed based on the Tropic State Index (TSI) values, computed adopting the land use and TP relationships. Therefore, the steps are needed to deal with the deterioration of water quality.

A Teknologi Lake in UTHM campus area located in Parit Raja, Batu Pahat is a man-made lake surrounded by palm oil plantation. The lake flows is connected with Sembrong Dam river. Parit Raja always occurs a flood when there are heavy rains. Thus, this lake reacts as a retention pond by slowing down the water flow. Lately, the lake serves the local people as an infrastructure for social activites such as fishing, jogging, fitness training (kayak) and for investigation process. Therefore, the study aims to characterize the pollutant in lake water either suitable for water activities or suitable for others purpose (irrigation).

2. Methodology

Twice sampling period were performed, the first sampling was in March 2018 and the second sampling in September 2018. Different sampling time was taken due to wet season and dry season. The samples were taken using polyethylene bottles, which had been soaked and cleaned before the sampling. After the collection, the samples were immediately placed into ice boxes and proceed for analysis in the laboratory. The water samples were collected on a range of time, from 8.00 to 10.00 am. Three replications of samples were taken randomly. The water quality analysis is divided into two measurements: In-situ measurement and in the lab measurement. Three water quality parameters (temperature, pH, dissolved oxygen) were measure in-situ using Eutech instruments DO 450. Seven other parameters are physico-chemical parameters such as the chemical oxygen demand (COD), biochemical oxygen demand (BOD), total suspended solids (TSS), ammoniacal nitrogen (AN), nitrite (NO2), nitrate (NO3) and turbidity were analysed in the laboratory. All the analyses were performed in accordance with the Standard Methods of Examination of Water and Wastewater [7] and HACH Water Quality Testing and Analytical Instruments procedure. The results were compared to USEPA water quality standards [8] and Interim National Water Quality Standard Malaysia [9]. The water quality index calculation [6] was used for the computation of observed data comprises six water quality parameters.

3. Results and Discussions

3.1. Lake water characteristics.

Lake water characteristics based on the geometry and environmental conditions of lakes is presented in Table 1. The Teknologi Lake which has a total area of 4067.18 m^2 consists of 343.77 m perimeter of the lake. The water level of lake is 1.136 m. The lake includes three different main land uses such as palm oil plantations, water bodies and industry. The range annual rainfall of the area is about 50 to 350 am and the average wind speed 5.56 m/s to 8.33m/s. According to [8], under water body survey and assessment, it is important to identify the several factors including physical factors that may be examined to determine whether an aquatic life protection is attainable for a given water body. The physical factors listed including in lake characteristics.

Table 1: Geometry and environmental conditions of lake

Parameter/ Geometry	
Area	4067.18 m^2 (observed data)
Perimeter	343.77 m (observed data)
Water level	1.136 m (observed data)
Wind speed	5.56 m/s to 8.33 m/s (Malaysian Meterological Station)
Mean Surface water temperature	28.8 °C
Total precipitation	50-350 mm (Malaysian Meterological Station)

3.2. Water Quality Assessment.

Lake water quality characteristics obtained from this study was compared to one of the recent previous studies on lake water [10] as same function with present study as retention lake and the National Water Quality Standards for Malaysia as shown in Table 2.

Table 2: Characteristics of Teknologi Lake water in campus area (Minimum, maximum and mean value) sampling was conducted on March 2018 and September 2018 (n=12).

Parameter	**Present study**			**Previous study Gasim *et al.*, 2015**	**National Water Quality Standards for Malaysia**	
	Minimum	**Maximum**	**Mean**		**Class I**	**Class IIA**
pH	6.42	6.61	6.51±0.095	6.9	6.5-8.5	6-9
DO (mg/L)	9.0	9.8	9.4±0.4	3.2	7	5-7
Temperature (°C)	28.7	28.9	28.8±0.1	28.4	28	28
BOD (mg/L)	105.2	106.0	105.7±0.42	0.97	1	3
COD(mg/L)	1	19	10.3±9.02	22.17	10	25
TSS (mg/L)	19	20	19.7±0.58	6.75±2.50	25	50
Turbidity (NTU)	12	13	12.3±0.58	NR	5	50
Ammoniacal nitrogen (mg/L)	3.6	4.3	3.95±0.38	2.46±0.32	0.1	0.3
Nitrite (mg/L)	0.13	0.11	0.12±0.01	0.19±0.12	0.04	0.4
Nitrate (mg/L)	1.4	1.5	1.36±0.15	1.17±0.10	7	7

*All parameters are expressed as mg L^{-1} except for pH and turbidity (NTU), NR= Not Reported

3.2.1. In-Situ Analysis

Three water quality parameters; pH, DO and temperature were measured in-situ (during sampling) to gain best appropriate result in this study. The range of pH value during the sampling period was 6.42 to 6.61, with a mean of 6.51±0.095. The mean pH value of the Teknologi Lake is indicated as slightly acidic. The low acidity of the lake could also be due to the free carbon dioxide as well as the various acids and alkalis which can penetrate the water bodies that came along with the industrial wastewater [10]. Low pH was probably due to the rainfall and runoffs from the nearby agricultural area (Palm oil plantation) and surrounding areas and attributed to the presence of high organic matter resulting from the discharge of organic matter into the lake. The pH value was ideal when compared to the NWQS and classified as class I. The pH values of the lake in this study are non different compared to the values that have been studied by [10] of Cempaka Lake, Bangi, Selangor, which was 6.9. Next, the range of dissolved oxygen value during the samplings was from 9.0 to 9.8 mg/L, with a mean of 9.4±0.4 mg/L. The DO value was higher during the first sampling period because due to raining. Sharma et al., (2010) reported lower values of dissolved oxygen in summer months due to a higher rate of decomposition of organic matter and limited flow of water in the low oxygen holding environment due to the high temperature [11]. Hence, higher temperature in surface water lower its oxygen content. Another factors increases of DO probably due to frequently of raining and occur high natural aeration. The DO values of the lake in present study are higher compared to the DO values at Cempaka Lake which was from 3.76 to 4.94 mg/L. According to NWQS, Teknologi Lake placed under class I. At that moment, The range of temperature values during samplings time was from 28.7 to 28.9 ^{o}C, with a mean of 28.8±0.1 ^{o}C. This reading may result from the high level of exposure of the sun and heat evaporation after raining. Several factors which are affected by weather variations and may influence the water temperature; sampling time and location were taken considerably. The temperature value was recorded at the Teknologi Lake was classified as normal based on the NWQS and classified as class I.

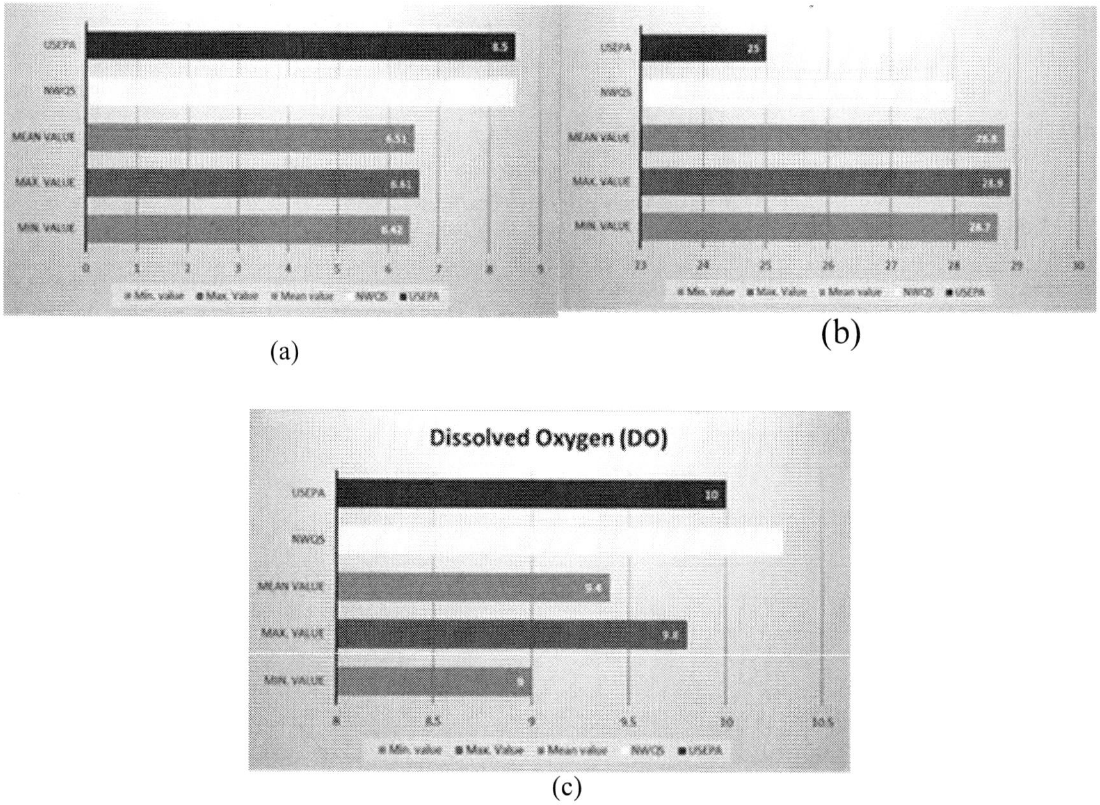

Figure 1. Observed data of a) pH, b) temperature and c) DO for Teknologi Lake compared to INWQS and USEPA standards.

3.2.2. Physico-Chemical Laboratory Analysis

The range of Chemical Oxygen Demand during the sampling period was from 1 to 19 mg/L, with a mean of 10.3±9.02 mg/L. In this study the value of COD take from lake in all samples was low. Increases of the COD level in the waters are due to the increase of organic matter and inorganic chemicals [10] runoff from the restaurant food waste and wastewater from areas around the lake. Based on the NWQS, the level of COD measured in this study in the lake was classified as class I. The COD values in this lake are not different compared to the COD values at Cempaka Lake in Bangi which was 14.25 mg/L. After that, The range value of biochemical oxygen demand during the sampling time was from 105.2 to 106.0 mg/L, with a mean of 105.7±0.42 mg/L. It is an empirical test to measure the amount of oxygen used by the microorganisms in the aerobic oxidation, or breakdown of organic matter in the lake, the higher BOD, the more organic matters in the water. According on the NWQS, the BOD value Teknologi Lake was classified as class IV. However, when comparing with a study which was done by [12] at Engineering Lake, UPM, the value of BOD was not much different, 78.33 mg/L. Then, the range of total suspended solids during the sampling time was from 10 to 20 mg/L, with a mean of 19.7±0.58 mg/L. The result of TSS values is high during the sampling, because there was raining and generated a strong rate of soil erosion and contributing more suspended solids in the area of flat land. According to [13], the solids may in fact consist of algal growths and hence be indicate be of severely eutrophic conditions. Based on the NWQS, level of TSS in the study area is classified as class I. However, the result was slightly lower compared to the TSS values at Sembrong Lake [14], which was 54.33 mg/L. At that time, The range of turbidity for the sampling was from 12 to 13 NTU and a mean of 12.3±0.58 NTU. The effect of raining increases the turbidity in water. However, referring to the NWQS it can be classified in class I which is suitable conversation of natural environment, water supply and fishery. Moreover, the comparison values with previous studies by [12] showed that the lakes turbidity still below the previous studies which was 13.86 NTU.

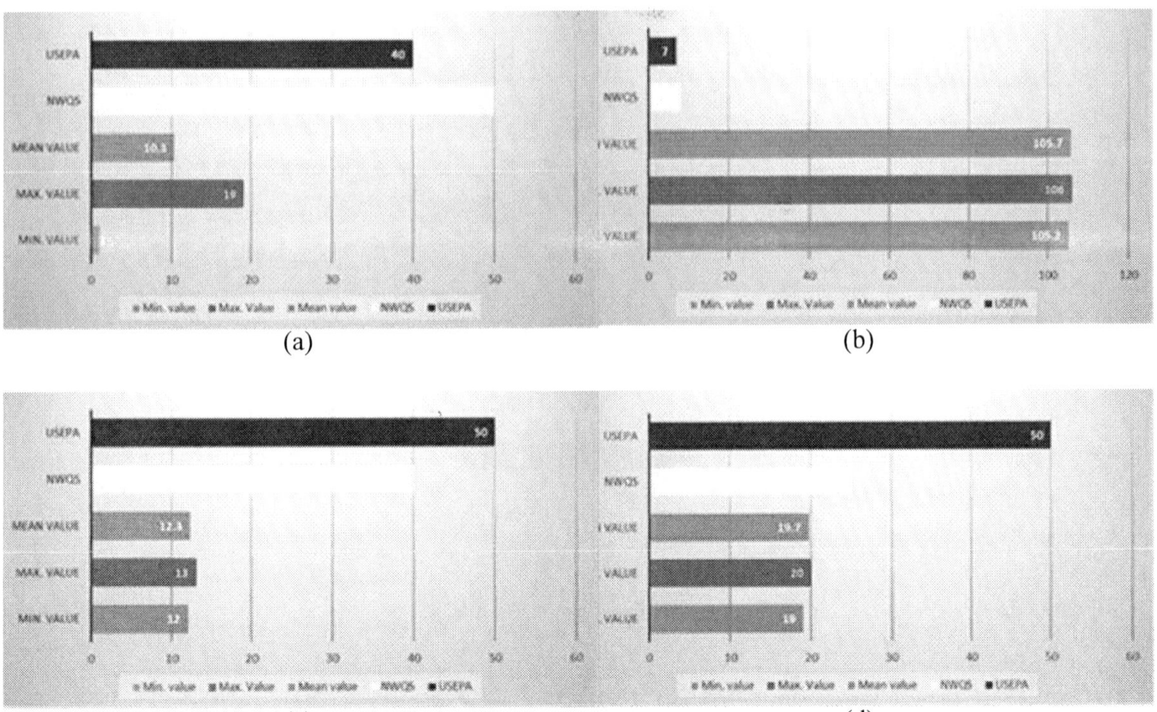

(a) (b)

(c) (d)

Figure 2. Observed data of a) BOD, b) COD, c) TSS and d) Turbidity compared to standards for Teknologi Lake

The range of ammoniacal nitrogen during the sampling was from 3.6 to 4.3 mg/L, with a mean of 3.95±0.38 mg/L. The pH of the water is less than 7 the ammonia is present as ammonium ion while pH increases above 7, more of the ammonia is present as ammonium hydroxide [15]. There have correlation between pH and AN in this study. The AN is increases because the pH is lower than 7 (refer Table 1).The high level of ammoniacal nitrogen present in lake water may cause eutrophication to arise [1]. Based on the NWQS, the AN in Teknologi Lake is classified as class V. However, AN values in this study not much different when comparing with AN value of sampling in [10], 2.46±0.32 mg/L. After that, the range of nitrite for the first sampling ranged from 0.13 to 0.11 mg/L or a mean of 0.12±0.01 mg/L. Nitrite exists normally in very low concentrations and even in waste treatment plant effluents levels are relatively low. According to [16] nitrite is an intermediate in the oxidisation of ammonia to nitrate. Principally, the nitrite will tend to exist in the more reduce of ammonia or more oxidised nitrate forms. The main source of NO2 was due to runoff from the restaurant food waste and wastewater from areas around the lake. Based on the NWQS showed NO2 of the Teknologi Lake, and classified as class I. This indicated the lake low of nitrite because the lake far from food industries. Next, the range of nitrate during the sampling was (minimum) from 1.4 to 1.5 mg/L (maximum), with a mean of 1.45±0.15 mg/L. Relatively little of the nitrate found in natural waters most coming from organic and inorganic sources (artificial fertilizers) [8]. The highest NO_3 values was received discharge from human activities such as sanitary waste and fertilizers waste. Lawniczak et al. (2016), stated that high levels of nitrate are more likely to indicate significant run-off from agricultural land [17]. Based on the NWQS, the NO_3 of the Teknologi Lake was within the normal range and classified as class I. However, when comparing with data from [10] at Cempaka Lake, it is considered high.

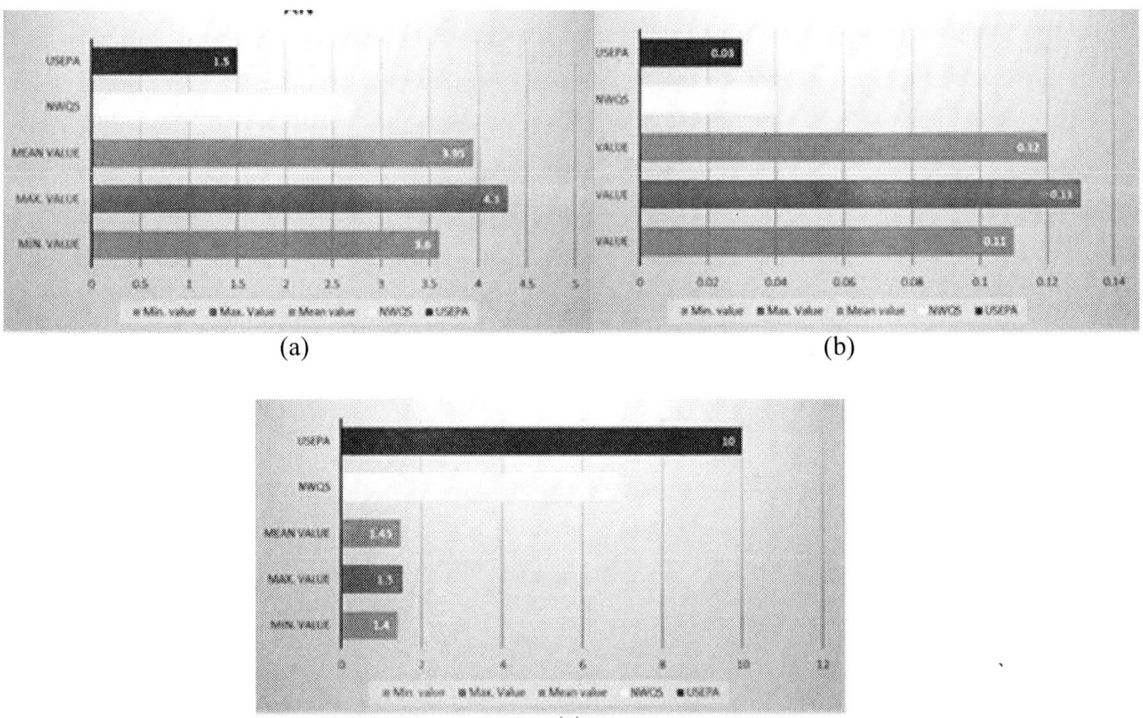

Figure 3. Observed data of a) Ammoniacal nitrogen, b) nitrite and c) nitrate for Teknologi Lake compared to standards

4. Conclusion

The results of the study indicated that the mean concentrations of some parameters such as pH, temperature, TSS,COD, DO and turbidity for two different samplings were within the normal range and are classified under Class I. The mean BOD and ammoniacal nitrogen under class IV and V, respectively. The results clearly shown the majority of the water quality parameter is more polluted in term of BOD and ammoniacal nitrogen. Water body in the study area was contaminated may probably originate from the agricultural activities, domestic sewage and industrial activities. These activities were generated both organic and inorganic waste and these wastes are ultimately contaminating the water bodies.

5. References

[1] Huang Y K, Ang S Y, Lee K M and Lee T S (2015) Quality of Water Resources in Malaysia. *Research and Practices in Water Quality*

[2] NAHRIM SM (2009) Pelan Strategi: Strategi Pembangunan dan Pengurusan Lestari bagi-Tasik dan Empangan Air di Malaysia Jilid 1: Laporan Utama

[3] Wieben C, Baker R J and Nicholson R S (2013) Nutrient Concentrations in Surface Water and Groundwater and Nitrate Source Identification Using Stable Isotope Analysis in the Barnegat Bay-Little Egg Harbor Watershed New Jersey 2010-11 US Department of the Interior US Geological Survey

[4] Dressing S A, Meals D W, Harcum J B, Spooner J, Stribling J B, Richards R P and O'Donnell J G (2016) *Monitoring and Evaluating Nonpoint Source Watershed Projects*. EPA 841-R-16-010. US Environmental Protection Agency Washington, DC. Accessed December 20 2016.https://www.epa.gov/nps/monitoring-and-evaluating-nonpoint-source-watershed-projects

[5] Naidoo S and Olaniran A O (2013) Treated wastewater effluent as a source of microbial pollution of surface water resources *International journal of environmental research and public health* **11**(1) 249-270

[6] DOE 2006 Department of Environment Malaysia Malaysia Environmental Quality Report (2006) In: Chapter 3: River Water Quality Department of Environment Ministry of Science Technology and Environment, Malaysia

[7] APHA (2012) Standard methods for study of examination of water and waste water analysis 20th Ed., Washington DC: American Public Health Association.

[8] USEPA (2014) Water Quality Standards Handbook – Chapter 2: Designation of Uses. Retrieved from website: http://water.epa.gov/scitech/swguidance/standards/handbook/chapter02

[9] DOE (2008) Department of Environment Malaysia, Interim National Water Quality Standards for Malaysia Department of Environment, Ministry of Science Technology and Environment, Malaysia

[10] Gasim M B, Toriman M E, Muftah S, Barggig A, Aziz N A A, Azaman F and Muhamad H (2015) Water quality degradation of Cempaka Lake, Bangi, Selangor, Malaysia as an impact of excessive E. coli and nutrient concentrations. *Malaysian Journal of Analytical Sciences* **19**(6) 1391-1404

[11] Sharma A, Ranga, M M and Sharma, P C (2010). Water quality status of historical Gundolav Lake at Kishangarh as a primary data for sustainable management. *South Asian Journal of Tourism and Heritage* **3**(2) 149-158

[12] Daud N N N, Abdulrahman A and Idrus S (2016) Preliminary assessment of lakes water quality status at campus area in selangor, malaysia *Malaysian Journal of Civil Engineering* ***28*** (Special issues 1) 42-49

[13] Rabbani M L and Sarker S (2018) Evaluation of Pollution Status of Turag River, Bangladesh *Journal of Water Resources and Pollution Studies* **3**(1)

[14] Sharip Z, Zaki A T, Shapai M A, Suratman S and Shaaban A J (2014) Lakes of Malaysia: Water quality, eutrophication and management. *Lakes & Reservoirs: Research & Management* **19**(2) 130-141
[15] Körner S, Das S K, Veenstra S and Vermaat J E (2001) The effect of pH variation at the ammonium/ammonia equilibrium in wastewater and its toxicity to Lemna gibba. *Aquatic botany* **71**(1) 71-78
[16] Ma J, Yao H, Yu H, Zuo L, Li H, Ma J and Li X. (2018) Hydrazine addition enhances the nitrogen removal capacity in an anaerobic ammonium oxidation system through accelerating ammonium and nitrite degradation and reducing nitrate production *Chemical Engineering Journal* **335** 401-408
[17] Lawniczak A E, Zbierska J, Nowak B, Achtenberg K, Grześkowiak A and Kanas K (2016) Impact of agriculture and land use on nitrate contamination in groundwater and running waters in central-west Poland. *Environmental monitoring and assessment* **188**(3) 172.

Acknowledgement
The authors would like to express their appreciation to the Micropollutant Research Centre (MPRC), Universiti Tun Hussein Onn Malaysia (UTHM) for the use of their research facilities and the authors also like to thank the Research Management Centre (RMC) Universiti Tun Hussein Onn Malaysia (UTHM) for its support under Grant GPPS (U957).

IOP Conf. Series: Materials Science and Engineering **601** (2019) 012014 doi:10.1088/1757-899X/601/1/012014

Assessment of Cause and Effect Factors of Poor Communication in Construction Industry

Ismail Abdul Rahman[1], **Yaser Gamil**[1*]

[1] Faculty of Civil and Environmental Engineering, University Tun Hussein Onn Malaysia, Johor, Malaysia
*Corresponding author: ismailar@uthm.edu.my

Abstract. Effective communication in construction industry is a core competency element to the success of the project. But it has been always a challenging mission to perform the communication effectively because the industry itself is characterized as fragmented, dynamic, and comprises of many parties (client, consultant, contractor, authorities) which leads to the presence of poor communication. There has been no publication to address the poor communication in a broad context especially identifying and assessing the causative factors and effects to this issue in the construction industry. However, this article aimed to identify and assess the significance of these factors through ranked them individually based on its level of significance/severity. Quantitative approach was adopted to conduct this study using structured questionnaire survey. A 5-point Likert's type scale was used to assess the factors gradually based on their significance. Descriptive analysis was used to analyse the data collected from the survey. Average index (AI) and standard deviation were used to rank the importance of the factors in descending order. The study found that, the most dominant cause factors of poor communication is fear to communicate in the project space however the most dominant effect factor is high stress in workplace. The findings of this study serve best to draw a platform for construction practitioners to introduce strategies and plans to avoid the cause and diminish the effect of poor communication in construction industry.

1. Introduction

Currently According to Oxford dictionary, communication is defined as "The imparting or exchanging of information through speaking, writing, or using some other medium" [1]. In a similar concept, Longman dictionary defined communication as "the process by which people exchange information or express their thoughts and feelings" [2]. Communication can have various different meanings, contexts, and forms which relatedly depend on the discipline referred accordingly. The key aspect of both mentioned definitions introduced communication as the way of sending or receiving information by different means [3]. Sinha [4] defined communication synonymously and equivalently as the process of coding one's thought, ideas and feeling into signals that are transmitted to others who receive, decode, understand and react to them on the basis of their own previous experiences and future expectations. However, in the construction industry, communication can be defined as the mutual exchange of project information and processes with the assurance of creating an understandable platform between receiver and sender [5]. An effective communication is one of the crucial factors contributing to the success of the project [6-8]. The Project Management Body of Knowledge guide [9] defined project communication management as "one of the main knowledge areas in project management which include the processes requires to ensure timely and appropriate generation, collection, distribution, storage, retrieval and ultimate disposition of project information". In addition, communication covers all task related to producing, compiling, sending, storing, distributing and managing project records. It also requires an accurate report on the project status, performance, change and earned value [10].

Furthermore, communication is considered one of the most important aspects of project management that pervades all others [11]. Emmitt and Gorse [12] described communication as a key management competency of leadership and decision-making are founded on good communication skills. In another word, communication is the ability to effectively communicate with employees, though not an easy skill to master is an obvious requirement for good leadership [13]. Emuze and James [14] reveals that communication is one of the fundamental aspects of the construction industry and it is the key aspect to project success. Communication affects the project implementation from the inception stage to the handover stage; as a result, it is necessary to understand its important role in the success of construction industry [8, 15].

A survey was conducted by Coopers [16] on the current state of project management revealed that 17% increase in finishing project within budget was associated due to effective communication. It was then stressed by PMI [5], the cost overrun in construction projects is also associated with improper communication and information management.

Communication is an essential element in the processes of construction and it is an augmentation factor of collaboration and cooperation among construction parties [17-18]. However, communication among the parties is not always concise and effective and, in many situations, it turns out to be unsuccessful and this is resulted in the phenomenon of poor communication [19]. Poor communication in construction industry is a repetitive issue in all construction projects and it is caused by many elements which of these, the unique characteristics of construction industry as a fragmented and involves many parties [6]. It also leads to many other issues such as cost, time overruns, delay, dispute and sometimes failure of the projects as a whole [20].

The reason of communication lack in the construction project is due to the complex and fragmented nature of the industry. In a single project several players are in charge of communication in the construction project which includes contractor (general contractor, design-builder, and construction manager), developer, sub-contractors, and owner/client: public (government), private and labour: unionized, non-unionized [21,26]. These characteristics of construction industry impact the methods and quality of communication which then need to be investigated in a broad context.

Poor communication issue has attracted many researchers and publication platforms to investigate its cause and introduces solutions and strategies. PMI [5] in its press release considered poor communication as one of the most causes of cost and time overruns in construction projects and it was vindicated that this issue causes overruns in the form of direct and indirect means [5,25]. Hoezen et al [22] declared that, most of the major issues in construction projects are caused by improper communication and it mostly reported between the demand side party and the supply side. In spite of the study mentioned before, this article aimed to assess degree of significance for causes and the degree of severity of effect factors which were predefined by [20] in article entitled "Identification of Causes and Effects of Poor Communication in Construction Industry: A Theoretical Review "and this paper also discussed the findings obtained from questionnaire survey.

2. Method of Study

Figure 1 show the process adopted to carry out this study in sequence order

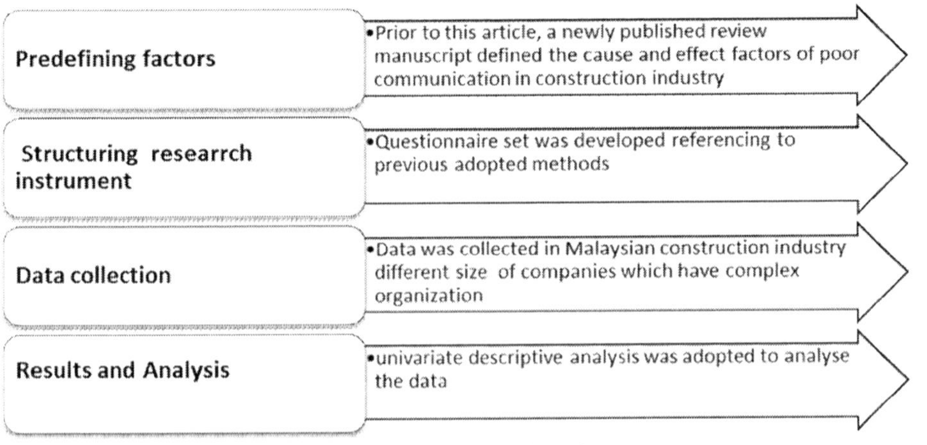

Figure 1. Methodologies Chart

According to figure 1, the first step is to identify the cause and effect factors of poor communication in construction industry worldwide. The second step to develop questionnaire survey and collect the data. Questionnaire consists of three parts which includes: demography of respondents causes factors and effects factors. The respondents were asked to rank the significance and severity of causes and effects respectively using 5 points Likert type score. The third step is to analyses the data using statistical software. It was found that the cause and effect factors of poor communication in construction industry are 41 and 27 respectively. The basic method to assess the importance of these factors is using 5-point Likert's type scales to give the respondent the space to evaluate the significance and severity of each factor. Questionnaire set consisted of two parts, the first part is about the demographic profile of targeted respondents and the second part is the assessments of predefined factors. To understand the sequences of this research, figure 1 illustrate the major steps followed in this research. After the completing data collection, the responses were validated and any response was incomplete considered valid. The internal consistency of the instrument was tested using Cronbach's alpha. The following section is to analyse the data and discusses the findings.

3. Results and Analysis

This part presents the results and analysis of data. The Level of significance and level of severity for causes and effects factors were determined based on the collected data through questionnaire tool. This part starts by introducing the demographic profile of participating respondents then analysing the ranking of the factors accordingly using AI and standard deviation.

3.1 Demography's of Respondents

This subsection presents the demographic attributes of participants who participated in this study. Table 1 demonstrates the demographic data of respondents which includes organization, category of organization, role of respondent in the organization, and highest level of qualification of respondents and number of years of experience in the construction industry.

Table 1. Participant's demography

Category	Items	Percentage of responses
Role of respondent in the organization	Company Director	13.3
	Project manager	3.3
	Architect	33.3
	Project engineer	10.0
	Clerk of work	6.7
	General worker	33.3
	Others	13.3
Respondents qualification	Diploma	13.3
	Degree	50.0
	Master	23.3
	PhD	6.7
	Others	6.7
Respondents' working Experience in construction (years)	0-10	50.0
	11-20	26.7
	21-30	16.7
	More than 31	6.7
Type of organization	Consultant	30.0
	Contractor	30.0
	Client	26.7
	Others	13.3
Category of organization	Government	20.0
	Private	76.7
	Others	3.3

Table 1 shows the respondents demography data elicited from the study. It is shown that respondents involved in this study are mostly in top management level of their companies. Regarding the role of respondents, directors are 4 (13.3%), project manager 1 (3.3%), architect 10 (33.3%), Project engineer 3 (10.0%), Clerk of work 2 (6.7%), general worker 10 (33.3%) and others 4(13.3%) It is also shown in table 2 that respondents mostly obtained high level of qualification and the number of degree holders is 15 (50%). In regards of number of years of experience, it is shown that all respondents are senior employees in their organization. Table 1 also presents the data of respondent's demography from organization perspectives. Consultant and contractors allocated 30% of the type of organization however client and others allocate 26.7% and 13.3% respectively. In term of category of organization, percentage of private companies carries the highest percentages of 76.7% however governmental carries 20% and others only 3.3%. This signifies that, the organization selected are acceptable to be considered in this study because the companies are considered to have many projects in place. The data elicited in table 1 confirms that, respondents participated in this study are qualified and hold an adequate knowledge of construction industry therefore, their responses and opinion regarding the issue of communication and evaluating the causes and effects factors are sufficient and satisfactory.

3.2. Reliability Analysis of Study Data

To conduct reliability test of questionnaire items in this study, it is essential to calculate Cronbach's alpha. Cronbach's alpha is used to estimate the internal consistency reliability associated with scores derived from Likert scale used in this study [23]. It also tests the reliability and validity of scores. In this study, Likert type scale was used in this study in the range of 1 to 5 to assess the factors based on given indications. SPSS is used to calculate Cronbach's alpha. Cronbach's alpha value for the five Likert scales which is (α =0.812). To interpret this data, a rule of thumb is derived by George [24] who stated any value between 0.8 and 0.9 is considered good internal consistency. For the study data, it is considered good and this indicates the scales hold sufficient degree of reliability, satisfied and consistent to be used for further analysis and investigations.

3.3. Assessment of Cause Factors of poor communication in construction industry

The average index (AI) score is referred for all causes and effects factors. It is employed to assess the factors and rank them from the most important to the less important.

Table 2. Causes factors of poor communication Sorted by Descending AI

Causative Factors	STDV	AI	Ranking
Fear to communicate	0.56	4.37	1
Delay notification of change	0.61	4.33	2
Lack of sector experience	0.53	4.30	3
Individual barrier (habits)	0.59	4.17	4
Poor progress measurement	0.83	4.17	5
Contractual barriers (restrictions)	0.97	4.13	6
Poor communication skills	0.71	4.10	7
Slow information flow among parties	0.66	4.10	8
Lack of communication among parties	0.80	4.10	9
Lack of effective communication system and platform	0.83	4.07	10
Usage of different terms and methods	0.78	4.07	11
Lack of confidence between stakeholders	0.72	4.03	12
High confidentiality	0.72	4.03	13
Informality	0.67	4.03	14
Improper communication channel	0.79	4.00	15
Lack of communication plan	0.98	4.00	16
Language barrier	0.61	3.97	17
Lack of mutual trust among construction teams	1.08	3.93	18
Lack of collaboration and representation between stakeholders	0.78	3.93	19
Negligence and casualness	0.71	3.90	20
Poor communication management	0.92	3.90	21
Inaccessibility of project information	0.68	3.87	22
Frequent change of project contract	0.82	3.87	23
Weak organizational structure for communication purpose	0.82	3.87	24
Incorrect delivery of instructions or technical info	0.83	3.83	25
Lack of communication procedure and training	0.85	3.80	26
Poor detailed drawings	0.90	3.77	27
Different level of skills among construction team	0.53	3.70	28
Different level of education among construction team	0.65	3.70	29
Lack of clear communication objectives	0.86	3.57	30
Improper communication time management	0.86	3.57	31
Lack of appropriate communication medium	0.94	3.50	32
Poor planning	0.94	3.50	33
Lack of support for advanced communication technology	0.90	3.47	34
Natural problem on site (Noise, rain)	1.00	3.40	35
Lack of understanding among parties	0.81	3.37	36
Complexity of construction industry	0.96	3.37	37
Unavailability of information in time of need	1.01	3.23	38
Diversity of culture and ethics among parties	0.89	3.20	39
Frequent technology malfunction	1.00	3.20	40
Gender difference	0.74	3.00	41
Overall average	**0.87**	**3.80**	

3.4. Assessment of effect Factors of poor communication in construction industry

In this subsection, the assessment of effect factors is introduced by evaluating the average index. Table 3 demonstrates the factors and their assessments.

Table 3. Effect factors of poor communication Sorted by Descending AI

Effect Factors	STDV	AI	Ranking
High stress in workplace	0.57	4.50	1
Cost overrun	0.84	4.30	2
Information overlapping	0.75	4.30	3
Conflict among construction parties (Dispute)	0.94	4.27	4
Worsening relationship among construction parties	0.78	4.27	5
Poor project planning	0.74	4.27	6
Wrong execution of activities	0.77	4.23	7
Poor risk management	0.68	4.23	8
Time overrun	0.66	4.20	9
Poor project information management	0.91	4.17	10
Late response to disaster	0.82	4.13	11
Unclear communication channel	0.94	4.13	12
Poor collaboration among construction parties	0.86	4.13	13
Design error	0.76	4.10	14
Poor team work	0.94	4.07	15
Failure of the project	1.00	4.03	16
Low level of satisfaction among parties	1.03	4.03	17
Rework occurrence	0.87	4.00	18
Demotivated workforce	1.10	3.97	19
Frequent remedies in design and planning schedule	0.87	3.93	20
Misinterpretation	0.98	3.93	21
Poor project documentation	0.91	3.93	22
Low productivity	0.99	3.90	23
High accident rate	1.18	3.83	24
Waste generation	1.10	3.77	25
Affects construction process	0.76	3.37	26
Misunderstanding	1.01	3.23	27
Average value	**0.92**	**4.00**	
Maximum Value	**1.18**	**4.50**	
Minimum Value	**0.57**	**3.23**	

Table 3 shows the analysis for effect factors of poor communication in construction industry. From the calculation, standard deviation for all the factors equals 0.92 which displayed good variations of the answers from targeted respondents. This is justified that; the background of respondents' matters in their opinions toward the effect factors of poor communication furthermore signifies good distribution of the mean values. Aside from that, the maximum and minimum average index (AI) is 4.50 and 3.23 respectively and the overall average score of AI is 4.00. The most ranked effect factor is high stress in workplace which carries an AI of 4.5 and this indicates poor communication leads to high stress in workplace. The second most ranked is cost overrun which carries an AI of 4.30. This confirmed to the findings by PMI in 2013 [5] press release which stated that, more than half of all project budget risk is due to ineffective communications and improper time management of project communication. It also stated ineffective communications among relevant parties contribute 56% of the total risk of a construction project. The report elaborated poor and substandard communications is the primary contributor to project failure one third of the time, and had a negative impact on project success more than half the time. Thus, it suggested a high-level, executable communications strategy to reduce financial risks.

4. Conclusion

This article presented the assessment of cause and effect factors of poor communication in construction industry. A total of 41 cause factors and 27 effects factors were assessed based on their significance and severity correspondingly. The study considered the factor of fear to communicate as the most dominant causative factors of poor communication in construction industry. In the other hand, stress in workplace as the most severe effect of poor communication in construction industry. The findings of this study help to define the most cause and effect factors of poor communication in construction industry which eventually assist construction parties to understand the causes and effects of poor communication and introduce effective resolution for these factors by evading the cause and reducing the effect. In conclusion, poor communication issue requires more attention from researchers and academicians to introduce effective methods and strategies to overcome the issue for the sake of augmenting the performance level and produce successful projects.

5. References

[1] Dictionary, O. E 1989 Oxford: *Oxford university press*
[2] Mayor, M 2009 Longman dictionary of contemporary English. (Pearson Education)
[3] Al-Reshaid K and Kartam 1999 Improving Construction Communication: The Impact of the On-line Technology. *Institute for Research in Construction* (Ottawa, Canada)
[4] Sinha J B 2009 Culture and organizational behaviour. (SAGE Publications India)
[5] Press Releases 2013 Project Management Institute (PMI)" more than half of all project budget risk is due to ineffective communications" (www.pmi.org)
[6] Murray M., Dainty A and Moore 2007 Communication in construction: Theory and practice. (Routledge Publishing)
[7] Fichet H and Giraud 2007 How the information flow is processed in project-based companies compared to others and how it affects strategic drift. Master thesis. (UMEA school of business. Sweden)
[8] Gamil, Yaser, Ismail Abdul Rahman, Sasitharan Nagapan, and Nashwan Alemad 2017 Qualitative Approach on Investigating Failure Factors of Yemeni Mega Construction Projects *in MATEC Web of Conferences*, vol. 103, p. 03002. (EDP Sciences)
[9] Guide A 2001 Project Management Body of Knowledge (PMBOK® GUIDE). (In Project Management Institute)
[10] Crowe, A 2005 The PMP exam: How to pass on your first try (Velociteach Press)
[11] Olaniran, H. F 2015 On the role of communication in construction projects in Nigeria *Journal of Environmental Science and Technology* **2**(5) 048-054
[12] Emmitt, S. and Gorse 2006 Communication in construction teams (Taylor & Francis. London)
[13] Dingsdag, Donald P. and Biggs, Herbert C., Sheahan, Vaughn L. and Cipolla 2006 Construction Safety Competency Framework: Improving OH&S Performance by Creating and Maintaining a Safety Culture. *CRC for Construction Innovation* (Brisbane, Qld)
[14] Emuze, F and James 2013 Exploring communication challenges due to language and cultural diversity on South African construction sites. *Acta Structilia*, **20**(1), 44-65
[15] Zulch, B G 2012 The Construction Project Manager as Communicator in the Property Development and Construction Industries. Doctoral dissertation (University of the Free State)
[16] Coopers P W 2012 Insights and Trends: Current Portfolio, Programme and Project Management Practices. Price Waterhouse Coopers. A report retrieved from: http://www.pmi.org/~/media/PDF/RCP/PwC_PPPM_Trends_2012.ashx

[17] Goh, K C Goh, H H, Toh, S H and Peniel Ang 2014 Enhancing communication in construction industry through BIM *Proceedings of the 11th International Conference on Innovation & Management*. PP (313-324)

[18] Gamil, Y and Abdul Rahman I 2018b Assessment of critical factors contributing to construction failure in Yemen *International Journal of Construction Management* 1-8.

[19] Wikforss Ö and Löfgren 2007 Rethinking communication in construction. *Journal of Information Technology in Construction* (ITcon) **12**(23) 337-346

[20] Gamil Y and Rahman I. A. 2018a Identification of Causes and Effects of Poor Communication in Construction Industry: A Theoretical Review *Emerging Science Journal* **1**(4) 239 - 247

[21] Ballan S and El-Diraby 2011 A value map for communication systems in construction. *Journal of Information Technology in Construction* **16**(44) 745-760

[22] Hoezen M E L, Reymen, I M M J and Dewulf, G P M R 2006 The problem of communication in construction. *In CIB W96 Adaptable Conference* (University of Twente)

[23] Gliem, Joseph A and Rosemary R Gliem 2003 Calculating, interpreting, and reporting Cronbach's alpha reliability coefficient for Likert-type scales." *Midwest Research-to-Practice Conference in Adult, Continuing, and Community Education*

[24] George D 2011 SPSS for windows step by step: A simple study guide and reference, 17.0 update, 10/e. (Pearson Education India)

[25] Assah-Kissiedu, M., Fugar, F. D. K., & Badu 2010 Triggers of disputes within the Ghanaian construction industry. *In Proceedings 5th Built Environment Conference* (Vol. 18, p. 20)

[26] Kasimu, M A and Abubakar 2012 Causes of delay in Nigeria construction industry. *The Journal of contemporary research in business* **4**(2) 785-794

Acknowledgement

Authors appreciatively acknowledge the financial support provided by BP Renalcare grant and also Malaysia Ministry of Higher Education (MOHE).

IOP Conf. Series: Materials Science and Engineering **601** (2019) 012015 doi:10.1088/1757-899X/601/1/012015

Influence of groundnut shell ash on the properties of cement pastes

Jamilu Usman [1*], Nasiru Yahaya[2], Mohammed Mazizah E[2]

[1] Building Department, Ahmadu Bello University Zaria, Nigeria
[2] Civil Engineering Department, University College of Technology Sarawak, Malaysia
*Corresponding author:jamilonline05@gmail.com

Abstract. This paper presents the effect of groundnut shell ash (GHA) on the properties of cement paste. Cement pastes containing GHA as cement replacement in different proportions up to 50% by weight were prepared. At fresh state, the consistency and setting times of the pastes were checked, while at hardened state, soundness and compressive strength of the pastes were determined. In addition, the microstructure of the hardened paste was investigated using the Fourier Transformed Infrared (FTIR) technique. The results show that GHA increased water demand and delayed setting times, but improved soundness of cement paste. Moreover, compressive strength enhancement of hardened cement paste due pozzolanic reaction as evidenced by the microstructure analysis (FTIR) was observed with up to 10% GHA replacing cement. Thus, GHA is a good potential material for cement replacement in cement paste.

1. Introduction

Currently and for the foreseeable future, Portland cement remains the binder of choice for concrete to develop infrastructures for our socio-economic needs [1]. However, cement production is characterized with both high energy consumption and the emission of enormous greenhouse gases to the atmosphere, which contribute to economic and environmental sustainability pressures in the construction industry. This calls for the utilization of supplementary cementitious materials (SCMs), such as; fly-ash, slag, and metakaolin, as partial replacement of Portland cement for the production of cement-based materials. Since, most of the SCMs are predominantly industrial by-products; their successful inclusion into cement-based material would equally guarantee an effective means of waste management for the industries from which the wastes are generated. Similar to other industries, agricultural industry also generates wastes, commonly known as agro-wastes. The typical agro-wastes among others are; palm kernel shell, bagasse, rice husk, and groundnut shell. Interestingly, because of the high calorific values of these wastes, they are often utilized as biofuels in boilers for energy production [2-4]. In addition, most of the ashes (residues) generated from burning the biofuels which are mostly disposed of in landfills can be used as cement replacement materials in concrete, owing to their pozzolanic characteristics [3].

Groundnut shell is an agro-waste generated from milling of groundnut. Groundnut production worldwide had reached about 42.31 million tons in 2014, and it was projected to continuously increase annually (FAO, 2014) [5]. China, India and Nigeria are the first three leading countries that contributed about 37.3%, 15.5% and 8.1% of the total global groundnut production, respectively. These countries are still leading in the production of groundnut. Approximately, groundnut constitutes about 25-40% shells. And in every one ton of the shells, 2.5% ash is produced [6, 7]. Therefore, the total global groundnut ash produce could be up to 0.42 million tons. So far, the ash has less economic value; instead, it becomes an environmental nuisance when disposed of in landfills, causing soil, groundwater and air pollutions.

Groundnut shell was used as fine aggregate replacement to produce lightweight concrete [4, 8]. The shell was equally utilized to produce insulation board and wardrobe [9, 10]. However, there are limited studies on the use of groundnut shell ash as cement replacement in cement based material. Therefore, the aim of this study is to investigate the effect of groundnut shell ash on the properties of cement paste. The properties examined include; consistency, setting times, soundness and compressive strength. Equally, the microstructure of the pastes was checked using the Fourier Transformed Infrared (FTIR).

2. Materials and Methods

2.1 Materials

Limestone Portland cement and groundnut shell ash (GSA) were used as binders in this study. The ash was obtained by burning groundnut shell to 700°C in an incinerator. The shell was sourced from a groundnut milling factory located in northern Nigeria. To improve its reactivity, the ash was ground to finer particles after heating. The grinding was conducted using a les Angeles grinding machine. After grinding, the medium particle size of the ash was found to be 11.42μm which is less than that of cement 15.90μm as shown in Table 1. The chemical compositions of the binders are shown in Table 1. Based on its chemical composition, GSA constitutes silica, calcium oxide and alumina as the major oxides. The summation of silica, iron oxide and alumina is 46.94%, which is less than the minimum of 50% stipulated by the ASTM 618 [11] for Class C pozzolan. However, due to its high calcium oxide content (21%), but low SO3 ($\leq$ 5%) and LOI ($\leq$6%), GSA may loosely be classified as Class C pozzolan.

Table 1. Chemical composition and physical properties of binders

Chemical composition (%)	OPC	GSA
Silicon dioxide (SiO2)	19.78	31.80
Aluminum oxide (Al2O3)	3.90	10.12
Iron oxide (Fe2O3)	3.00	5.03
Calcium oxide (CaO)	63.38	20.79
Magnesium oxide (MgO)	2.00	8.77
Sulfur trioxide (SO3)	2.85	2.46
Sodium oxide (Na2O)	0.75	0.09
Phosphorus pentoxide (P2O5)	0.12	4.82
Chloride (Cl)	0.01	0.35
Potassium oxide (K2O)	0.18	12.51
Titanium oxide (TiO2)	-	2.07
Zinc oxide (ZnO)	-	0.19
Mn2O3	-	0.81
SrO	-	0.16
Loss on Ignition (LOI)	1.90	7.95
SiO2 + Fe2O3 + Al2O3	-	46.94
Physical properties		
Specific gravity	3.15	2.22
Strength activity index (%)	-	82
Median particle size, D50 (μm)	15.90	11.42

2.2 Preparation of Samples

Six different paste mixes constituting GHA in different proportions were produced. Constant water to binder ratio of 0.36 was used for all the mixes. The mix proportions are shown in Table 2. The mix without GHA (100% cement) represents the control sample upon which the other mixes were compared to. 50mm cube moulds were used to cast the samples for compressive strength test. After 24 hours of casting, the samples were demoulded and then cured in water for 7 and 28 days before testing. The result reported is the average of three samples.

Table 2. Mix proportion of binders by weight

Paste type	Cement (%)	GHA (%)
Control	100	0
10GHA	90	10
20GHA	80	20
30GHA	70	30
40GHA	60	40
50GHA	50	50

2.3 Methods

The consistency and setting times by Vicat needle of the mixes were determined in accordance with the ASTM C 187 [12] and ASTM C191 standards respectively. Soundness of binders by Le-Chateliers method (IS: 4031- Part 3- 1988)[13] and compressive strength based on ASTM C 109 [14] were also determined. The compressive strength test was done after 7 and 28 days of curing.

3. Results and Discussion

3.1 Consistency

The consistencies (water demands) of the different mixes are shown in Figure 1. It can be observed that the specimens with GHA required more water than the control to achieve the desired consistency recommended by ASTM C 618 [11]. And the rate of water demand increased with the increasing GHA content. In fact, the water demand for the blended specimens with 10%GHA, 30%GHA and 50%GHA was higher than that of the control by 46%, 100% and 146% respectively. The spongy nature of GHA particles which is typical to most agro-waste ashes, leading to their high surface area may be responsible for the high water requirement [15].

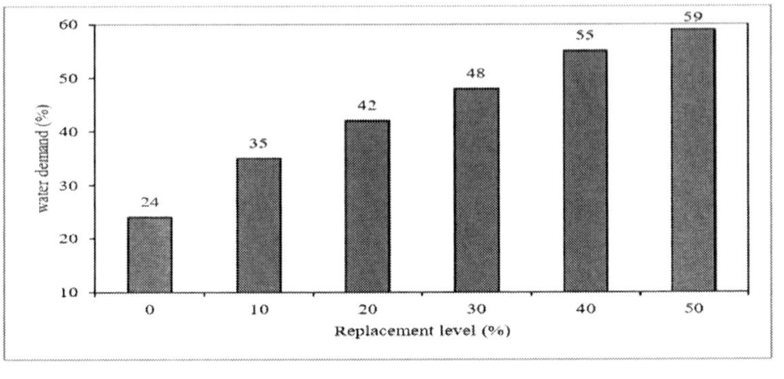

Figure 1. Consistencies of mixes

3.2. Setting Times

Figure 2 presents the initial and final setting times of the control and blended pastes. It can be seen that GHA retarded both the initial and final setting times of cement paste, and the retarding effect increased with the increasing ash content. However, it is noteworthy that at all the replacement levels, the minimum initial setting time (60mins.) and the maximum final setting times (600mins.) recommended by the ASTM standard [16] were not exceeded. This implied that, GHA may be a suitable material for use where retardation of setting time is significant. The retarding effect exhibited by GHA is similar to those of the other ashes such as, rice husk ash, fly ash and palm oil fuel ash which are characterized with a low reactivity [3, 17].

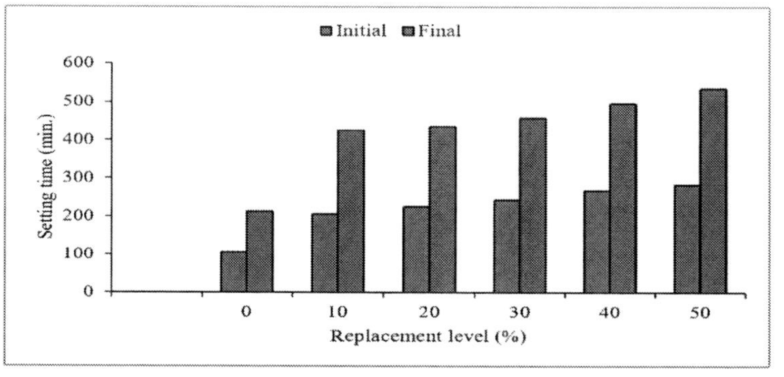

Figure 2. Setting Times

3.3. Soundness

Soundness indicates the ability of hardened binder paste to resist expansion due to excessive content of magnesia or free lime content in the binder [18]. The expansion of the control and blended specimens are depicted in Figure 3. Remarkably, none of the specimens exceeded the maximum allowable expansion of 10mm stipulated by the Indian Standard (IS: 1489-1991: Part 1) for sound Portland cement. Apparently, despite its high magnesia content as shown in Table 1(chemical composition), GHA improved the soundness of paste and the extent of the improvement increased with the ash content. Compared to the control, the expansion of blended specimens with 10%, 30% and 50% GHA were less by 8%, 28% and 56% respectively. The pattern of soundness improvement exhibited by GHA is similar to that shown by fly ash reported in [3].

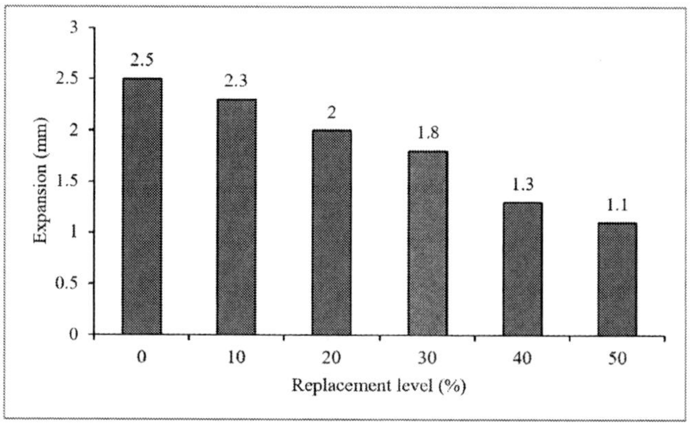

Figure 3. Expansion

3.4. Compressive strength

The influence of GHA on the compressive strengths of the cement paste specimens at 7 and 28 days are shown in Figure 4. At early age, 7 days, all the specimens with GHA showed compressive strength

IOP Conf. Series: Materials Science and Engineering **601** (2019) 012015 doi:10.1088/1757-899X/601/1/012015

lower than that of the control. The strength reduction effect of the blended specimens increased with the ash content. However, at 28 days, strength comparable to that of the control was achieved when 10% GHA was used. Therefore, the optimum replacement level of GHA to improve compressive strength of cement is 10%. The strength reduction effect of GHA on the paste at early age may be unconnected with the low pozzolanic characteristic of agro waste ashes. However, dilution effect may be attributed to the reduction of strength for the specimens containing GHA above 10% levels at 28 days. The trend of strength development of GHA observed in this study is similar to that when fly ash was used to replace cement in concrete [3].

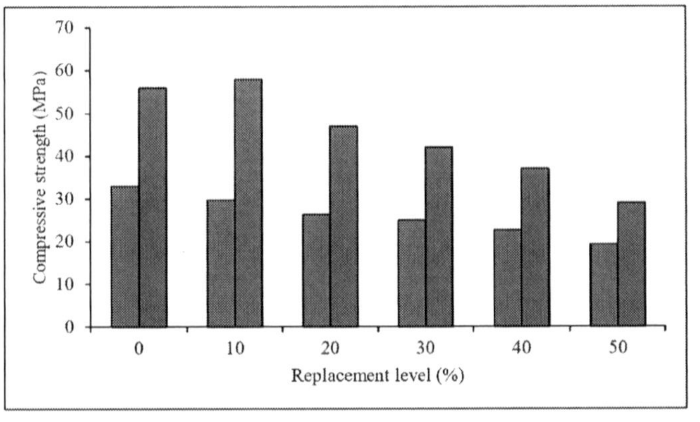

Figure 4. Compressive strength

3.5. FTIR

The IR-spectra of the control and blended samples (with 10% GHA) are shown in Figure 5. It can be observed that $Ca(OH)_2$ is present in both the samples as evidenced by the appearance of the OH-stretching band at 3642 cm^{-1}. Similarly, the detection of the asymmetric Si-O stretching band (v_3) at 954 cm^{-1} confirmed the formation of calcium silicate hydrate (C-S-H) in the samples. The C-O stretching (v_3) at 875cm-1 and bending at 1424 cm-1 indicated the existence of calcite. However, compared to the control, the blended sample has less $Ca(OH)_2$ but more CSH content as reflected by their respective intensities. This highlighted the evidence of pozzolanic reaction between GHA and cement hydrate. Thus, the improvement of compressive strength shown by the blended sample can be attributed to the pozzolanic reaction.

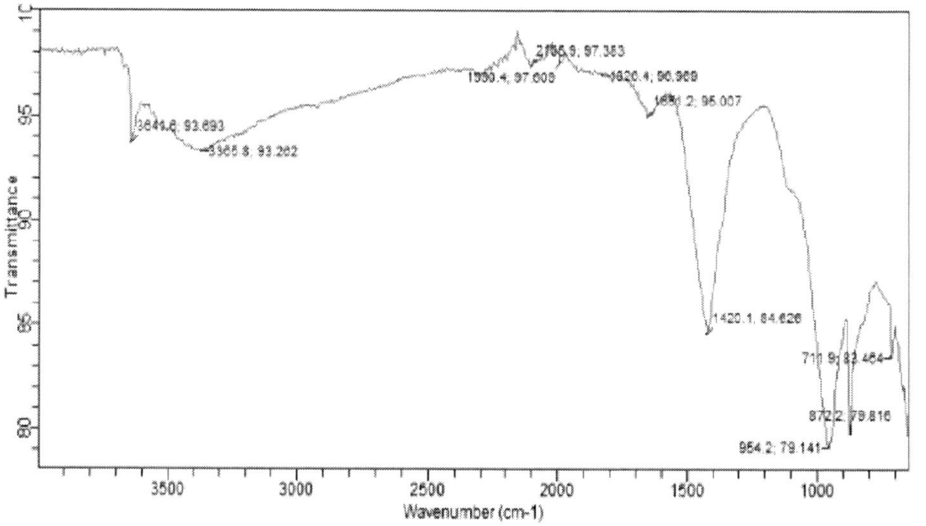

(a)

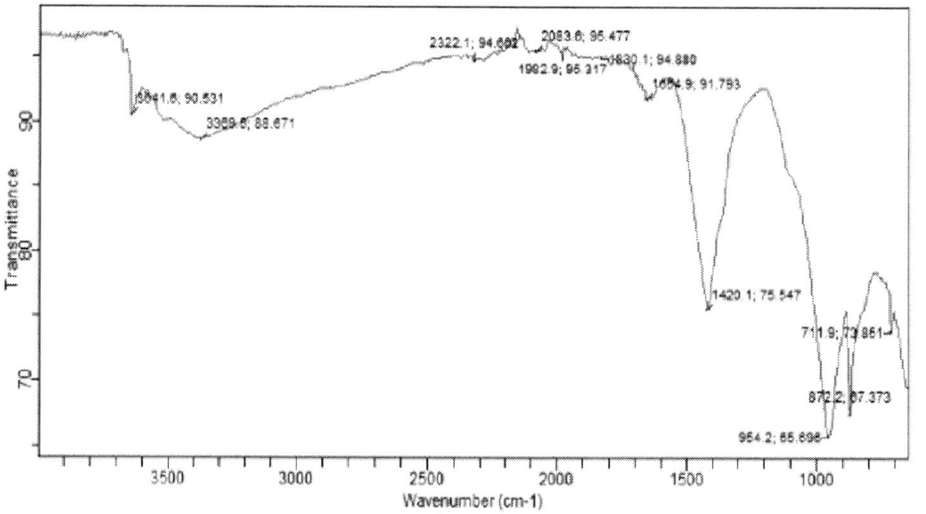

(b)

Figure 5. IR-spectra of samples; (a) 100% PC (control) (b) 10%GHA

4. Conclusion

Based on the results obtained in this study, the following conclusions were drawn:

a) Groundnut shell ash (GHA) increased water requirement, delayed both initial and final setting times, but improved soundness of cement paste.

b) Compressive strength of hardened cement paste improved with 10% GHA replacing cement. Therefore, 10% is the optimum replacement level of GHA in cement paste

5. References

[1] Aïtcin P C and Mindess S 2011 *Sustainability of concrete* (London: Spon Press) pp 1-10

[2] Tay J H and Show K Y 1995 Use of ash derived from oil-palm waste incineration as a cement replacement material. *Res. Conserv. and Recycl.* **13** 27-36

[3] Ahmaruzzaman M A 2010 Review on the utilization of fly ash. *Progrs. in Energ. and Combs. Sci.* **36(3)** 327-363.
[4] Prusty J K, Patro S K and Basarkar S 2016 Concrete using agro-waste as fine aggregate for sustainable built environment- A review. *Int. Journ. of Sust. Built Env.* **5(2)** 312-333.
[5] ASTMC348 2014 *Standard Test Method for Flexural Strength of Hydraulic-Cement Mortars* (West Conshohocken: ASTM International) pp 1-6
[6] Singh F and Diwakar B 1993 *Nutritive value and uses of pigeonpea and groundnut*, Skill Development Series No. 14 (India: International Crops Research Institute for the Semi-Arid Tropics) pp 1-34
[7] Abdulrazak S, Otie D and Oniwapele Y 2014 Proximate analysis and anti-nutritional factors of groundnut and melon husk. *Online Journ. of Anim. and Feed Res.* **4(2)** pp 25-28
[8] Sada B, Amartey Y and Bakoc S 2013 An Investigation into the use of groundnut as fine aggregate replacement. *Nigerian. Journ. of Tech.* **32(1)** pp 54-60
[9] Gunasekaran K and Kumar P 2008 Lightweight concrete using coconut shell as aggregate. Proc. *Int. conf. on advances in concrete and construction.* (Hyderabad) vol 1, (India) pp 450-459
[10] Raju G, Gaitonde V and Kumarappa S 2012 Experimental study on optimization of thermal properties of groundnut shell particle reinforced polymer composites. *Internal. Journ. of Emerging Sci.* **2(3)** pp 433-454
[11] ASTMC618 2012 *Standard Specification for Coal Fly Ash and Raw or Calcined Natural Pozzolan for Use in Concrete* (West Conshohocken: ASTM International) pp 1-5
[12] ASTMC187 2011 *Standard Test Method for Amount of Water Required for Normal Consistency of Hydraulic Cement Paste* (West Conshohocken: ASTM International) pp 1-3
[13] IS:4031 Part 3, 1988 *Methods of Physical Tests for Hydraulic Cement: Determination of Soundness*, (New Delhi: Bureau of Indian Standard) pp 1-10
[14] ASTMC109C109M 2013 *Standard Test Method for Compressive Strength of Hydraulic Cement Mortars (Using 2-in. or [50-mm] Cube Specimens* (West Conshohocken: ASTM International) pp 1-10
[15] Veeresh, S.J. and J. Narayana, Assessment of Agro-Industrial Wastes Proximate, Ultimate, SEM and FTIR analysis for Feasibility of Solid Bio-Fuel Production. *Universal Journal of Environmental Research & Technology*. 2012. 2(6).
[16] ASTMC150C150M 2012 *Standard Specification for Portland Cement* (West Conshohocken: ASTM International) pp 1-9
[17] Kroehong W, Theerawat S, Chai J and Prinya C 2011 Effect of palm oil fuel ash fineness on the microstructure of blended cement paste. *Constr. and Build. Mater.* 25 pp 4095-4104
[18] Nevillie A 2002 *Properties of Concrete–Fourth and Final Edition* (Essex: Pearson Education Limited) pp 51

Acknowledgement
The authors are indebted to the technicians of the Concrete Laboratory, Department of Building, Ahmadu Bello University Zaria, for their assistance towards conducting the experimental works

Preliminary Study on Properties of Oil Palm Shell Lightweight Concrete with Cockle Shell as Mixing Ingredient

Mat Aris Shariza[1*], Dick Carlos Philip[1], Syed Mohsin Sharifah Maszura[1]

[1] Faculty of Civil Engineering and Earth Resources, Universiti Malaysia Pahang, Lebuhraya Tun Razak, 26300 Gambang, Kuantan, Malaysia
*Corresponding author: sharizaaris@ump.edu.my.

Abstract. Environmental pollution caused by the dumping of oil palm shell (OPS) from palm oil mill and cockle shell from cockle trade has initiated early exploration to discover the potential of these waste incorporated in concrete production. The present research investigates the effect of integrating crushed cockle shell as partial fine aggregate replacement on compressive strength and flexural strength of OPS lightweight aggregate concrete. A total of five mixes were prepared. OPS lightweight aggregate concrete containing 100% river sand was used as control specimen. Other mixes were produced by varying the percentage of crushed cockle shell by weight of sand. All specimens were water cured for 28 days before subjecting it to compressive strength and flexural strength. The finding shows the concrete exhibit strength increment when crushed cockle shell is added as partial fine aggregate replacement. Conclusively, crushed cockle shell has the potential to be used as mixing ingredient in OPS lightweight concrete production.

1. Introduction

Sand and gravel are two important ingredients required to make concrete in the building construction. As the amount of population is increasing, the needs for development and facilities also raises inducing growths in the construction industry. In Malaysia, most of the buildings and infrastructures are made from reinforced concrete as the materials are cheaper, easy to construct, stronger, durable and more sustainable. About 60-70% of the concrete components are made from fine and coarse aggregate, which are normally sand and granite, respectively. The aggregates are from natural resources obtained through mining the river and rock quarries. The consumption of these natural resources should be handled cautiously to avoid disruption on the environmental and topographical of river [1] and scarcity of the resources. Therefore, others alternative should be considered to overcome this situation. Using wastes as aggregates replacement is one of the potential solutions.

Malaysia is a country that have good soil for agricultural and plantation. Due to this, Malaysia becomes a major player in agricultural industries such as palm oil, paddy, and rubber. The agricultural industries also produce agricultural waste that can pollute the environment and requires significant amount of space for landfill disposal. Incorporating this waste into construction industry to replace natural resources will impact a win-win situation for both industries. The problem arises due to shortage of dumping areas can be eliminated if the waste can be recycled and reuse in the construction industries. Furthermore, the usage of natural resources such as sand and granite, or even cement can be replaced with this waste to produce sustainable concrete and minimize pollution. At the same time, the environmental problems due to the river mining, waste disposal and even air pollution can be controlled. Thus, previous work has studied and tries to improve the performance of the concrete by using the continuously generated palm oil wastes such as oil palm shell [2-7], palm oil clinker [8-11] and palm oil fuel ash [12-17] which results in production of more environmental friendly concrete.

In the meantime, the expanding of fisheries industry such as cockle trade also poses environmental issue. Cockles that exist in muddy area is harvested, processed to remove the shell before the edible meat is sold. The shells are dumped at open area causing pollution to the environment [12] and emit unwelcomed stinking smell. The growing cockle industry generates larger amount of shell which dumped as waste. Turning this thrash into beneficial material would be rewarding both in terms of reduced land used for waste disposal and cleaner environment. The availability of free shells has inspired researchers [19-22] to utilize it as mixing ingredient partially replacing certain natural resource in construction material production. However, very limited research is available on the utilization of two types of wastes namely oil palm shell and cockle shell as mixing ingredient in concrete production. Thus, this research investigates the effect of crushed cockle shell content as partial fine aggregate replacement on the strength performance of oil palm shell lightweight aggregate concrete.

2. Experimental Details

2.1. Materials

Ordinary Portland cement (OPC) complying to ASTM C150 [23] was used as the binder. River sand from a single source was used as fine aggregate. Supplied tap water was used for concrete preparation and curing. Oil palm shells were collected from dumping area in the area of palm oil mill as illustrated in Fig. 1. The oil palm shells used is produced from palm oil tree of Dura species. Oil palm shells were cleaned to ensure it is free from debris. Then it is oven dried before kept in a closed container. Cockle shells shown in Fig. 2 were obtained from the dumping area nearby cockle processing location. Then, the shells were washed to remove the sands and dirt on its surface. Then it was dried before crushed into fine particles.

Figure 1. Oil palm shell disposal area.

Figure 2. Cockle shell.

2.2. Mix Proportion

Two types of oil palm shell lightweight aggregate concrete mixes were used in this experiment. A Grade 20, plain OPS lightweight aggregate concrete consisting 100% river sand was used as control specimen, CS0. The rest of the mixes CS5, CS10, CS15 and CS20 were prepared by varying the percentage of crushed cockle shell by weight of fine aggregate replacement. The quantities of crushed cockle shell used are 5%, 10%, 15% and 20% by weight of sand. Other mixing ingredients were kept constant. Table 1 presents the mix proportion of the mixes used.

Table 1. Mix proportion of concrete samples.

Materials	**kg/m³**				
	CS0	**CS5**	**CS10**	**CS15**	**CS20**
Cement	550	550	550	550	550
Oil palm shell	360	360	360	360	360
Fine aggregate	870	826	783	739	696
Crushed cockle shell	0	44	87	131	174
Superplasticizer	5	5	5	5	5
Water	225	225	225	225	225

2.3. Specimens preparation and testing

Mixes were prepared in form of cubes and beams. All the specimens were cured up to 28 days using water curing method in the laboratory. Then, the specimens were subjected to mechanical properties testing at 7, 14 and 28 days. The cubes compressive strength test was conducted following the procedure in BS EN 12390-3 [24]. The flexural strength of the beam specimens was measured in accordance with BS EN 12390-5: [25].

3. Results and Discussion

The Figure 3 and 4 illustrate the effect of crushed cockle shell content as partial sand replacement on compressive strength and flexural strength of oil palm shell lightweight concrete. All specimens exhibit strength increment throughout the curing age. Leaving the specimens immersed in water at all time, ensures undisturbed hydration process. Presence of moisture allows chemical reaction by cement that generates larger amount of binding gel known as calcium silicate hydrate gel. These gels fill in the voids of internal structure making the hardened concrete denser. The formation of compact structure increases the capacity of concrete to sustain larger load. Calcium silicate hydrate gel plays significant role in increasing concrete strength through densification of the microstructure [26]. Therefore, providing a suitable environment for uninterrupted hydration process is vital for strength development of concrete containing cement as a sole binder.

Looking at the effect of adding crushed cockle shell, this preliminary study shows that inclusion of crushed cockle shell waste up to 20% contributes positively in enhancing the compressive strength and flexural strength of concrete. It is believed that role of fine crushed particle acts as filler creates a more compacted concrete internal structure. This assists the concrete to carry higher load. The encouraging contribution of other waste material when used as fine aggregate replacement at suitable amount in enhancing concrete strength has been reported in past studies [27-29]. Further study needs to be carried out to determine the optimum percentage of crushed cockle shell which contains high calcium oxide that can be used without detrimental effect to the oil palm shell lightweight aggregate concrete strength and durability aspect. The role of crushed cockle shell in improving oil palm shell lightweight aggregate concrete performance remains to be ventured. Conclusively, the success in combining both agricultural wastes for production of construction material saves the use of the precious natural resources, ensures environmental sustainability and reduces amount of waste thrown as landfill.

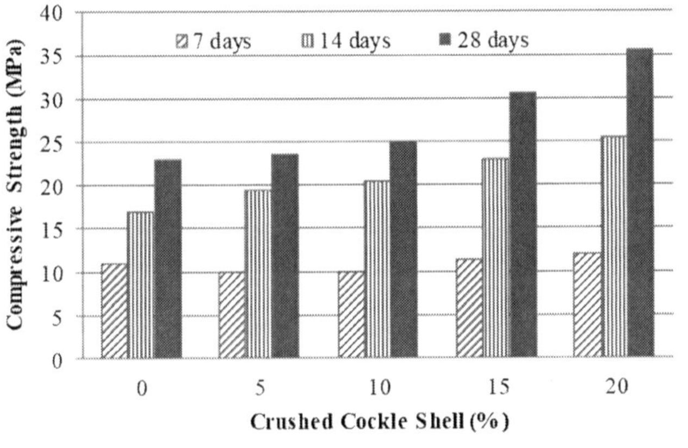

Figure 3. Compressive strength of specimens containing various percentages of crushed cockle shell

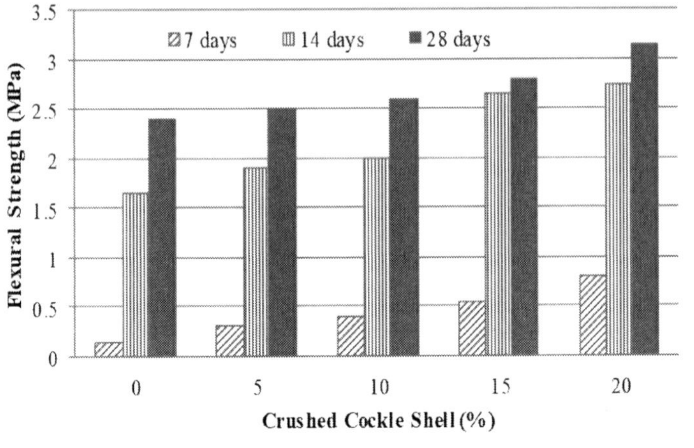

Figure 4. Flexural strength of specimens containing various percentages of crushed cockle shell.

4. Conclusion

This investigation founds that integration up to 20% of crushed cockle shell contributes towards strength enhancement of oil palm shell lightweight aggregate concrete. Further investigation on the effect of integrating crushed cockle shell as partial fine aggregate replacement towards the long term

IOP Conf. Series: Materials Science and Engineering **601** (2019) 012016 doi:10.1088/1757-899X/601/1/012016

mechanical, durability and thermal conductivity properties of oil palm shell lightweight aggregate concrete is among the area that remains to be ventured.

5. References

[1] Sathiparan and HTSM. De Zoysa 2018 The effects of using agricultural waste as partial substitute for sand in cement blocks *Journal of Building Engineering* **19** 216-227

[2] Shafigh P, Johnson Alengaram U, Mahmud H B and Jumaat M Z. 2013 Engineering properties of oil palm shell lightweight concrete containing fly ash, *Materials and Design* **49** 613-621

[3] Azimi S J, Syed Mohsin S M, Yahaya F and Namdar A 2014 An investigation on engineering properties of composite beam *Research Journal of Applied Sciences, Engineering and Technology* **8**(6) 702-705

[4] Syed Mohsin S M, Azimi S J and Namdar A 2014 Behaviour of oil palm shell reinforced concrete beam added with kenaf fiber, *Applied Mechanics and Materials* **567** 351-355

[5] Muthusamy K, Zamri N A, Zubir M A, Kusbiantoro A and Wan Ahmad S. 2015 Effect of mixing ingredient on compressive strength of oil palm shell lightweight aggregate concrete containing palm oil fuel ash *The 5th Int. Conf. of Euro Asia Gul. Engineering Forum (Surabaya, Indonesia)* vol 125 pp 804-810

[6] Muthusamy K, Zulkepli N A and Mat Yahaya F 2013 Exploratory study of oil palm shell as partial sand replacement in concrete *Research Journal of Applied Sciences, Engineering and Technology* **5**(7) 2372-2375

[7] Muhammad Nazrin Akmal A Z, Muthusamy K, Mat Yahaya F, Mohd Hanafi H and Nur Azzimah Z 2018 Utilization of fly ash as partial sand replacement in oil palm shell lightweight aggregate concrete *Proceeding of Global Congress on Construction, Material & Structural Engineering (Johor Baru, Johor)* vol 271(1) (UK: IOP Publishing Ltd.) pp 012003

[8] Mohammed B S, Foo W L, Hossain K M A and Abdullahi M 2013 Shear strength of palm oil clinker concrete beams, *Materials and Design* **46** 270–276

[9] Abutaha F, Razak H A and Kanadasan J 2016 Effect of palm oil clinker (POC) aggregates on fresh and hardened properties of concrete *Journal of Construction and Building Materials* **112** 416-423

[10] Ahmmad R, Alengaram U J, Jumaat M Z, Sulong N H R, Yusuf M O and Rehman M A 2017 Feasibility study on the use of high volume palm oil clinker waste in environmental friendly lightweight concrete *Journal of Construction and Building Materials* **135** 94-103

[11] Muthusamy K, Zamri N A and Kusbiantoro A 2018 Effect of palm oil fuel ash on compressive strength of palm oil boiler stone lightweight aggregate concrete *Proceeding of 1st Int. Conf. on Innovative, Technology, Engineering & Sciences (Pekan, Pahang)* vol 342 (UK: IOP Publishing Ltd.) pp 012098

[12] Chindaprasirt P, Homwuttiwong S and Jaturapitakkul C 2007 Strength and water permeability of concrete containing palm oil fuel ash and rice husk–bark ash *Journal of Construction and Building Materials* **21** 1492-1499

[13] Ismail M A, Budiea A M A, Hussin M W and Muthusamy K B 2010 Effect of POFA fineness on durability of high strength concrete *Indian Concrete Journal* **84**(11) 21-28

[14] Abdul Awal A S M and Shehu I 2015 Performance evaluation of concrete containing high volume palm oil fuel ash exposed to elevated temperature *Journal of Construction and Building Materials* **76** 214-220

[15] Muthusamy K and Zamri N A 2016 Mechanical properties of oil palm shell lightweight aggregate concrete containing palm oil fuel ash as partial cement replacement *KSCE Journal of Civil Engineering* **20**(4) 1473-1481

[16] Saffuan W A, Muthusamy K, Salleh N A M and Nordin N 2017 Properties of concrete containing ground palm oil fuel ash as fine aggregate replacement *Int. Conf. on Building*

Materials and Material Engineering (Lyon France) vol 264(1) (UK: IOP Publishing Ltd.) pp 012008

[17] Sulaiman M A, Muthusamy K, Mat Aris S, Paramasivam R and Othman R 2018 Effect of unground oil palm ash as mixing ingredient towards properties of concrete *Proc. of 4th Int. Con. On Civil & Environmental Engineering for Sustainability (ICONCESS 2017) (Langkawi Kedah)* vol 140(1) (UK: IOP Publishing Ltd.) pp 012150

[18] Boey L, Maniam G P, Abd Hamid S and Ali D M H 2011 Utilization of waste cockle shell (Anadaragranosa) in biodiesel production from palm olein: Optimization using response surface methodology *Fuel* **90** 2353-2358

[19] Mijan N A, Sahari F and Yusoff S B 2011 Cockle shell in artificial reef construction *Proceedings of 3rd International Conference on Applied and Creative Arts (Unimas, Kuching Sarawak)* pp 1-13

[20] Muthusamy K and Sabri N A 2012 Cockle shell: a potential partial coarse aggregate replacement in concrete *International Journal of Science, Environment and Technology* **1**(4) 260-267

[21] Nor Hazurina O, Badrul Hisham A B, Mashitah M D and Megat Johari M A 2013 Cockle shell ash replacement for cement and filler in concrete, *Malaysian Journal of Civil Engineering,* **25**(2) 201-211

[22] Mat Aris S, Muthusamy K, Uzer A and Wan Ahmad S 2018 Properties of palm oil fuel ash cement sand brick containing pulverized cockle shell as partial sand *Proc. of 4th Int. Con. On Civil & Environmental Engineering for Sustainability (ICONCESS 2017) (Langkawi Kedah)* vol 140(1) (UK: IOP Publishing Ltd.) pp 012145

[23] ASTM C150-18 2018 *Standard Specification for Portland Cement* (West Conshohocken, PA: ASTM International)

[24] BS EN 12390-3: 2009 *Testing hardened concrete Part 3: Compressive strength of test specimens* (London: British Standard)

[25] BS EN 12390-5: 2009 *Testing hardened concrete Part 5: Flexural strength of test specimens* (London: British Standard)

[26] Safiuddin Md, Raman S N, and Zain M F M 2007 Effect of different curing methods on the properties of micro silica concrete *Australian Journal of Basic and Applied Science* **1**(2) 87–95

[27] Bisht K and Ramana P V 2018 Sustainable production of concrete containing discarded beverage glass as fine aggregate *Journal of Construction and Building Materials* **177** 116-124

[28] Shettima A U, Hussin M W, Ahmad Y and Mirza J 2016 Evaluation of iron ore tailing as replacement for fine aggregate in concrete *Journal of Construction and Building Materials* **120** 72-79

[29] Muthusamy K, Mohamad Hafizuddin R, Mat Yahaya F, Sulaiman M A, Syed Mohsin S M, Tukimat N N, Omar R and Chin S C 2018 Compressive strength performance of OPS lightweight aggregate concrete containing coal bottom ash as partial fine aggregate replacement *Proceeding of 1st Int. Conf. on Innovative, Technology, Engineering & Sciences (Pekan, Pahang)* vol 342(1) (UK: IOP Publishing Ltd.) pp 012099

Acknowledgement
The authors would like to acknowledge the financial support provided by Universiti Malaysia Pahang through internal grant RDU170337.

High Strength Concrete Incorporating Oil-Palm-Boiler Clinker as Coarse Lightweight Aggregate

Jin Chai Lee[1*], Payam Shafigh[2,3] and See Kim Lee[1]

[1]Department of Civil Engineering, Faculty of Engineering, UCSI University, Cheras, 56000 Kuala Lumpur, Malaysia
[2]Department of Building Surveying, Faculty of Built Environment, University of Malaya, 50603 Kuala Lumpur, Malaysia
[3]Center for Building, Construction & Tropical Architecture (BuCTA), Faculty of Built Environment, University of Malaya, 50603 Kuala Lumpur, Malaysia
*Corresponding author: leejc@ucsiuniversity.edu.my

Abstract. Construction industry demands for high amount of raw material to produce concrete. The constant reduction of natural resources results in negative impact to the environment and lack of raw material. Promotion of utilization of agricultural wastes in production of concrete can minimize the environmental impact towards the sustainable development. In this study, the oil-palm-boiler clinker (OPBC) which is an agricultural waste from palm oil industry was used to substitute the crush granite in conventional concrete from 0 to 50% by volume in increments of 12.5%. Tests conducted in this study were compressive strength in different curing regimes, flexural strength and water absorption. The test results showed that concretes containing 37.5 to 50% OPBC can be categorized as semi-lightweight aggregate concrete with high strength. This aggregate has considerable potential for performing internal curing in concrete. Using OPBC in concrete reduced the flexural strength; however, the reduction was not significant. Although water absorption showed that concretes containing OPBC have slightly higher water absorption, however, all these concrete could be considered as good quality concrete.

1. Introduction

Due to the rapidly growing world population, the demand for and cost of building materials is tremendously ascending [1]. In order to cater to market demand and economic factors, the constant reduction of natural resources has resulted in the negative impact to the environment and shortage of raw materials. Some excessive wastes were disposed to dumping sites and this led to environmental pollution. Nowadays, many researchers focus on utilization of solid waste discharge from agriculture to full or partial replacement of raw material in the construction industry. Therefore, many researchers are interested to explore the use of waste and utilize it as full or partial replacement of raw material for sustainable development [2]. In the past, the unwanted materials that has been utilised in construction sector, such as oil-palm-boiler clinker (OPBC), oil palm shell (OPS), corncob rice husk, coconut shell, tobacco waste and coconut shell [3].

Malaysia is well known for its palm oil industry and is considered the second largest country in the world to produce palm oil [4]. However, the palm oil industry generates million tons of agricultural wastes yearly and is classified as the main source to contribute the environmental pollution in the country. This kind of waste is known as oil-palm- boiler clinker (OPBC) and is obtained from the

IOP Conf. Series: Materials Science and Engineering **601** (2019) 012017 doi:10.1088/1757-899X/601/1/012017

boiler of every palm oil mill through the process of incineration of oil palm shell and fiber. In Malaysia, OPBC is abundant with little selling value and easily to obtain from palm oil mill. Other than using OPBC as a landfill material, this agro waste has high potential to convert into construction materials. OPBC that are crushed and sieved into required sizes is found suitable to replace normal gravel aggregate in concrete mixture as OPBC is highly porosity, irregular shape with thorn. It also possesses good lightweight nature [5].

In construction industry, there is always special attention to the use of lightweight structures due to their significant advantages. For example, cable-stayed bridge with the length of long span and small scale structure is considered as lightweight structures [6]. The weight of concrete structures can be reduced significantly if structural lightweight concrete is used instead of conventional concrete. The common method for manufacturing structural lightweight concrete is to replace coarse aggregate with lightweight aggregate and the fine aggregate with normal sand. The OPBC as lightweight aggregate can be used to produce structural lightweight aggregate concrete due to OPBC possessing lightweight characteristics [5].

The main objective of this study is to evaluate the effect of the partial substitution of crushed granite with OPBC from 0 to 50% in increments of 12.5% in high strength normal weight concrete on the compressive strength with different curing regimes, flexural strength and water absorption.

2. Experiment

2.1. Materials and properties

The binder used was ordinary Portland cement (OPC) with the Blaine specific surface area and specific gravity of 3510 cm^2/g and 3.14, respectively.

The coarse aggregate was OPBC and crushed granite. The OPBC in lump was sourced from palm oil producing mill in Dengkil. Figure 1 shows the OPBC in a lump. Prior to using the OPBC, it need to be washed for removing the dust, then crushed into 9.5 mm maximum nominal size by the crushing machine.

The fine aggregate was normal sand which sourced from a local mine. All the aggregates were used in dry condition. The physical properties of aggregates was shown in Table 1. The water sourced from normal tap water. To facilitate the dispersion of cement particles and improvement of concrete workability, a super-plasticizer (SP) was used in all mixes.

Figure 1. OPBC in a lump.

Table 1. Physical properties of aggregates.

Aggregate	Specific gravity	Compacted bulk density (kg/m^3)	Water absorption (%)	
			30min	24h
OPBC	1.9	1409	2.31	4.11
Granite	2.67	1453	0.53	0.72
Sand	2.65	1615	1.12	1.21

2.2. Mix proportions

All mixes were containing of OPC with 500 kg/m^3. The crushed granite was replaced with OPBC from 0 to 50% by volume of total crushed granite. The percentages of replacements were from 0 to 50% with an increment of 12.5%. Table 2 shows the mix proportions for all mixes.

Table 2. Mix proportions of concretes (kg/m^3).

Material	Mix code				
	C	**C12.5**	**C25**	**C37.5**	**C50**
OPBC	0	81	162	243	324
Granite	910	796	683	569	455
Sand	760	760	760	760	760
Cement	500	500	500	500	500
Water	170	170	170	170	170
SP	5	5	5	5	5

2.3. Test method

The casting ingredients were well prepared. First, all aggregates were placed into rotary drum mixer for two minutes. Secondly, OPC cement was placed into mixer with aggregate mixing for further three minutes. Thirdly, 70% water mixing with SP was poured into mix until well blended. Lastly, the balance water was added into the mix for five minutes. Then the mixture was to precede the slump test to determine the workability of the concrete.

In order to investigate the effect of different curing regimes on the compressive strength at the age of 28 days, the 100 mm cubes were performed under different curing regimes after 24 hours of casting. The different curing regimes are described in Table 3. Besides, the prisms of 100 mm x 100 mm x 500 mm beam for flexural strength and 100 mm cubes for water absorption tests were conducted after 27 days of water curing.

Table 3. Curing regimes.

Curing code	Description
AC	Specimen to be kept in air in a laboratory condition with RH% of 73 ± 5 and temperature of 29 ± 3 °C after demoulding
3W	Specimen kept in water for 2 days after demoulding and then kept in air under laboratory condition
5W	Specimen kept in water for 4 days after demoulding and then kept in air under laboratory condition
7W	Specimen kept in water for 6 days after 1 day demoulding and then kept in air under laboratory condition
FW	Specimen kept in water for 27 days after demoulding

3. Results and discussion

3.1. Compressive strength

Table 4 shows compressive strength at the age of 28 days under continuous moist curing and different curing regimes of all mixes. It was reported by Khoshkenari et al. [7] that the density of semi-lightweight aggregate concrete is between 1840 and 2240 kg/m^3. Therefore, the mixes of C37.5 and C50 in this study are considered as high strength semi-lightweight aggregate concrete. However, the mixes of C, C12.5 and C25 are under category of high strength conventional concrete. The high strength conventional concrete inclusion of 50% OPBC (mix of C50) showed the satisfactory workability without reducing compressive strength compared to mix of C37.5 under continuous moist curing. The control concrete mix (mix of C) and concrete containing low volume OPBC (mixes of

IOP Conf. Series: Materials Science and Engineering **601** (2019) 012017 doi:10.1088/1757-899X/601/1/012017

C12.5 and C25) shows a slight decrease on the compressive strength at the age of 28 days under AC condition. However, at higher OPBC contents (mixes of C37.5 and C50), the compressive strength has the same (or higher than) compressive strength at the age of 28 days as for the full water curing condition. In addition, these two mixes have equivalent strength to the control mix in AC condition. The test results clearly show the potential for using high volume OPBC in concrete mixture for improving compressive strength when proper curing is not provided for concrete.

All partial early curing exhibits a considerable effects on the compressive strength at the age of 28 days. The effectiveness is more significant for concrete containing a higher volume OPBC. Increasing the period of water curing gives improvement of concrete strength. Similar to AC curing conditions, improvement in the compressive strength of C37.5 and C50 concrete mixes is more than for the other concretes. This shows that OPBC aggregates provides internal curing leading to further improvement of the compressive strength.

Compared to full water curing, the highest improvement of compressive strength is exhibited in the 7W curing condition. Therefore, 7-day wet curing in concrete is recommended in most codes of practice [8]. However, it can be seen from Table 4 that for the 3W curing regime, the compressive strength is the same as the standard curing for control mix and for mixes containing low volume OPBC, while it is even higher for concretes containing high volume OPBC.

In general, it can be concluded from Table 4 that when crushed granite as coarse aggregate in the control mix is substituted by OPBC in high volume (more than 37.5%), water curing is not necessary. This may be a great advantage for using these wastes as coarse aggregate in concrete in that we know that the curing cost is about 0.1 to 0.5% of the concrete members in construction expenses [9].

Table 4. Slump, density and compressive strength of concretes under different curing regimes.

Mix No.	Slump (mm)	Oven dry density (kg/m^3)	The 28-day compressive strength (MPa)				
			FW	Partial early curing			
				AC	3W	5S	7W
C	160	2308	71.7	68.8	71.7	80.6	80.6
C12.5	140	2298	74.4	72.9	76.2	74.7	84.9
C25	130	2242	72.0	72.1	72.1	76.3	78.0
C37.5	85	2225	68.1	67.7	72.0	75.0	77.1
C50	75	2202	68.1	71.9	72.2	73.3	77.0

3.2. Flexural strength

The flexural strength at the age of 28 days of all concrete mixes is shown in Table 5. Compared to control mix, the concrete mixes containing OPBC show a lower flexural strength. For the compressive strength of concrete exceeding 25MPa, the flexural strength generally appears about 8 to 10% of 28-day compressive strength [10]. The ratio of flexural strength to compressive strength of concrete containing OPBC in this study is in the range of 9.3-10% within the usual range. Previous studies have revealed that this ratio for high strength lightweight concrete is about 9-11% [11]. Therefore, the ratio for concrete containing OPBC in this study is within this range.

Table 5. The flexural strength and compressive strength of concretes (MPa) at age of 28 days.

Mix No.	f_{cu}	f_r	$\frac{f_r}{f_{cu}}$%
C	71.7	7.87	11
C12.5	74.4	6.94	9.3
C25	72	6.98	9.7
C37.5	68.1	6.83	10
C50	*68.1*	*6.66*	*9.8*

3.3. Water absorption

Measurement for the quality of concrete based on initial water absorption (water absorption in 30 minutes) has been recommended by the CEB-FIP as poor, average and good for the ranges of 5% and above, 3-5% and 0-3%, respectively [12].

Figure 2 showed the initial and 24-hour water absorption for concrete mixes at the age of 28 days. This figure shows that increasing substitution of OPBC coarse aggregate in the mixture shows slightly increasing water absorption. This is due to water absorption of OPBC is much higher than that of crushed granite by about six times. The texture surface of OPBC aggregate is very porous compared to crushed granite. The water is absorbed into porosity of OPBC grain and hence increases the water absorption. However, it is noted that the initial absorption for concrete mixes is lower than 3%, which can be considered as good quality concrete by CEB-FIP [12]. The results show that even 24 hours water absorption of OPBC mixes are still less than 3%.

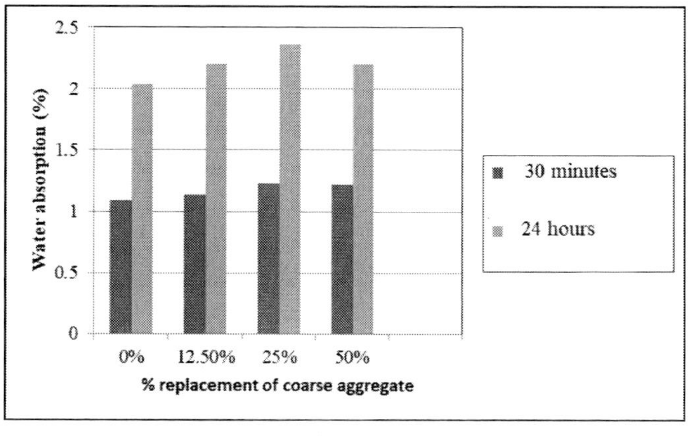

Figure 2. Relationship between OPBC volume and water absorption.

4. Conclusions

From the results obtained in this study, the following conclusion can be drown:

1. High strength semi-lightweight aggregate concrete with good workability could be manufactured by inclusion of 37.5 to 50% lightweight OPBC as coarse into normal weight concrete.
2. Concrete containing 12.5% OPBC has the highest compressive strength among the different curing conditions compared to the other concrete mixes. Therefore, it is recommended that at least 12.5% of conventional coarse aggregate to be replaced with OPBC in concretes.
3. Two days wet curing is sufficient for OPBC concretes to obtain the equivalent compressive strength of the concrete under full water curing. However, increasing the wet curing time to 4 and 6 days results in a higher compressive strength than the standard curing. Therefore, it is expected that OPBC coarse aggregate has great potential for internal curing.
4. Inclusion of OPBC in conventional concrete decreases the flexural strength.
5. With increasing the replacement of normal coarse aggregates with OPBC, the water absorption of mixture increases. A concrete containing OPBC up to 50% of the total volume of coarse aggregate can still be categorized as good concrete.

5. References

[1] Bashar S M, Foo W L and Abdullahi M 2014 Flexural strength of palm oil clinker concrete beams *Materials and Design* **53** 325-331

[2] Mannan M A and Neglo K 2010 Mix design for oil-palm-boiler clinker (OPBC) concrete *Science and Technology* **30**(1) 111-118

[3] Shafigh P, Mahmud H B, Jumaat M Z and Zargar M 2014 Agricultural wastes as aggregate in concrete mixtures – A review *Construction and Building Materials* **53** 110-117
[4] Basri H B, Mannan M A and Zain M F M 1999 Concrete using waste oil palm shells as aggregate *Cement Concrete Research* **29** 619–622
[5] Chai L J, Shafigh P and Bin Mahmud, H. 2017. Production of high-strength lightweight concrete using waste lightweight oil-palm-boiler-clinker and limestone powder *European Journal of Environmental and Civil Engineering* 1-20.
[6] Javanmardi A, Ibrahim Z, Ghaedi K, Jameel M, Khatibi H and Suhatril M 2017 Seismic response characteristics of a base isolated cable-stayed bridge under moderate and strong ground motions *Archives of Civil and Mechanical Engineering* **17** 419-432
[7] Khoshkenari A G, Shafigh P, Moghimi M and Mahmud H B 2014 The role of 0–2 mm fine recycled concrete aggregate on the compressive and splitting tensile strengths of recycled concrete aggregate concrete *Materials and Design* **64** 345 – 354
[8] Haque M N 1990 Some concretes need 7 days initial curing *Concrete International* **12**(2) 42–46
[9] Aitcin P C, Haddad G and Morin R 2004 Controlling plastic and autogenous shrinkage in high-performance concrete *ACI Special Publication* **220** 69-83
[10] Shetty M S 2006 *Concrete technology theory and practice* (New Delhi: S. Chand and Company) 624
[11] Omar W and Mohamed R N 2002 The performance of pre-tensioned concrete beams made with lightweight concrete, *Journal of Civil Engineering* **14**(1) 60-70
[12] CEB-FIP 1989 *Diagnosis and assessment of concrete structures - state of the art report* (192: CEB Bull) 83–85

Shear strength of FRP-Externally strengthened High Strength RC T-Beams

Basil Ibrahim[1*], Moussa Leblouba[1], Salah Al-Toubat[1] and Samer Barakat[1]

[1]University of Sharjah, Department of Civil & Environmental Engineering, P.O. Box 27272, University City, Sharjah, United Arab Emirates
*Corresponding author: basil@sharjah.ac.ae

Abstract. This paper presents preliminary results of an extensive experimental program on the behavior of high-strength (HS) reinforced concrete (RC) T-beams strengthened with Carbon Fiber Reinforced Polymer (CFRP) sheets. The test program involves the fabrication and testing of 56 RC beams and many objectives to achieve. The objective of this paper is to investigate the effect of the number of CFRP layers on the shear strength of HS RC beams. To this end, three full-scale specimens were prepared and instrumented with displacement transducers and strain gauges: one control beam, one single-layer of CFRP sheets, and one double-layer of CFRP sheets. All beams were simply supported and subjected to a monotonically increasing load until failure. Preliminary results showed that in the absence of transverse steel reinforcements, the number of CFRP layers does not affect the ultimate strength of the HSC beams.

1. Introduction

Generally, RC beams may undergo two types of failure: flexural and shear. The flexural type of failure is ductile as it occurs gradually, giving a warning before the failure occurs. Contrarily, the shear failure is brittle and catastrophic since it does not provide any kind of warning before it happens. Shear failure of concrete beams is commonly attributed to the insufficient shear reinforcement. However, it can be prevented by using different strengthening techniques. Different shear strengthening or repairing of concrete structures are available depending on the type of structural element, material, etc., these include: prestressing, shotcreting, and steel plate bonding. An alternative to these methods is the Externally Bonded Fiber Reinforced Polymers (EB-FRP). EB-FRP has been the subject of many studies [1-5].

The EB-FRP composites are used as shear reinforcements to improve the load-bearing capacity of beams and to control the deflection at the failure by constraining the growth of concrete cracks and delaying the loss of the aggregate interlock. The FRP can be bonded in S, U, or W configurations. It's called S configuration when the FRP is bonded only to the sides of the beam. When the FRP is bonded to both sides and the tension face of the beam, it's called U configuration. The W configuration, is the case where the FRP is fully wrapped around the beam section [6].

The most significant two modes of failure of beams strengthened in shear by FRP are the debonding of FRP sheets and FRP rapture. There are other modes however, such as peeling-off of the concrete cover, shear failure of the beam without the FRP being raptured, anchorage failure, etc.

In the S and U configurations, the FRP usually fail by debonding or peeling off of the concrete cover. The adhesive layer between the concrete surface and FRP sheets transmits the stress from the concrete to the FRP, thus, producing in-plane shearing between them. Therefore, debonding, which is mainly the loss of adhesion, is the main control of the ultimate strength in the S and U configurations.

Many studies on the use of FRP in shear strengthening of RC beams were carried out in recent years [1-3,7-9, among others). Although numerous studies were performed on the behavior of the bond between the concrete and the FRP, it still not well investigated yet and needs more research. The main objective of this work is to report preliminary results of a test program to study the effect of the number of CFRP layers in the absence of transverse steel reinforcements on HS RC beams.

2 Experimental Program

2.1 Material Properties

Three high strength (HS) concrete beams were casted with a concrete mix having an average of 70 MPa compressive strength (f'_c) at 28th day according to ASTM C39. The longitudinal bars used for the flexural reinforcement of the beams had a yield strength (f_y) of 560 MPa.

The EB-CFRP discrete strips used in this study are unidirectional woven carbon fiber flexible sheets. The CFRP sheets have a thickness (t_{frp}) of 0.168 mm, an ultimate elongation of 1.8%, a tensile modulus and tensile strength of 230 GPa and 4.8 GPa respectively. The CFRP sheets were applied to each beam specimen using a polymeric encapsulation resin.

2.2 Test specimens

All the three beams are large scale, HS RC T-beams without transverse reinforcement. The control specimen (Specimen 1) is virgin (i.e., without EB-CFRP). Specimen 2 is a beam that is strengthened with a layer of EB-CFRP strips in U-jacketing configuration placed at 90 mm from center to center. Specimen 3 similar to Specimen 2 except that the CFRP sheets are applied twice (double-layer). Table 1 summarizes the geometry and material properties of test specimens. Figure 1 shows the beams' section properties and Figure 2 illustrates the geometry of the beams and test setup.

As shown in Table 1, the beams have a T-section with a web width of 150mm (b_w), flange width of 450mm (b_f), flange height of 100mm (h_f), total height of 420mm (h), effective depth of 340mm (d), length of 3700mm and test span of 3300mm. The effective depth d_{frp} and the thickness t_{frp} of the CFRP are 240 mm and 0.166 mm/layer, respectively. All the 3 beams are designed to fail in shear. The longitudinal reinforcements consist of 4 25mm diameter bars as tensile reinforcements and 4 12mm diameter bars as compression reinforcements. Figure 1 shows the section details of the beams.

Table 1. Geometry and material properties of test specimens.

beam	h	h_f	b_w	b_f	Span	a/d	f'_c	ρ_s	d_{frp}	t_{frp}	w_{frp}	S_{frp}	ρ_{frp}
Units	mm	mm	mm	mm	m	-	MPa	-	mm	mm	mm	mm	-
1	420	100	150	450	3.3	3.2	70	0.0385	0	0	0	0	0
2	420	100	150	450	3.3	3.2	70	0.0385	240	0.166	50	90	0.0012
3	420	100	150	450	3.3	3.2	70	0.0385	240	0.332	50	90	0.0049

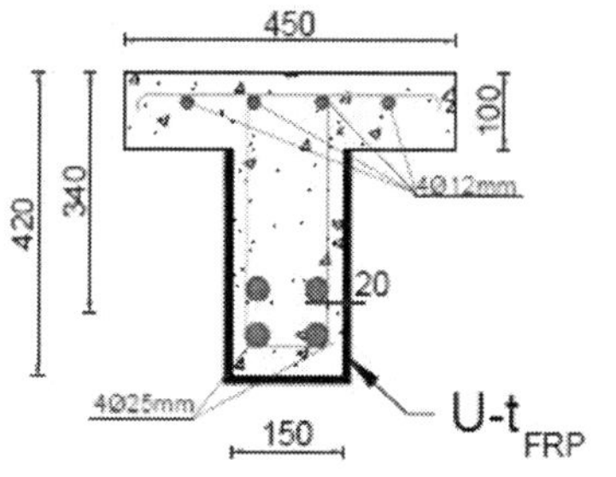

Figure 1. Beams section details.

2.3 Test setup and instrumentation

As shown in Figure 2 below, the test setup consisted of a simply supported condition using rollers. To control the shear failure location to be in the shorter span, the load has been applied at 1/3 of the beam's span. Thus the strain gauges were assigned at the expected shear failure location. Each beam specimen was tested under monotonically increasing point load in a displacement-controlled manner with a rate of 0.01mm/sec. along the test, beams deflection and cracking sequence were monitored during testing. Linear voltage displacement transducers (LVDTs) at three different locations along the span were used to recorded the beams deflection as shown in the figure.

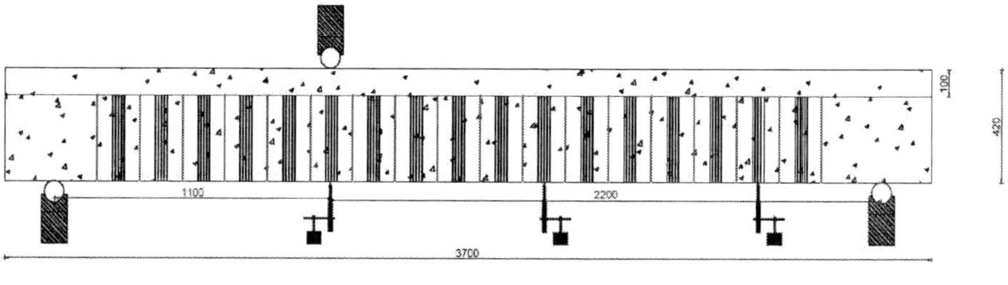

Figure 2. Geometry of the beam and test setup.

3. Results and discussion

As discussed previously, all 3 beams were designed to fail in shear. In what follows we discuss the preliminary results obtained from testing these beams.

3.1 Control Beam

Specimen 1 failed due to shear as shown in Figure 3. After the appearance of flexural cracks in the tension zone of the beam at a load level of 70 kN, the shear cracks started to appear at a load level of 130 kN in the shear span from the point of loading until the support, and since there's no transverse reinforcement to control the shear cracks, the cracks widened until the sudden failure at a load level of 196 kN.

Figure 3. Control beam after failure.

3.2 Single-Layer
The behavior of this beam, as shown in Figure 4, was not much different from the control beam. In the beginning, shear cracks appeared, at a similar load level of 130 kN in the shear span after the formation of flexural cracks. The shear cracks started to get wider from the point of load application until the support level at a load of 170 kN. After the beam reached its ultimate load capacity of 220 kN, the main shear crack increased abruptly. The contribution of the CFRP strips in resisting the shear cracks started beyond the load capacity of the beam then dropped sharply. Subsequently, the CFRP strips started to debond from the concrete surface without reaching the rupture of the fibers.

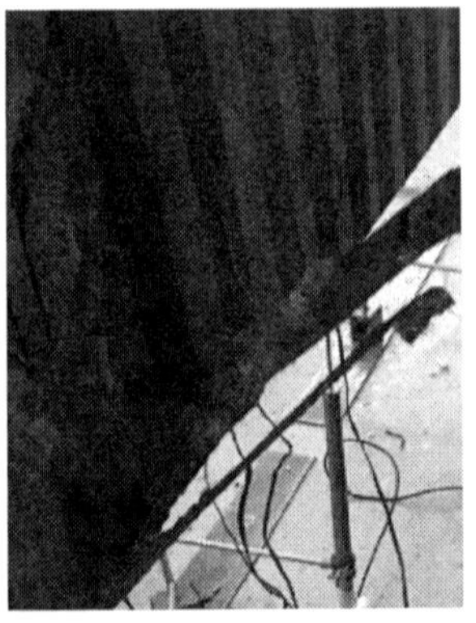

Figure 4. Single layer beam after failure.

3.3 Double-Layer
The performance of this beam was actually close to the behavior of the single layer. In the absence of the transverse steel reinforcement, adding another layer of CFRP strips was not effective as anticipated, which was mainly due to the debonding failure without rupture of the CFRP sheets (Figure 5).

Figure 5. Double layers beam after failure.

Figure 6 shows the load-mid-span deflection curves of the three beam specimens. Clearly, the CFRP had the effect of enhancing the post failure behavior of the strengthened beams. However, the CFRP layers had a negligible influence on the overall load-carrying capacity of Specimens 2 and 3; CFRP shear reinforcement increased the load capacity of the beams by around 20 kN only, shown in Table 2 below, which represents about 10% of the capacity of the control beam (i.e., Specimen 1). A noticeable effect of EB-CFRP was the residual strength which was about 23% of the ultimate load-capacity of the beams. The control beam failed without recording any residual strength.

The observed debonding failure of the EB-CFRP in both, Specimen 2 and 3, suggests that the performance of the single-layer and double-layer specimens are no different, which means adding more layer was not effective in enhancing the shear behavior of the HS RC beam in the absence of transverse steel reinforcements.

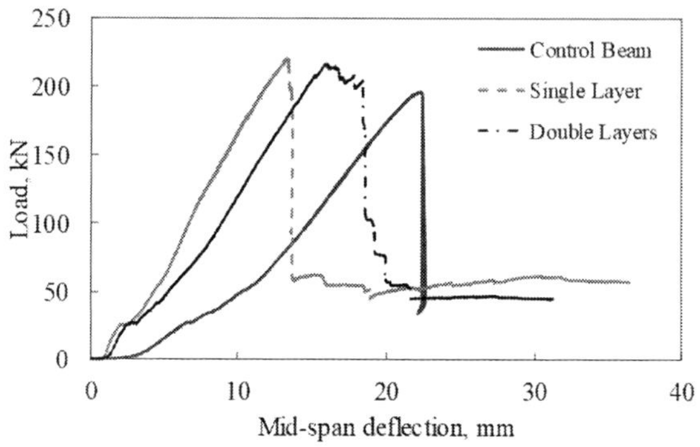

Figure 6. Load-Deflection curve.

Table 2. FRP contribution of the shear load capacity.

Beam	**Shear capacity load (kN)**	**FRP Contribution (kN)**
Control	196.60	-
Single Layer	220.95	24.35
Double Layers	216.97	20.37

4. Conclusion

The aim of this study was to investigate the behavior of the HS RC beams strengthened with single and double layers of EB-CFRP. This paper reported preliminary results pertaining to 3 specimens. Based on test results, the following conclusions were drawn:

- The shear failure of the externally U-bonded CFRP sheets to the T-section HS RC beams was of the debonding type, where the CFRP strips debonded before reaching rupture of CFRP sheets;
- The CFRP shear reinforcement enhanced the post-failure behavior of the HS RC beams. However, it increased the load capacity by no more than 10% of its control capacity only;
- CFRP sheets (be it single or double-layer) provided a residual strength of about 23% of the load-capacity of strengthened beams;
- Adding another layer of CFRP sheets to the HS RC beams was not efficient in the absence of the transverse steel reinforcement due to the debonding failure behavior.

5. References

[1] Uji K 1992 Improving shear capacity of existing reinforced concrete member by applying carbon fibers sheet *Trans. Japan Concr. Inst.* **14** 253-266

[2] Triantafillou T C 1998 Shear Strengthening of Reinforced Concrete Beams Using Epoxy-Bonded FRP Composites *ACI Structural J.* **95** 107–115

[3] Bousselham A and Chaallal O 2006 Behavior of Reinforced Concrete T-Beams Strengthened in Shear with Carbon Fiber-Reinforced Polymer — an Experimental Study *Structural J.* **103** 339–347

[4] Menegotto M, Monti G and Antonio M 2009 Modelling Shear Mechanisms in FRP-Strengthened R/C Beams *Arch. Civ. Eng. Env.* **3** 57–68

[5] Gonzalez-Libreros J H, Sneed L H, D'Antino T and Pellegrino C 2017 Behavior of RC beams strengthened in shear with FRP and FRCM composites *Eng. Structures* **150** 830-842

[6] Mostofinejad D and Kashani A T 2013 Experimental study on effect of EBR and EBROG methods on debonding of FRP sheets used for shear strengthening of RC beams *Composites Part B: Engineering* **45**(1) 1704-1713

[7] Bousselham A and Chaallal O 2004 Shear strengthening reinforced concrete beams with fiber-reinforced polymer: assessment of influencing parameters and required research *ACI Struc. J.* **101**(2) 219–227
[8] Bousselham A and Chaallal O 2008 Mechanisms of shear resistance of concrete beams strengthened in shear with externally bonded FRP *J. Compos. Constr.* **12**(5) 499–512
[9] Pellegrino and Modena 2002 Fiber Reinforced Polymer Shear Strengthening of Reinforced Concrete Beams with Transverse Steel Reinforcement *J. Compos. Constr.* **6**(2) 104–111

Acknowledgment
Special thanks for the Sustanable Construction Materials and Structural Syetems group in the Research Institute of Sience and Engineering (RISE), University of Sharjah. We would also like to acknowledge CONMIX TLD for their support in providing the strenghening materials, and supervising on the FRP application process.

Post-disaster permanent housing: user participation and long-term satisfaction

Bee Lan Oo[1*]

[1] Faculty of Built Environment, University of New South Wales Sydney, Kensington, NSW 2052, Australia
*Corresponding author: bee.oo@unsw.edu.au

Abstract. There is empirical evidence on varying levels of user satisfaction with the provided post-disaster permanent housing based on their levels of participation. This study compares non-participating and participating users' long-term satisfaction with permanent housing constructed under the REKOMPAK (a community-based reconstruction program) after the 2010 Merapi volcano eruption. The results show that both groups have recorded highly similar levels of satisfaction with their permanent housing attributes with only four (out of 38) attributes that are statistically significantly different between the groups. These findings provide suggestive evidence on the maturity of the REKOMPAK program since it was created in 2005. This has implications for local authorities, humanitarian and other agencies in their decisions to adapt the REKOMPAK or other similar community-based programs for post-disaster reconstruction.

1. Introduction

The context of post-disaster housing reconstruction differs markedly from that which of routine construction as it is characterized by the need for speedy construction, the bureaucratic and community participation issues, the extraordinary financial requirements, and the resource challenges [1]. In a large collection of case studies on post-disaster permanent housing, authors have attempted to classify the project delivery methods in their evaluation of project successes and failures. For example, Bareastein [2] have documented four different approaches, namely: contractor-driven, subsidiary, participatory and owner-driven approaches. Dikmen et al. [3] classified the contractor, funding agency and owner-driven projects based on two divergent top-down and bottom-up principles. Although there are some variations in defining these approaches between authors, it appears there is a consensus among authors to broadly classify the respective project delivery methods into donor- and owner-driven approaches, as evidenced in a handful of literature that aimed to compare the strengths and limitations of these two approaches in delivering post-disaster permanent housing (e.g., [4, 5]). The other area of focus in previous case studies is the user, community, recipient, or beneficiary participation in the housing reconstruction processes. Davidson et al. [6] classified the level of community participation in post-disaster housing reconstruction into five levels, namely: manipulate, inform, consult, collaborate and empower. He indicated that an empowered community would be happy and satisfied with the housing and resettlement. Fanany [7] reported that the residents' high level of involvement in the housing design and selection of materials, and the construction processes in the reconstruction helped them to deal with the devastation and loss caused by the tsunami in Aceh,

Indonesia. Mazza et al. [8], on the other hand, found that citizens were dissatisfied with the post-earthquake reconstruction in the city of L'Aquila because they have not been well informed about the reconstruction phases. In comparing donor- and owner-driven approaches, Karunasena and Rameezdeen [5] found that beneficiary satisfaction is higher in owner-driven approach. In terms of psychological needs and quality of interior spaces, Erinsel Önder et al. [9] found that those who participated in the design and construction processes were more satisfied than those who had not participated. Previous findings are mixed concerning the effect of user participation in post-disaster housing reconstruction, and this can partly be explained by the time frame when the evaluations were performed; and the different mechanisms in the respective project delivery methods. While authors have reported on changes in user satisfaction and permanency at post-disaster housing sites over time (e.g., [10, 11]), there is limited studies on objective evaluation of user satisfaction from a long-term perspective. This study aims to fill this literature gap by reporting the findings on user participation and long-term satisfaction in the REKOMPAK program.

2. The REKOMPAK program in Merapi

The October 2010 Merapi volcano eruption was an unusually large magnitude explosive eruption that caused over 350 fatalities and more than 300,000 people evacuated [12]. Close to 3900 houses were damaged in the affected areas, and more than 75% of all damaged housing were in the Slemen Regency in Yogyakarta Province [13]. Cangkringan district is one of the fourteen districts in the Slemen Regency, and was worst affected by the 2010 Merapi Eruption with over 2500 houses destroyed [14]. The post-disaster housing reconstruction was implemented through a scheme called REKOMPAK with funding from the Java Reconstruction Fund contributed by donors after Central Java earthquake in 2006 [15]. Under the REKOMPAK program, there are fifteen resettlement sites (called *huntap*, permanent settlements) with 1596 houses built in five villages that make up the Cangkringan district. These resettlement sites can be classified into five typologies based on the housing and infrastructure provisions and sources of fund, namely [13]: (A) collective resettlement sites; (B) independent collective resettlement, (C) combined support collective resettlement; (D) independent individual reconstruction; and (E) independently non-REKOMPAK supporter rebuilding. Type A was the typical type in most resettlement sites where the planning and construction of permanent housing were carried out by resident group representatives with the support from facilitators provided by REKOMPAK. The funding was in the amount of 30 million Rupiah for building a 6 x 6 meter reinforced concrete block house on $100m^2$ lot of village-owned land for each household. With the mechanism of having resident group representatives in the reconstruction processes, this means not all the affected households had directly participated in the decision making.

3. Research method

This paper is part of a project on testing a conceptual framework on users' long-term satisfaction with post-disaster permanent housing in Oo et al. [16-17]. Data was collected using a structured survey questionnaire written in Bahasa Indonesia. The results reported here are related to the users' long-term satisfaction with permanent housing attributes. They were asked to indicate their level of satisfaction with 38 attributes based on a five-point Likert scale (1 = very dissatisfied and 5 = very satisfied). These attributes were adopted from a small collection of similar studies that investigated users' satisfaction with post-disaster permanent housing on medium- to long-term perspectives (for e.g., [4], [10-11]). The field survey took place between Dec 2017 and Feb 2018. The resettlement sites in this study: (i) Batur, (ii) Pagerjurang; (iii) Gondang 2; and (iv) Dongelsari are all Type A collective resettlement sites. They are major resettlement sites as the total number of houses (738) built represents 46% of the 1596 houses built in fifteen resettlement sites in Cangkringan District [13]. This forms a strong basis to comparatively study the users' long-term satisfaction with post-disaster permanent housing according to their levels of participation. With the focus on users' long-term satisfaction, 285 (out of 297) sets of completed surveys were included in the data analysis. This selection criterion was based on the requirement that a household must have stayed in the provided housing for at least 5 years (i.e., moved in before year 2014). This represents an average response rate of 38.62% (285/738) for the four resettlement areas, which is considered high and representative. For the data analysis, the respondents (or households) were classified into two groups according to their levels of participation in the reconstruction processes, namely: non-participating and participating groups. A non-parametric Mann-Whitney U test was used to test the equality of average satisfaction scores between the two groups.

4. Results and discussion

Table 1 shows the profile of the 285 respondents and about 67% of them were from Batur and Pagerjurang resettlement sites. The biggest age group (51%) is between 36 and 55 years old and about 80% of the respondents are either employed or self-employed. About 70% of the households have a typical family size of three to four members living in the housing permanently.

Table 1. The respondents' profile.

Characteristics		Frequency	%
Resettlement site (Huntap)	Batur	96	33.68
	Pagerjurang	96	33.68
	Gondang 2	48	16.85
	Dongelsari	45	15.79
Gender	Male	172	60.35
	Female	113	39.65
Age	< 25	33	11.58
	26 to 35	44	15.44
	36 to 45	74	25.96
	46 to 55	72	25.26
	56 to 65	39	13.68
	> 65	23	8.07
Employment status	Employed/ self-employed	226	79.30
	Housewife	42	14.74
	Student	12	4.21
	Unemployed	5	1.75
No. of people live permanently	1	15	5.26
	2	44	15.44
	3-4	198	69.47
	5-8	28	9.82

Table 2 shows the respondents' level of participation in the reconstruction processes. There are 94 respondents in the non-participating group, indicating that about one-third (i.e., 94/285) of the respondents had not directly participated in any of the reconstruction processes. While the remaining 191 (67%) households had participated in various processes, only about 10% of them were involved in more than three processes. The largest group (27%) had participated in at least two processes. The seven processes are: (i) selection of site location; (ii) selection of the housing type (traditional vs. modern); (iii) design of the housing (size, space, layout); (iv) selection of construction materials used; (v) selection of material suppliers; (vi) selection of subcontractors or labours; and (vii) supervision during construction stage.

Table 2. The respondents' level of participation in the housing reconstruction processes.

Level of participation		Frequency	%
No participation		94	32.98
Participated in different processes	1 process	50	17.54
	2 processes	76	26.67
	3 processes	37	12.98
	4 processes	14	4.91
	5 processes	4	1.40
	6 processes	3	1.05
	7 processes	7	2.46

Table 3 shows the respondents' average satisfaction scores and the Mann-Whitney U test results for comparing equality of means between the two groups. With the exception of one or two attributes, the average satisfaction scores for the remaining attributes for both groups are above 3 (i.e., the neutral level on 5-point Likert scale). However, there is a rather wide spread in the respondents' responses for some attributes in both groups as demonstrated by the high standard deviation values. Attributes related to the resettlement site locations scored low for both groups, including proximity to city centre (SC1) and public transportation (IS6) with mean values below or slightly above 3. The 'neutral' perceptions among the respondents could possibly be explained by the fact they were relocating within their own villages under Type A topology. Nonetheless, for both groups, the social environment (SE) attributes scored rather high mean values with most attributes ranked top on the list. These provide suggestive evidence that the resettlement programs had addressed the community social environment needs.

The test results show that the mean values of only four (out of 38) attributes are statistical different between the two groups at $p < 0.05$ level. These attributes are: size of bedrooms (PC7); shopping and conveniences (IS1); health facilities (IS3); and availability of job or business opportunities (SE5). Also, the participating group's mean satisfaction scores for the first three attributes (i.e., PC7, IS1 and IS3) are lower than that of the non-participating group. For these observed differences between the groups, there are three possible explanations based on the present data. First, the small differences in mean scores for most attributes between the two groups can be partly due to the differences in perceptions among respondents from four different resettlement areas that have been reflected in the means. This phenomenon of differences in perceptions are not uncommon in similar past studies (e.g., [9], [11]). Indeed, Mei et al. [18] found that there are considerably differences in perceptions of the REKOMPAK program among respondents from another two Type A resettlement sites. Next, the lower mean scores from the participating group could also possibly be explained because those involved were more critical of the housing requirements or had higher expectations that cannot be met during the reconstruction processes. Thirdly, the findings, in general, provide suggestive evidence that the resident group representatives and the REKOMPAK facilitators were successfully in informing community during the reconstruction processes, although not all the households had directly participated. The third possible explanation is further supported by the average overall satisfaction scores of the two groups that is not statistically different at $p < 0.05$ level. The average overall satisfaction scores of the non-participating and participating groups are considerably high at 3.953 and 4.021 (close to 4 on a 5-point Likert scale), respectively. Above all, the highly similar levels of satisfaction between the two groups provide evidence on the maturity of the REKOMPAK program since it was created in 2015. The established mechanisms and expertise in the program contributed to its successful implementation in Merapi. Indeed, authors have attributed the success of the REKOMPAK program in Merapi to past experiences and lessons learnt from its implementation in Aceh post-tsunami and Central Java earthquake in 2006 ([13, 14]). The World Bank [15] has shared this community-based reconstruction model and relayed the experiences and lessons learned in Aceh and Java due to its tremendous success.

Table 3. The respondents' satisfaction scores and test statistics for comparing two groups

Attribute		**Level of participation**						**Sig. (Mann-Whitney U)**
		No participation			Participated in various processes			
		Mean	Std. Dev.	Rank	Mean	Std. Dev.	Rank	
Site characteristics								
SC1	Proximity to city centre	2.968	0.921	37	3.168	0.873	36	0.105
SC2	Proximity to workplace	3.564	0.665	27	3.508	0.787	27	0.603
SC3	Convenience to livelihood	3.787	0.637	12	3.832	0.756	6	0.645
SC4	Space between houses	3.840	0.627	8	3.764	0.822	12	0.464
SC5	General cleanliness	3.819	0.548	10	3.770	0.710	11	0.572
SC6	Landscaping (trees, garden areas)	3.574	0.726	26	3.623	0.692	20	0.951
SC7	Resilient for future disasters	3.787	0.620	12	3.670	0.681	16	0.106
Physical conditions of housing								
PC1	Size of the house	3.287	0.887	36	3.157	0.850	37	0.170
PC2	Durability of material used	3.766	0.679	15	3.780	0.770	10	0.745
PC3	Quality of workmanship	3.734	0.706	16	3.796	0.751	9	0.546
PC4	External appearance of the house	3.628	0.639	20	3.634	0.769	19	0.761
PC5	Internal layout of the house	3.628	0.656	20	3.513	0.717	26	0.117
PC6	Number of bedrooms	3.479	0.729	33	3.372	0.770	35	0.228
PC7 *	Size of bedrooms	3.606	0.609	23	3.419	0.763	33	*0.038*
PC8	Toilets/ bath	3.564	0.712	27	3.435	0.729	32	0.112
PC9	Kitchen/ cooking space	3.511	0.684	31	3.450	0.772	31	0.373
PC10	Access to utilities supply (water, electricity)	3.777	0.642	14	3.707	0.731	14	0.462
PC11	Natural lighting	3.681	0.626	17	3.654	0.685	17	0.828
PC12	Natural ventilation (fresh air)	3.596	0.661	24	3.613	0.678	22	0.986
PC13	Availability of space to meet your household needs	3.670	0.709	18	3.586	0.741	23	0.455
PC14	Visual privacy	3.415	0.679	35	3.471	0.753	29	0.491
PC15	Aesthetic value	3.436	0.649	34	3.571	0.714	24	0.226
PC16	Flexibility to incorporate changes (modification or improvement)	3.500	0.699	32	3.503	0.774	28	0.958
PC17	Fire and other safety measures	3.660	0.648	19	3.571	0.684	24	0.269
PC18	Easy to upkeep	3.628	0.604	20	3.649	0.724	18	0.951
PC19	Feeling of home	3.851	0.621	7	3.806	0.732	8	0.415

Institutional services								
IS1 *	Shopping and conveniences (market)	3.809	0.723	11	3.618	0.831	21	*0.050*
IS2	Educational services	3.840	0.493	8	3.764	0.650	12	0.302
IS3 *	Health facilities	3.894	0.497	6	3.686	0.722	15	*0.007*
IS4	Facilities for religious practice (church, mosque)	4.053	0.516	1	3.958	0.639	5	0.227
IS5	Leisure and sport facilities	3.553	0.784	29	3.414	0.841	34	0.117
IS6	Public transportation	2.713	1.064	38	2.675	0.978	38	0.951
IS7	Spaces for social gathering (functions, parties, cultural events)	3.532	0.786	30	3.471	0.753	29	0.203
Social environment								
SE1	Interaction between neighbours	4.043	0.506	3	4.110	0.675	1	0.255
SE2	Nearness to relatives and friends	3.989	0.475	5	4.026	0.721	2	0.484
SE3	Personal safety	4.043	0.484	3	3.984	0.714	4	0.629
SE4	Neighbourhood or public safety	4.053	0.472	1	4.010	0.718	3	0.857
SE5 *	Availability of job or business opportunities	3.596	0.794	24	3.817	0.770	7	*0.041*
Overall satisfaction		4.021	0.439		3.953	0.555		*0.353*

* Significant at $p < 0.05$ level

5. Conclusions

This study examines users' long-term satisfaction with post-disaster permanent housing attributes according to their levels of participation in REKOMPAK - a community-based post-disaster rehabilitation and reconstruction settlement program in Merapi, Indonesia. The results show that both non-participating and participating groups have recorded highly similar levels of satisfaction with their permanent housing attributes. The non-participating groups has not been disadvantaged in terms of the provided permanent housing. The REKOMPAK program was very successfully in addressing the users' long-term social environment needs. These findings provide suggestive evidence on the maturity of the REKOMPAK program with established mechanisms and expertise since it was created in 2005. This empirical evidence has implications for local authorities, humanitarian and other agencies in their decisions to adapt the REKOMPAK or other similar community-based programs for post-disaster reconstruction. Nonetheless, it is recognized that local authorities and agencies should consider the cultural and local conditions as well as users' needs in developing their respective models, guidelines and policies of community-based post-disaster reconstruction program. This forms the scope of the next stage of this project.

6. References

[1] Bilau A and Witt E 2016 An analysis of issues for the management of post-disaster housing reconstruction *Int. J. of Strategic Property Management* **20**(3) 265-276
[2] Barenstein J D 2006 *Housing Reconstruction in Post-Earthquake Gujarat: A Comparative Analysis* (London: Humanitarian Practice Network at Overseas Development Institute) pp 54
[3] Dikmen N, Elias-Ozkan S T and Davidson C 2012 Comparison of post-disaster housing procurement methods in rural areas of Turkey *Open House Int.* **37**(1) 28-39

[4] Andrew S A, Arlikatti S, Long L C and Kendra J M 2013 The effect of housing assistance arrangements on household recovery: an empirical test of donor-assisted and owner-driven approaches *J. of Housing and the Built Environment* **28**(1) 17-34
[5] Karunasena G and Rameezdeen R 2010 Post-disaster housing reconstruction: Comparative study of donor vs owner-driven approaches *Int. J. of Disaster Resilience in the Built Environment* **1**(2) 173-191
[6] Davidson C H, Johnson C, Lizarralde G, Dikmen N and Sliwinski A 2007 Truths and myths about community participation in post-disaster housing projects *Habitat International.* **31**(1) 100-115
[7] Fanany I 2010 Towards a model of constructive interaction between aid donors and recipients in a disaster context: the case of Lampuuk *Post-disaster reconstruction: Lessons from Aceh* eds M Clarke and I Fanany (London: Taylor and Francis) pp 107-125
[8] Mazza M, Chiara Pino M, Peretti S, Scolta K and Mazzarelli E 2014 Satisfaction level on quality of life post-earthquake rebuilding *Int. J. of Disaster Resilience in the Built Environment.* **5**(1) 6-22
[9] Erinsel Önder D, Köseoğlu E, Bilen O and Der V 2010 The effect of user participation in satisfaction: Beyciler after-earthquake houses in Düzce *ITU A|Z.* **7**(1) 18-37
[10] Manatunge J M A and Abeysinghe U 2017 Factors affecting the satisfaction of post-disaster resettlers in the long term: A case study on the resettlement sites of tsunami-affected communities in Sri Lanka *J. of Asian Development* **3**(1) 94-124
[11] Snarr D N and Brown E L 1994 Post-disaster housing reconstruction: A longitudinal study of resident satisfaction, *Disasters* **18**(1) 76-80
[12] Smithsonian Institution and US Geological Survey 2016 *Global Volcanism Program, Report on Merapi (Indonesia)* in S.K. Sennert (ed.) Weekly Volcanic Activity Report 26 October-1 November 2016
[13] Maly E, Iuchi K and Nareswari A 2015 Community-based housing reconstruction and relocation: REKOMPAK program after the 2010 eruption of Mt. Merapi *Indonesia Institute of Social Safety Science Journal* **27** 205-214
[14] Samekto C B D and Nuh M 2017 Evaluation of community-based settlement reconstruction program: case study in post-disaster recovery of 2010 Merapi volcano eruption in Cangkringan district, Sleman Regency, Yogyakarta Province *J. of Public Administration Studies* **1**(3) 64-70
[15] The World Bank 2012 *REKOMPAK: Rebuilding Indonesia's Communities After Disasters. Jakarta, Indonesia* The Secretariat of the Multi Donor Fund for Aceh and Nias and the Java Reconstruction Fund
[16] Oo B L, Sunindijo R and Lestari F 2018 Users' long-term satisfaction with post-disaster permanent housing programs: a conceptual model, *Int. J. of Innovation, Management and Technology* **9** 28-32
[17] Oo B L, Sunindijo R and Lestari F 2018 Users' long-term satisfaction with post-disaster permanent housing: a case study of 2010 Merapi Eruption, Indonesia *Proc. of the Int. Conference on Engineering, Applied Sciences and Technology (Phuket)* **192** (Paris:MATEC Web of Conference) 02066
[18] Mei E T W, Fajarwati A, Hasanati S and Sari I M 2016 Resettlement following the 2010 Merapi Volcano eruption *Procedia-Social and Behavioral Sciences* **227** 361-369

Acknowledgment
This project was supported by the UNSW-Indonesia Seed Fund (Grant no.: RG162919). The author gratefully acknowledges the efforts of Dr. Riza Sunindijo in the questionnaire translation and research ethics application process, and Dr. Fatma Lestari and her research team for their invaluable help in completing the data collection and entry.

Green Design for the Comfort Environment of Kindergarten Building in Malaysia: A review

Fitrynadia Mohd Shahli[1] and Zainal Abidin Akasah[2*]

[1] Faculty of Civil and Environmental Engineering, Universiti Tun Hussein Onn Malaysia, 86400 Parit Raja, Batu Pahat, Johor, Malaysia
[2] Faculty of Civil Engineering, Universiti Teknologi Malaysia, 81310 Johor Bahru, Johor, Malaysia
*Corresponding author: zainal59@uthm.edu.my

Abstract. Public buildings such as kindergarten can be used as a potential educational spot in promoting environmental sustainability for the children. Most of the public kindergartens in Malaysia are facing uncomfortable issues, such as due to the poor air ventilation and inefficient design at the respective building and facilities. By promoting green design and sustainability, this paper presents the synthesis of other studies, which consist of the factors and aspect approaches towards the application of green design in developing development of kindergarten building. This paper will review different articles and studies, which have published within 2009 until recently from the selected online database. Through green design at kindergarten, it is hoped to be able to promote sustainability for the children and foster designers to design more friendly buildings for them.

1. Introduction

A comfortable kindergarten building is important for the development of learning process for children in preschool environment. Since, children spend their time and play mostly within the building, therefore the building needs to be designed based on the specification of children's needs in order for them to feel the comfort while learning. In Malaysia, most kindergarten buildings not originally built as a function of kindergarten and need to renovate to fit the purpose of learning space for children [1]. Therefore, the buildings are facing several problems such as sick building syndrome, poor air quality [1] and inefficient physical design [2] due to the renovating activity. This may result uncomfortable environment, limited space for children to move around, to receive stack air within building and have potential to expose to poor air quality that lead to bad health [3].

Moving towards green and sustainability invention, the development of kindergarten building also need to be involved with the specification of green designs in order to achieve the best convenient level and become more children friendly. Since the awareness and application of green design in kindergarten have become attention for most developed country [4]. This paper tends to overlook the aspect of green design that influencing the development of comfortable environment kindergarten from the current literature review. The specific concerns of this paper are: (1) to identify the kindergarten development in context of Malaysia (2) to review the comfort factors and green design applications worldwide as a guideline for green and sustainable kindergarten buildings in Malaysia.

IOP Conf. Series: Materials Science and Engineering **601** (2019) 012020 doi:10.1088/1757-899X/601/1/012020

2. Methods

Literatures were selected based on different disciplines including green design, building design, environmental education, children's developments, and health. The combination of the keywords were used to guide the computerized search from online database consists of different articles and studies. The papers were selected from the literatures that have been published within a ten years period, starting from 2009 until 2018 including theoretical review and empirical article that consists of qualitative and quantitative of research. The aspects that consider include the factors, criteria, preferences, and importance of green designs in influencing the comfort environment of kindergarten buildings. As results, this paper will review and discuss the aspects of kindergarten development and green design applications for comfort environment of kindergarten buildings.

3. Kindergarten Developments

The development of kindergarten building and its facilities is important to ensure the quality of learning for the children. According to the Education Act 1996, Act 550 [5], kindergarten is defined as any places where pre-school education (children with age 4 years to 6 years old) is provided to ten or more pupils. This can be understood that kindergarten become as a place that is set up in order to facilitate the preschool education in Malaysia.

In developing kindergarten building and its facilities, *Jabatan Perancangan Bandar dan Desa Semenanjung* Malaysia has introduce specific outline in order to starting set up of the kindergarten. According to *Garis Panduan Penubuhan Taska dan Tadika* [6] several criteria has been outlined which are site planning, building design, establishment process of kindergarten, implementation and payment rate for establishment of kindergarten.

A good environment of kindergarten were designed based on four important environmental stimuli, which are movement, comfort, competence, and control [7]. Research in early childhood environment found that, children require comfort environment, facilities and space to foster their quality of learning [2]. This shows that, the fulfillment of comfort parameter is crucial in developing physical-sensory, socio-emotional and mental development of children toward progress of adulthood [8]. This becomes focus for this study.

4. Comfort Environments for Children in Kindergarten

Comfort defines as a reaction to the environment that is strongly influenced by cognitive and behavioral processes that is controlled by habit and sensation of human [9]. Since children spend most of their time inside buildings, it is very important to maintain the comfort level during learning activity. Therefore, the design and planning of the building and educational facilities need to be considered in the aspect of comfort in order to achieve the conducive and effective learning for students. Since kindergarten becomes a place for children to explore and learn, the comfort environment of kindergarten is served when the children feel comfortable with their physical surrounding through its convenient setting [7]. The consideration of comfortable setting can be felt in variation of architecture elements such as scale, height, lighting, and finishes material that involve systematic manipulation of space to achieve comfort environment [2].

4.1. Factors Influencing Comfort Environment

There are several comfort parameters in creating appropriate space for children to feel comfort during stay in kindergarten building. The identified comfort parameters are air comfort, space comfort, light comfort, and aesthetic comfort specifically for indoor environment of the building [8-10].

4.1.1. Air and thermal comfort

The indoor and outdoor air quality of kindergarten is important for children's health and development. Children require good indoor air quality to influence their growth and wellbeing development [3]. Therefore, the implementation of good and natural ventilation system is important for children. In order to enhance the comfort, the quality of fresh air regulates the air humidity, removes odor, lower dust content and keep the thermal comfort of the buildings [8]. Special attentions are needed when selecting proper finishing materials for interior surfaces such as floors, walls, ceilings, and other surfaces of kindergarten [11].

It might be harmful for the children health and performance if the selected material contain unsafe ingredient. Thus, the use variety type of natural and man-made materials such as wood, brick, concrete, steel, and plastic such as polycarbonate board can become more eco-friendly for the children [4]. The materials used to construct the kindergarten need to have pleasant texture in order to keep the comfort temperature of the kindergarten building, although without the use of any mechanical cooler. Besides that, the use of natural elements such as deciduous trees and indoor plants can keep the environmental quality of the kindergarten. Plants act as filter from direct sun, keep room temperature, and promote sustainability solution for the building.

4.1.2. Space comfort

The space and size of surrounding kindergarten may different based on design guideline of the country. As for context of Malaysia, the provided overall space for kindergarten is divided based on the guideline from *Jabatan Perancangan Bandar dan Desa Semenanjung* Malaysia. The space for the kindergarten is divided by two types of space which are floor space and open space (total space $=166m^2$). As example from other developed countries the floor area or space of kindergarten depending on the capacity of the building (number of children) and based on the aged of the children [4]. The space of the kindergarten building is divided into three main spaces which are child area, staff area, and service area (total space $= 162m^2$).

To ensure the comfortable spaces, the organization, circulation, and shape of the space need to take into consideration so that children can move around and do activities [8]. The daily activities of children (such as reading, writing, listening, playing, and etc.) require optimization of space and function by providing subdividing zone to improve the comfort of the children movement. Although most of kindergarten space is controlled by the standard, designers need to be smart in arranging the space by promoting innovative solutions to make the space more children-friendly. It is to ensure the safety and comfortable environment for the children.

Since, most kindergartens in Malaysia are developed in the existing built up dwelling type of building [2]. The provided spaces are limited and the arrangements of children's space need to be designed in minimize area. Therefore, the important of circulation in helping children to move around and become separate area between children activity area and others spaces [4]. Moreover, the selection of furniture and facilities such as toilet needs to be in proportion and have suitable height for children for easy access without assistance from teacher [7].

4.1.3. Light Comfort

Light comfort associates with the existing daylight and artificial lighting, which are built in the building. As for kindergarten building, the specific parameter such as window position, window area, glazing type, intensity of color and type of artificial light can influence the light comfort of kindergarten [8]. In supporting the idea of sustainability, the use of efficient window and skylight can increase the application on natural daylight in school [12]. As a result, the sufficient light comfort is important to achieve more pleasant environment and visual comfort effect of kindergarten.

In term of color and lighting, most of the walls are painted and decorated with appropriate color and displayed mural to attract children's interest in learning. The sufficient of natural sunlight is encouraged for most kindergarten designs to encourage children movement. However, most of kindergarten in Malaysia facing limited daylight effecting from existing building designs such ad dwelling and commercial building [2]. Due to that reason, children need to use mechanical lighting during learning session, as a result children is exposed to the excessive artificial light and limit children's sensory motor [13].

4.1.4. Visual and aesthetic comfort
The visual appearance and aesthetic comfort also can become aid for children in developing their physical and mental development [2]. Therefore, the places where children spend their time need to be aesthetically pleasant and beautiful to attract children attention. The interior space elements, texture, shapes, colors and size play significant roles in creating aesthetic comfort of kindergarten buildings. Somehow, aesthetic comfort may be affected by arrangement of space that utilize the senses such as eye contact, feel to touch, and experience for interaction [8].

4.1.5. Acoustic comfort
The noise can affect children's academic performance for two reasons, which are noise distraction and make them unable to understand the teacher due to the inappropriate level of signal-to-noise ratio and articulation loss of consonants [14]. Children that chronically exposed to the noise distraction will affect in lower reading scores rather than those that live in quieter surrounding [15]. This shows that, the acoustic conditions of a classroom are particularly important for children as younger learners [16]. The distraction of noise usually come from the noise from traffic, building system, exterior and adjacent space sound transmission, noise from HVAC, and plumbing system [17].

5. Green design applications for comfort environment of kindergarten building
The use of green application is one of strategies in modern sustainable architecture. Green design involves a holistic approach in designing, which include the green material that contributes to the need of the user to promote the sustainability [18]. In the context of school, the green application can involve in various aspects such as improvements is additional daylight, improved indoor air quality, enhanced classroom acoustics, and comfortable and consistent indoor temperatures [19] . Previous researchers, particularly from Malaysia and other countries have proven that, the application of natural elements such as vegetation, integrated design and technology can become a solution to maintain the comfortable environment of the building. Table 1 indicates the implementation of green design can be applied in influencing comfort environment of kindergarten building in Malaysia.

Since Malaysia is still looking towards the development of green design in the country, the establishment of green design is more focusing towards the non-residential building [35], residential building, the township [36], and the lack consideration of school building [17]. Unlike, other western and developed countries, the green design specifically introduces for school building including kindergarten [37]. As Malaysia is situated in the tropics and hot climate region, the application of green design aspect needs to be considered in the design of the building with the respond towards the climate and culture to maintain the comfortable environment of kindergarten.

Table 1. Implementation of green design based on comfort factors

Comfort factors	Green design application	References
Air and thermal	**Natural elements** - Vegetation influenced microclimate surrounding - Green structure: covering vegetation, isolated trees, groves and lines of trees - Indoor plants - Passive design: natural ventilation, courtyard	 [20] [21] [22] [23] [24] [25] [26]
	Integrated design and technology Rooftop garden and living wall	 [27] [28]
Light	**Natural elements** - Matured tree: shading and solar filtration	 [29]
	Integrated design and technology - Solar energy system - Skylight	 [30] [4]
Space, visual and aesthetic	**Natural elements** - Landscape design: plants arrangement - Biophilic design	 [31] [32] [33]
	Integrated design and technology Eco building material : timber, double glass façade solar, light steel structure	[34]
Acoustic	**Natural elements** Plants: reduce pollution and become barrier for noise	 [21]

6. Conclusion

Literature reviews indicated that there are four important comfort factors influence the development of kindergarten buildings which are thermal, light, space, aesthetic and acoustic comfort in Malaysia and other countries. Through consideration of green design, several examples and study of existing kindergarten have been developed in most developed country, which resulted in several ecological and integrated approaches in the implementation of green design for kindergarten building. Thus, the diverse solution through green design can change the formal physical kindergarten building in Malaysia toward more sustainable buildings in the future.

7. References

[1] Kamaruzzaman S N, Norhanim Z and Yau A 2011 The effect of indoor air quality towards students performance in refurbished private kindergarten in Malaysia, *Proc. International Engineering Education Conference 2011(Madinah al-munawarah)* pp 25-27

[2] Mohidin H H B, Ismail A S and Ramli H 2015 Effectiveness of Kindergarten Design in Malaysia, *Procedia - Soc. Behav. Sci.* **202** 47-57

[3] Kamaruzzaman S N and Razak R A 2011 Measuring Indoor Air Quality Performance in Malaysian Government Kindergarten, *J. Build. Perform.* **2**(1) 70-79

[4] Kotnik J 2014 *New Design in Kindergartens: Design guide + 31 case studies* (Spain: LinksBooks) pp 10-19

[5] Laws of Malaysia 2012 *Act 550 Education Act 1996* (Kuala Lumpur: The Commissioner of Law Revision Malaysia) pp 17

[6] GP 027 2012 *Garis Panduan Penubuhan Tadika dan Taska* (Kuala Lumpur: Jabatan Perancangan Bandar dan Desa Semenanjung Malaysia) pp 2-3

[7] Leinonen J and Venninen T 2012 Designing learning experiences together with children, *Procedia - Soc. Behav. Sci.* **45** 466-474

[8] Stankovic D, Tanic M, Kostic A, and Timotijevic M 2015 Revitalization of Preschool Buildings : A Methodological Approach, *Procedia Eng.* **117** 723-736

[9] Ortiz M A, Kurvers S R, and Bluyssen P M 2017 A review of comfort , health , and energy use : Understanding daily energy use and wellbeing for the development of a new approach to study comfort, *Energy Build.* **152** 323-335

[10] Shaari M F and Ahmad S S 2016 Physical Learning Environment : Impact on Children School Readiness in Malaysian Preschools, *Procedia - Soc. Behav. Sci.* **222** 9-18

[11] Salleh N M, Kamaruzzaman S N, Sulaiman R and Mahbob N S 2011 Indoor Air Quality at School : Ventilation Rates and It Impacts Towards Children- A review *Proc. 2nd International Conference on Environmental Science and Technology* vol 6 (Singapore: IACSIT Press) pp 418-422

[12] Schneider M 2002 *Do school facilities affect academic outcomes?* National Clearinghouse for Educational Facilities (Washington: National Institute of Building Sciences) pp 1-24

[13] West T L 2011 *Environments for Young Children : A Qualitative Study and Design of Healthy and Nuturing Preschool Environments* Master Thesis (USA: Florida State University) pp 169-178

[14] Singer M J 2003 *Acoustics in Schools* (Maryland: Educational Resources Information Center (ERIC)) pp 1-22

[15] Bronzaft A L 2007 *A Quieter School: An Enriched Learning Environment* Retrieved on August 20, 2018 from http://www.musicmotion.com/images/quieter_school.pdf

[16] Van Reenen C and Karusseit C 2017 Classroom acoustics as a consideration for inclusive education in South Africa, *South African J. Commun. Disord.* **64**(1)

[17] Ramli N H, Masri M H, Zafrullah M, Mohd H and Hamid N A 2012 A Comparative Study of Green School Guidelines, *Procedia - Soc. Behav. Sci.* **50** (7) 462-471

[18] Ragheb A, El-shimy H, and Ragheb G 2016 Green Architecture: A Concept of Sustainability, *Procedia - Soc. Behav. Sci.* **216** 778-787

[19] Gordon D E 2010 *Green Schools as High Performance Learning Facilities* National Clearinghouse for Educational Facilities (Washington: National Institute of Building Sciences) pp 1-16

[20] Panagopoulos T 2008 Using Microclimatic Landscape Design to Create Thermal Comfort and Energy Efficiency Retrieved on November 19, 2018 from https://pdfs.semanticscholar.org/e0ac/6268e33f624fc2155774044dd6203b42eeab.pdf

[21] Kumar S R, Arumugam T, Anandakumar C R, Balakrishnan S and Rajavel D S 2013 Use of Plant Species in Controlling Environmental Pollution- A Review, *Acad. Environ. Life Sci.* **2**(1) 52-63

[22] Perini K 2013 Retrofitting with vegetation recent building heritage applying a design tool-the case study of a school building, *Front. Archit. Res.* **2**(3) 267-277

[23] Abd Wahab I and Ismail L H 2012 A study on plant selection for green building design, *J. Sci. Technol.* **4**(2) 65-78

[24] de Selincourt K 2015 *Building a Better Passive School* Retrieved on December 24, 2018 from https://passivehouseplus.ie/magazine/new-build/building-a-better-passive-school

[25] Chan S C, Che-Ani A I and Nik Ibrahim N L 2013 Passive designs in sustaining natural ventilation in school office buildings in Seremban, Malaysia, *Int. J. Sustain. Built Environ.* **2**(2) 172-182

[26] Salameh M and Taleb H 2017 Courtyard as Passive Design Solution for School Buildings in Hot Area, in *World Congress on Civil, Structural, and Environmental Engineering (CSEE'17) Barcelona* **141**(2002), 2371-5294

[27] Fauzi M A, Malek N A and Othman J 2013 Evaluation of green roof system for green building projects in Malaysia, *Int. J. Environ. Geol. Min. Eng.* **7**(2) 124-130

[28] Mohammed Ahmed R and Halil Alibaba A Z 2016 An Evaluation of Green roofing in Buildings *Int. J. Sci. Res. Publ.* **6**(1) 366-2250
[29] Moreno A, Tangenberg J, Hilton B and Hilton J 2015 An Environmental Assessment of School Shade Tree Canopy and Implications for Sun Safety Policies: The Los Angeles Unified School District, *ISPRS Int. J. Geo-Information* **4**(2) 607-625
[30] Vatin N I, Nemova D V, Kazimirova A S and Gureev K N 2014 Increase of Energy Efficiency of the Building of Kindergarten, *Adv. Mater. Res.* **953-954**(6) 1537-1544
[31] El-Bardisy W M, Fahmy M, and El-Gohary G F 2016 Climatic Sensitive Landscape Design: Towards a Better Microclimate through Plantation in Public Schools, Cairo, Egypt, *Procedia - Soc. Behav. Sci.* **216**(1) 206-216
[32] Nagasawa S 2010 *A Collection of Exemplary Design of Kindergarten Facilities* (Japan: Ministry of Education, Culture, Sports, Science and Technology) pp 7-10
[33] Knodel C 2011 *Exploratory Case Study: How the Inclusion of Nature in the Design of Learning Environments Affects Learning Among Children* Master Thesis (Lincoln: University of Nebraska) pp 98-99
[34] Kamal M and Gani M O 2016 A Critical Review on Importance of Eco-structure Building or Green Building in Bangladesh, *Int. J. of Business Administration.* **7**(3) 1-15
[35] Ng B H and Akasah Z A 2011 An Overview of Malaysia Green Technology Corporation Office Building: A Showcase Energy-Efficient Building Project in Malaysia *J. Sustain. Dev.* **4**(5) 212-228
[36] Green Building Index 2009 *GBI Assessment Criteria for Non-Residential New Construction (NRNC)* (Kuala Lumpur: Greenbuildingindex Sdn Bhd) pp 1-18
[37] Johnson P D and Kritsonis W A 2010 Greener Schools , Greater Learning , and the LEED Value *Dr. Forum* **7**(1) 1-8

Acknowledgement
This study supported by Geran Penyelidikan Pascasiswazah (GPPS) UTHM (Vote H015).

Early career women in construction: career choice and barriers

Bee Lan Oo[1*], Xiyu Feng[1] and Benson Teck-Heng Lim[1]

[1] Faculty of Built Environment, University of New South Wales Sydney, Kensington, NSW 2052, Australia
*Corresponding author: bee.oo@unsw.edu.au

Abstract. Retaining female students through graduation and sending them out into the construction industry remain as challenges for greater women participation in the industry. This study examines the factors affecting early career women's career choice in construction and barriers faced by them at this career stage. Data was collected from female graduates of a construction management degree in an Australian university using an online questionnaire survey. The results show that significant factors affecting their career choice decisions are: (i) career opportunities; (ii) belief of getting better pay; and (iii) self-efficacy and high level of confidence of performing construction tasks. In terms of barriers, four statistically significant barriers are: (i) difficult to integrate into masculine culture in the industry; (ii) stressful and competitive working culture; (iii) long and inflexible working hours; and (iv) lack of informal networks for career opportunities. The results also show that most respondents were satisfied or very satisfied with their jobs. These findings have implications for the development of initiatives aimed at attracting and retaining female graduates in the construction industry.

1. Introduction

The construction industry facing skills shortages as a result of numerous causes including: aging and retirement of male workforce, difficulty to attract and retain female workforce in the industry, difficulty to recruit and retain students in construction related education programs to produce sufficient graduates, and graduates choose not to enter the industry at graduation. (e.g., [1-3]) Women are seen as an untapped resource to meet the workforce demand in the industry([3-4]. Nonetheless, and unfortunately, the presence of women in the industry remains low. For example, the Australian construction industry has a total workforce of approximately 1.16 million workers, representing nearly 9.3% of total employed workforce in Australia [5]. Of these, only 10.6% of the total workforce is female. In Europe, women only represented 10.6% of the total construction workforce [6]. While there is some evidence to suggest women are more willing to pursue construction management (CM) related degrees that would produce more female graduates (for e.g., [7-8]), the percentages of female student enrolments in CM remain low. Previous studies suggest that the female enrolment percentages recorded in the US are generally below 10% [9]. In addition, retaining female students through graduation and sending them out into the industry remain as challenges [2-3]. While the attraction, retention and work experiences of women in the construction industry have drawn significant attention from researchers, most previous studies have often focused on senior professional women (for e.g., [3], [10-11]). Little attention has been paid to early career women in construction who just commit their career and try to sustain their positions. Ling and Poh [8] referred them as 'fence-sitter' and they seem more likely to leave the industry. This study examines the factors affecting early career women's career choice in construction and barriers faced by them at this career stage. The findings have implications for initiatives by CM education providers and employers aimed at attracting and retaining female graduates in the construction industry.

2. Women career choice and barriers in construction

Career choice decision process is complex and affected by many inter-related factors [12]. With a group of 24 female CM graduates, Moore and Gloeckner [13] classified the identified factors into three categories, namely: (i) family background variables that include gender role specialization and parental influence (mainly from fathers); (ii) individual or psychological variables that include high school academic skills and interests, personality, and strength of self-efficacy; and (iii) environmental or sociological variables including role models, mentors and significant others (i.e., boyfriends, husbands, and friends) that were found to have greatest positive influence. Father's influence was pointed out as providing strong impact on women's lives especially when seeking out a non-traditional major [11], [14]. In examining the role of career counsellors, Francis and Prosser [15] found that career counsellors perceived construction as a less suitable career for females, and therefore may not actively direct them to consider careers in construction. In terms of strength of self-efficacy, women who prefer a non-traditional career usually demonstrated a high level of self-efficacy or self-confidence [6], [16). The other top factors that positively affecting female students' choice of CM careers include: career opportunities and internships [17]; students' interest in construction, and hands-on type work activities [14].

Most graduates commit their career with expectations about well-paying, secure workplace conditions and satisfied employment [18]. However, these initial expectations may change immediately once they have actual work experience [10], [18]. This phenomenon could be the results of barriers faced by women in the construction industry, the over-high expectations of employers and women's lack of knowledge of the industry [18, 19]. Facing such a situation, women tend to leave and move to other sectors [4]. Based on a systematic review of 60 publications between 2000 and 2015 on women's career barriers in construction, Navarro-Astor et al. [20] found that the top barriers include: difficulty of balancing work and family, gender stereotypes in the industry, and the lack of professionalism in construction organisations' human resource management. Other top barriers include: poor industry image as dirty and dangerous; sexual harassment; long and inflexible working hours; high level of work stress; and lack of informal networks for career opportunities (for e.g., [3, 4], [8], [10-11], [19]). It is noted that these previous studies have mainly focussed on professional women. Indeed, it is vital to identify barriers faced by both professional and tradeswomen in all career stages to attract women entering the industry and to improve their working lives. This study focusses on early career women in construction.

3. Research Method

Adopting a survey research design that permits timely data collection from a large population, the data was collected using a structured survey questionnaire distributed online in August 2017. The targeted population was fresh female graduates of a CM undergraduate program in an Australian university – University of New South Wales Sydney (UNSW). UNSW is one of the four universities in the New South Wales state of Australia that offer CM program. The female graduates from other institutes had not been invited to participate in this study due to research ethics considerations. The selection criteria for the targeted respondents are female graduates: (i) completed the UNSW CM program in the last five years, i.e., between 2012 and 2016; and (ii) had at least one-year of full-time working experience in the industry post-graduation. For the development of the questionnaire, factors affecting career choices and list of career barriers were adopted from similar studies on women in construction. The respondents were asked to indicate their perceptions and their overall level of job satisfaction based on a five-point Likert scale (for e.g., 1 = very dissatisfied and 5 = very satisfied). The online survey link was distributed to all 46 female CM students graduated between 2012 and 2016 via the UNSW Alumni office. No random sampling was conducted as the population size is considered small. For the data analysis, as the sample size hits $n = 30$, a one sample t-test was used to test the mean values of the examined variables based on a test value = 3 (i.e., the mid-point or neutral score on a five-point Likert scale) in identifying statistically significant variables.

4. Results and discussion

Thirty usable questionnaires were returned, representing a considerably high and representative response rate of 65%. Table 4.1 shows the profile of all respondents. Among the 30 respondents, almost half of them (46.7%) graduated in 2016 and had less than 3 years of construction-related job experience. Most (73.3%) of them were aged from 21 to 25, and the remaining were aged from 26 to 30. Considering the employment types, full-time employment (86.7%) outweighed part-time (6.7%) and casual employments (6.7%) significantly. Of these, about one-third (30%) of the respondents had an annual income above AUD 100k. When the respondents were asked whether the current job was their first job in the construction industry, half of them stated that they had left their previous job mainly for better career development opportunities and/or for accomplishing other life goals. This considerable high turnover rate (50%) suggests that women in the construction industry retain high turnover rates and are more likely to leave their current job to seek for alternatives [10].

Table 1. The respondents' profile.

	Frequency	**Percentage (%)**
Age		
21-25	22	73.3
26-30	8	26.7
Employment		
Full-time	26	86.7
Part-time	2	6.7
Casual	2	6.7
Salary		
Below 40k	5	16.7
41k-60k	3	10
61k-80k	8	26.7
81k-100k	5	16.7
Above 100k	9	30
Education Level		
Undergraduate	28	93.3
Postgraduate	2	6.7
Graduation Year		
2012	4	13.3
2013	2	6.7
2014	5	16.7
2015	5	16.7
2016	14	46.7
Experience		
Below 3 Years	15	50
3-5 Years	9	30
Above 5 Years	6	20

Table 2 shows the descriptive and t-test results on the nine factors affecting the respondents' career choice. The recorded average scores range from the lowest of 2.700 to the highest of 4.344. There is a rather wide spread in the respondents' responses for some factors as demonstrated by the high standard deviations. The three factors that are statistically significantly different from the test value are: (i) career opportunities (F1, mean = 4.344); (ii) belief of getting better pay (F2, mean = 3.767); and (iii) self-efficacy and high level of confidence of performing construction tasks (F7, mean = 3.433). As the mean values of these top three factors are statistically significantly higher than three, this indicates that the early career respondents gave greater importance to these factors in their career choice. These findings support authors' claims that career opportunities and opportunities of getting generous, lucrative wages with good earning potential in the industry should be fully introduced to female graduates in order to attract more women to the industry [15], [17]. Next, the third top factor suggests that female graduates with high level of self-efficacy and confidence in their abilities to perform construction tasks are more likely to consider a career in construction. This reinforces that self-efficacy or one's degree of confidence is related to career options [6], [13]. On the other hand, 'previous hands-on experiences related to construction (F8, mean = 2.70)' has recorded the lowest mean value. This is inconsistent with Koch's [14] study that hands-on work experience is a key motivator in career decision-making. This can be partly explained because some respondents had started working in the industry while studying (as evidenced in number of years of experience they had, see Table 1), and this respective group did not give greater importance to this factor that resulted in low mean value.

Table 2. Factors affecting the respondents' career choice.

	Factor	**Mean**	**Std. Dev.**	**Sig. (*t*-test, test value =3)**
F1	Career opportunities	4.344	0.897	**0.000***
F2	Belief of getting better pay	3.767	0.728	**0.000***
F3	Family influence and support	3.133	1.042	0.489
F4	Role models or mentors	3.200	1.031	0.297
F5	Encouragement from college advisors, teachers or counsellors	2.867	0.973	0.459
F6	Strong science ability and high academic achievements	2.900	1.062	0.610
F7	Self-efficacy and high level of confidence of performing construction tasks	3.433	1.006	**0.025***
F8	Previous hands-on experiences related to construction	2.700	1.208	0.184
F9	Education climate that encourages women to pursue non-traditional careers	2.933	1.143	0.752
	* Significant at $p < 0.05$ level			

Next, Table 3 shows the descriptive and *t*-test results on twelve career barriers faced by the respondents. The high variability in responses among the respondents are well-demonstrated by the considerably high standard deviations. This variability provides suggestive evidence that the respondents have had to address different challenges associated with these barriers in their early career. However, in overall term, ten barriers have recorded mean values above three, and four of them are statistically significantly higher than the test-value of three, namely: (i) difficult to integrate into masculine culture in the industry (B2, mean = 4.000); (ii) stressful and competitive working culture (B3, mean = 4.067); (iii) long and inflexible working hours (B5, mean = 3.933); and (iv) lack of informal networks for career opportunities (B8, mean = 3.600). The results suggest that these barriers are somewhat to certain extent significant in their early career in construction. Indeed, these barriers have all been identified as top barriers in previous studies [2], [10], [12], [19]. However, it is interesting to note that 'sex harassment (B12, mean 2.533)'- a top barrier in the literature has recorded the lowest mean value that is statistically significantly below three at $p < 0.10$ level. This seemingly suggests that this barrier did not apply to some of the respondents, or that some respondents never experienced this barrier and/or they have the skills to deal with this barrier in their early career. It is indeed encouraging to see this barrier ranks at the bottom of the list, it helps to take the stress out of the respondents in dealing with it while striving to move up in their early career. Above all, despite the high variability in responses, the results provide suggestive evidence the respondents as early career women have had not face barriers that are considerably different from other women's groups in construction in general.

Table 3. Barriers faced by the respondents.

	Barrier	**Mean**	**Std. Dev.**	**Sig. (*t*-test, test value =3)**
B1	The construction industry has poor images as "dirty" and "dangerous"	3.033	1.273	0.887
B2	Difficult to integrate into masculine culture in the industry	4.000	1.438	**0.001***
B3	Stressful and competitive working culture	4.067	1.437	**0.000***
B4	Bad working conditions for women especially on site	3.067	1.388	0.794
B5	Long and inflexible working hours	3.933	1.617	**0.004***
B6	Lack of role models in the industry	3.400	1.476	0.149
B7	Lack of mentoring	3.367	1.520	0.197
B8	Lack of informal networks for career opportunities	3.600	1.589	**0.048***
B9	Career glass ceiling for women	3.100	1.539	0.725
B10	Double standard performance evaluation for achievement	3.067	1.507	0.810
B11	Remuneration discrimination	2.733	1.285	0.265
B12	Sexual harassment	2.533	1.479	0.095
* Significant at $p < 0.05$ level				

Lastly, the respondents were asked to indicate their overall job satisfaction in their early career. The results show a considerably high mean value of 4.033 (out of 5), which is statistically significantly greater than the neutral point of three ($t = 8.464$, $p < 0.05$). The recorded standard deviation (0.669) is indeed the lowest compared to those reported in Tables 2 and 3, signifying high consistency in their overall job satisfaction responses. Encouragingly, none of the respondents were either very dissatisfied (score 1) or dissatisfied (score 2) with their jobs, and 80% (24 out of 30) of them were satisfied or very satisfied with their jobs. As early career women, this suggests there is a high possibility that they will remain in the industry, at least for some considerable time.

5. Conclusion

In examining the factors affecting early career women's career choice in construction and barriers faced by them at this career stage, the results show that the three statistically significant factors affecting their career choice decisions are: (i) career opportunities; (ii) belief of getting better pay; and (iii) self-efficacy and high level of confidence of performing construction tasks. Education providers and employers could reinforce their efforts on aspects related to these factors in attracting and retaining female graduates in the construction industry. For barriers, it was found that the early career women have had not faced barriers that are considerably different from other woman's groups in construction in general. However, it is recognized that early career women would need supports from mentors in sustaining their positions given the empirical evidence on the high possibilities of young women moving to another industry sector. Future studies could consider exploring in-depth the types of support for early career women received or would like to have received in their career development in construction. A focus group study that involves respondents from different tertiary institutions offering CM programs could be a possible way for this in-depth exploration.

6. References

[1] Bigelow B F, Saseendran A and Elliott J W 2018 Attracting students to construction education programs: An exploration of perceptions by gender *Int. J. of Construction Education and Research* **14**(3) 179-197
[2] Ling F Y Y, Leow X X , and Lee K C 2016 Strategies for attracting more construction-trained graduates to take professional jobs in the construction industry *J. of Professional Issues in Engineering Education and Practice* **142**(1)
[3] Worrall L, Harris K, Stewart R, Thomas A and McDermott P 2010 Barriers to women in the UK construction industry *Engineering, Construction and Architectural Management* **17**(3) 268-281
[4] Fielden S L, Davidson M J, Gale A W and Davey C L 2000 Women in construction: the untapped resource *Construction Management and Economics* **18**(1) 113-121
[5] Australian Bureau of Statistics 2017 *Labour force, Australia Table EQ09 Employed persons by Industry division (ANZSIC) and Occupation major group (ANZSCO) of main job and sex, August 1986 onwards, data cubes: excel spreadsheet 2017* cat. no. 6291.0.55.003
[6] Whitmarsh L, Brown D, Cooper J, Hawkins-Rodgers Y and Wentworth D K 2007 Choices and challenges: A qualitative exploration of professional women's career patterns *The Career Development Quarterly* **55**(3) 225-236
[7] Oo B L and Widjaja E C 2016 Female student enrolments in construction management programs, *In CRIOCM 21st International Conference on Advancement of Construction Management and Real Estate Proceedings* (Singapore: Springer) 789-798
[8] Ling F Y Y and Poh Y P 2004 Encouraging more female graduates to enter the construction industry *Women in Management Review* **19**(8) 431-436
[9] Oo B L, Li S, and Zhang L 2018 Understanding female students' choice of a construction management undergraduate degree program: Case study at an Australian university *J. of Professional Issues in Engineering Education and Practice* **144**(3)
[10] Barreto U, Pellicer E, Carrión A and Torres-Machí C 2017 Barriers to the professional development of qualified women in the Peruvian construction industry *J. of Professional Issues in Engineering Education and Practice* **143**(4)
[11] Francis V 2017 What influences professional women's career advancement in construction? *Construction Management and Economics* **35**(5) 254-275
[12] Dainty A R, and Lingard H 2006 Indirect discrimination in construction organizations and the impact on women's careers *J. of Management in Engineering* **22**(3) 108-118
[13] Moore J D and Gloeckner G W 2007 A theory of women's career choice in construction management: recommendations for academia *Int. J. of Construction Education and Research* **3**(2) 123-139
[14] Koch D C, Greenan J and Newton K 2009 Factors that influence students' choice of careers in construction management *Int. J. of Construction Education and Research* **5**(4) 293-307
[15] Francis V and Prosser A 2014 Exploring vocational guidance and gender in construction *Int. J. of Construction Education and Research* **10**(1) 39-57
[16] Gayani Fernando N, Amaratunga D and Haigh R 2014 The career advancement of the professional women in the UK construction industry: The career success factors *J. of Engineering, Design and Technology* **12**(1) 53-70
[17] Bigelow B F, Bilbo D, Mathew M, Ritter L and Elliott J W 2015 Identifying the most effective factors in attracting female undergraduate students to construction management *Int J. of Construction Education and Research* **11**(3) 179-195

IOP Conf. Series: Materials Science and Engineering **601** (2019) 012021 doi:10.1088/1757-899X/601/1/012021

[18] Jusoh M, Simun M and Chong S C 2011 Expectation gaps, job satisfaction, and organizational commitment of fresh graduates: Roles of graduates, higher learning institutions and employers. *Education+ Training* **53**(6) 515-530
[19] Dainty A R, Bagilhole B M and Neale R H 2000 A grounded theory of women's career under-achievement in large UK construction companies *Construction Management and Economics* **18**(2) 239-250
[20] Navarro-Astor E, Román-Onsalo M and Infante-Perea 2017 Women's career development in the construction industry across 15 years: main barriers *J. of Engineering, Design and Technology* **15**(2) 199-221

IOP Conf. Series: Materials Science and Engineering **601** (2019) 012022 doi:10.1088/1757-899X/601/1/012022

Dimensions of Tenants' Office Leasing Behaviours: An Introductory Study

Sumin Kim[1*] and Benson T.H. Lim[1]

[1] Faculty of Built Environment, UNSW Sydney, NSW, 2052, Australia
*Corresponding author: su.kim@unsw.edu.au

Abstract. The greenness of office buildings is becoming a more important issue as it contributes to the sustainability of the building sector. Moreover, by leasing greener offices, tenants can be benefitted by the improved indoor environmental quality and better corporate branding attributed to its positive image. These benefits, however, often come with relatively expensive prices that tenants need to pay. Nevertheless, the sources of the price premium have not been fully explored especially from the aspect of tenants' leasing behaviours. To address this, a questionnaire survey was conducted. Then, exploratory and confirmatory factor analysis were performed to identify the underlying dimensions of tenants' leasing behaviours and to ensure their reliability and validity. The results show that tenants' office leasing behaviours must be explained by several dimensions. Also, it is found that the obtained results have a sufficient reliability and validity. These allow proceeding to the next stage of the research to explain if and to what extent the greenness level of tenants' offices and the rents they pay are influenced by various motivational variables based on the established analytical model.

1. Introduction

Since green building certification schemes such as NABERS and Green Star were introduced in Australia in the late 90's–2000's, the number of green office buildings, buildings rated over 4 Stars from either of the certification schemes, has been increased steadily nationwide. Sydney Commercial Building District (CBD), the nation's biggest metropolitan area, has led the incremental trend as it is reflected by the most significant number of both NABERS and Green Star certified buildings in its local governing area [1]. From the tenants' perspective, the increasing number of green buildings provide more opportunity to enjoy the superior indoor environmental quality (IEQ) and energy efficiency which may lead to productivity and financial gains [2]. In addition, the positive image to customers and employees allow them to brand themselves as environmentally-conscious organisations [3].

However, these benefits often come with a hefty price tag. Studies show that tenants need to pay higher prices for greener office buildings with a certification compared to the non-green counterpart [4]. This is no exception to the tenants of the Sydney CBD office market as NABERS 5 Stars certified buildings command extra rents up to 2.7 percent compared to the buildings with lower ratings [5].

Meanwhile, many of these studies also pointed out that it may not only the greenness of the buildings that contribute to the price premium. This is supported by several studies suggesting that prices of buildings may be better explained as the results of collective effects of their tangible (e.g. building quality) and symbolic (e.g. the green image) aspect [6-7]. In particular, Kim et al. [4, 8]

suggested that tenants' organisational identity, as well as values and expectations that they place on their buildings, may be helpful to explain why they are paying the premium price for the greenness of their office buildings. Nevertheless, investigation on this is very limited, especially in the Australian office market context.

2. Research Aim and Methodology

The aim of this paper is to perform an introductory study on the impact of tenants' leasing behaviours on the greenness level of their offices and rents. To this, an online questionnaire survey was designed based on the research framework [8], results of a systematic literature review and meta-analysis [4], and preliminary interviews [9]. The survey is mostly composed of the 7-point Likert-scale questionnaires asking respondents to self-report their views on a range of different questions. The survey was distributed to the key leasing decision makers (e.g. CEO, office leasing director, sustainability director, national director) of the tenants from the top three sectors occupying the Sydney CBD office market (i.e. finance and insurance sector, real estate sector, and professional, scientific and technical sector) [9]. The key leasing decisions makers were contacted because they are either directly involved in their organisation's office leasing decision or have a sufficient knowledge in the leasing decision. Eventually, 51 responses containing expert opinions on their organisation's leasing motivations were obtained. The obtained responses were analysed using exploratory and confirmatory factor analysis (EFA & CFA). To this, IBM SPSS Statistics 3.0 and SmartPLS 3 software were used.

3. Results and Discussions

3.1. Exploratory Factor Analysis (EFA)

Firstly, EFA on the obtained data was undertaken using SPSS. The primary purpose of EFA was two-folds; (i) construct extraction and (ii) variable reduction. These are to identify the underlying dimensions (or factors) of tenants' leasing behaviours and to ensure their reliability and validity. A total of 82 variables included in the questionnaire survey were assessed based on multiple criteria (e.g. KMO and Bartlett's test of sphericity, item-to-total correlations, Eigenvalue, communalities, loadings, cross-loadings, Cronbach's alpha) [10–13]. This eventually led to a reduction in the number of variables from 82 to 50. These variables are composed of total 13 constructs including the single construct that showing the level of the greenness of tenants' office buildings and their rents (Table 1).

Table 1. Categorisation of the constructs

Tangible aspect of office buildings (X1)	Building sustainability ($X1_{BLDS}$)
	Lease contract features ($X1_{LEAS}$)
	Building quality ($X1_{BLDQ}$)
	Proximity ($X1_{PROX}$)
Symbolic aspect of office buildings (X2)	Occupant related symbolic values ($X2_{OCCU}$)
	Organisation related symbolic values ($X2_{ORGS}$)
Tenants' expectations (X3)	Become a socially conscious organisation ($X3_{SOCO}$)
	Become a sustainable organisation ($X3_{SUSO}$)
Tenants' identity (X4)	People-focused organisation ($X4_{PEOP}$)
	Sustainability-focused organisation ($X4_{SUST}$)
	Unique and approachable organisation ($X4_{UNIA}$)
The greenness of office buildings (X5)	Greenness (X5)
Office building rents (Y)	Rent (Y)

Table 2 shows the four constructs associated with the tangible aspect of office buildings (X1) and their associated variables. The first construct ($X1_{BLDS}$) consists of 5 variables of building lighting quality, indoor air quality (IAQ), acoustic quality, water efficiency, and environmental performance (e.g. CO2 emissions). All these variables are closely related to the sustainability aspect of an office building as it is reflected by the NABERS certification for office buildings [14]. The next construct ($X1_{LEAS}$) consists of four variables which are all closely related to lease contract features. The third construct ($X1_{BLDQ}$) contains three variables of office space size, building amenities (e.g. provision of end of the trip facilities, bicycle racks). and building grades (e.g. Premium, A, B and C) which all reflect building quality. The importance of this construct could be highlighted by the responses obtained from the preliminary interview that tenants leased their current offices because they 'tick the boxes'–reflecting that green buildings in Sydney CBD are also the high-quality buildings [9]. Lastly, the fourth construct ($X1_{PROX}$) only consists of two variables of proximity to major roads and to major competitors. This highlights the importance of accessibility and may indicate the existence of the 'urban agglomeration' that similar businesses within the CBD tend to cluster together. Collectively, the results show that the variables used to assess tenants' motivation towards the tangible aspect of office buildings (X1) could be explained based on the above-mentioned four constructs.

Table 2. Factor matrix for the tangible aspect of office buildings (X1)

	$X1_{BLDS}$	$X1_{LEAS}$	$X1_{BLDQ}$	$X1_{PROX}$
TA14_Building's lighting quality	.812			
TA15_Building's indoor air quality	.801			
TA13_Building's acoustic quality	.747			
TA17_Building's water efficiency	.718			
TA18_Building's environmental performance	.712			
TA23_Lease pre-commitment		.799		
TA22_Lease type		.769		
TA24_Build-out		.757		
TA21_Lease term		.748		
TA6_Office space size			.786	
TA11_Building's amenities			.785	
TA9_Building's grade			.763	
TA3_Proximity to major roads				.874
TA5_Proximity to major competitors				.833

Table 3 shows the symbolic aspect of office buildings (X2) which consists of two constructs, namely, occupant related symbolic values ($X2_{OCCU}$) and organisation related symbolic values ($X2_{ORGS}$). The variables associated with the first construct ($X2_{OCCU}$) well-reflect the interests of building occupants (e.g. employees) as these often influence their job satisfaction and performance [15]. This is further supported by the findings of preliminary interviews that the provision of employee benefits by finding more occupant-friendly offices is one of the key considerations for many organisations. Meanwhile, the second construct ($X2_{ORGS}$) is rather closely related to the interests and the demand of organisations themselves. For instance, the cutting-edge image of an organisation's office building may help them to be perceived as an innovator in their business, not a follower.

Table 3. Factor matrix for the symbolic aspect of office buildings (X2)

	$X2_{OCCU}$	$X2_{ORGS}$
SY6_Aesthetically pleasing office environment	.839	
SY7_Employee-friendly office environment	.832	
SY1_Comfortable office environment	.810	
SY10_Healthy office environment	.669	
SY8_Collaborative office environment	.594	
SY11_Office environment with cutting-edge technologies		.821
SY9_Environmentally-friendly office environment		.811
SY14_Office environment that well-reflects the corporate identity		.710
SY13_Office environment to attract future employees		.681

It is found that two constructs were extracted from the variables measuring tenants' expectations (X3) (Table 4). The first construct ($X3_{SOCO}$) represents the tenants' expectation to become a socially conscious organisation as these variables are well aligned with the Corporate Social Responsibility (CSR) of organisations and its impacts on stakeholder relations [16]. The other construct ($X3_{SUSO}$) refers to the tenants' expectations to become a sustainable organisation. This construct covers a range of expectations related to Corporate Sustainability (CS) which could be described as meeting the needs and interests of direct, indirect, and future stakeholders while contributing to the three domains of sustainability; i.e. environmental, economic, and social sustainability [17-18]. This is also supported by the findings of the preliminary interview that 'Keeping up with the Joneses' was one of the leasing motivators for tenants in a competitive industry [9].

Table 4. Factor matrix for the tenants' expectations for their offices (X3)

	$X3_{SOCO}$	$X3_{SUSO}$
EP9_Innovative organisation	.938	
EP10_People-first organisation	.893	
EP12_Flexible organisation	.888	
EP8_Trustworthy organisation	.883	
EP13_Publically well-known organisation	.851	
EP14_Approachable organisation	.818	
EP6_Collaborative organisation	.789	
EP7_Ethical organisation	.767	
EP11_Fair organisation	.728	
EP5_Environmentally friendly organisation		.897
EP2_Fast follower organisation		.861
EP3_Socially responsible organisation		.815

Table 5 shows the matrix for tenants' organisational values that reflecting their identity (X4). The first construct ($X4_{PEOP}$) includes a range of variables related to the values promoted by people-friendly organisations. These include the values associated with the organisation's current employees as well as their customers and potential employees. The second construct ($X4_{SUST}$) explains the tenants' identity as a sustainable organisation. All associated four variables of this construct are well-aligned with the triple bottom line of sustainability. Moreover, a variable of acknowledgement is well-supported by the interview findings that one of the biggest motivators for leasing a 'certified' green building is to be acknowledged by others as a sustainability-focused organisation. Meanwhile, the last construct ($X4_{UNIA}$) represents the tenants' identity as a unique and approachable organisation. It should be noted that the community variable is cross-loaded across the first and third construct. However, considering the gap between the two loadings are over 0.2 [12], the cross-loading can be disregarded.

Table 5. Factor matrix for tenants' organisational values (X4)

	$X4_{PEOP}$	$X4_{SUST}$	$X4_{UNIA}$
OV13_Customer satisfaction	.875		
OV1_Leadership	.805		
OV14_People (Employee)	.798		
OV8_Innovation	.777		
OV5_Openness	.776		
OV4_Teamwork	.751		.485
OV15_Community (Corporate Citizenship)		.898	
OV10_Social responsibility		.894	
OV12_Environmental sustainability		.791	
OV16_Acknowledgement		.741	
OV17_Uniqueness			.865
OV18_Approachability			.803

Lastly, the level of the greenness of tenants' office buildings (X5) extracted only a single construct (Table 6). This is because only two variables (i.e. NABERS and Green Star ratings) were used to assess the level of the greenness. This is consistent with the dependent variable of this research (Y), rent, that used a single variable. Nevertheless, all these two variables meet the requirements to be used for the subsequent stages of the analysis.

Table 6. Factor matrix for the greenness of tenants' offices (X5)

	X5
NABERS ratings	.884
Green Star ratings	.884

3.2. Confirmatory Factor Analysis (CFA)

Upon the successful removal of irrelevant (or inconsistent) variables and extraction of the constructs through EFA, CFA was undertaken based on the analytical model of the research (Figure 1). The established analytical model depicts the hypothesised relationships among the constructs identified through the EFA procedure. As a part of CFA, internal consistency reliability and construct validity of the retained data were assessed as it is recommended by Hair et al. [19]. Firstly, the results show that the retained data have a sufficient level of internal consistency reliability as all the constructs meet the required value of Cronbach's alpha and composite reliability of 0.7 or above. This confirms that variables loaded as the same construct measure the same factor (or dimension). Then, construct validity of data was evaluated. To this, assessments of convergent validity and discriminant validity were undertaken. Convergent validity indicates which two measures are related to each other, thus capture a common construct [20]. Firstly, it is found that Average Variance Extract (AVE) of all the constructs involved in the model meet the recommended thresholds of 0.5 or above. Moreover, the size of loadings, commonly known as indicator reliability, are all above the recommended 0.7 thresholds. These all confirm the establishment of convergent validity. Lastly, discriminant validity was assessed. To establish discriminant validity, correlations between variables related to each other must be higher than variables from other constructs that are theoretically meant not to correlate [21]. Assessments on discriminant validity were undertaken based on several approaches including cross-loadings, Fornell-Larcker criterion, and Heterotraid-Monotrait (HTMT) ratio of correlations. The results show that the variables used in the analytical model are not cross-loaded across different constructs. Moreover, the results of both Fornell-Larcker criterion and HTMT ratio of correlations deemed as satisfactory. Therefore, it can be confirmed that the model and its variables have no discriminant validity issues.

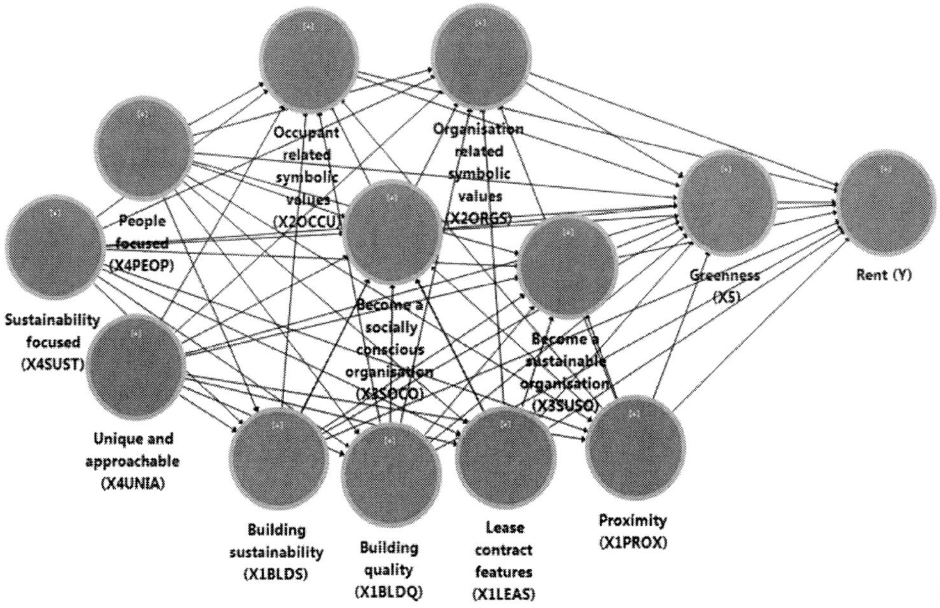

Figure 1. Analytical model of the research

4. Conclusion

The results of the analysis identified several underlying dimensions of tenants' office leasing motivators. Moreover, the internal consistency reliability and the construct validity of data were established. Collectively, these allow proceeding to the subsequent stages of the research to analyse the impact of tenants' behaviours on the level of the greenness of their offices and rents they pay using the developed analytical model.

5. References

[1] Kim S, Lim B T H and Kim J 2016 The Effect of Building Sustainability Regulation on the Green Office Building Stock in Australia *Proceeding of International Conference on Sustainable Built Environment Actions for the Built Environment of Post-Carbon era Complying with COP21 (Seoul)* (Seoul: Sustainable Built Environment 16 Seoul) pp 470–473

[2] Kats G H 2003 *Green Building Costs and Financial Benefits* (Massachusetts: Massachusetts Technology Collaborative) pp 1–8

[3] Edwards B W and Naboni E. 2013 *Green Buildings Pay: Design, Productivity and Ecology* 3rd edn. (Abingdon: Routledge) pp 1–296

[4] Kim S, Lim B T H and Kim J 2017 Green Features, Symbolic Values and Rental Premium. Systematic Review and Meta-analysis, *Procedia Engineering* **180** 41–48

[5] Newell G, MacFarlane J and Kok N 2011 *Building better returns. A Study of the Financial Performance of Green Office Buildings in Australia* (Sydney: Australian Property Institute and Property Funds Australia) pp 1–48

[6] Ledgerwood A, Liviatan I and Carnevale P J 2007 Group-Identity Completion and the Symbolic Value of Property *Psychological science* **18**(10) 873–878

[7] Healey P 1992 An institutional model of the development process *Journal of Property Research* **9**(1) 33–44
[8] Kim S, Lim B T H and Kim J 2017 Tenants' Decision to or not to Lease Green & Non-green Buildings: A Conceptual Framework *Procedia Engineering* **180** 1551–1557
[9] Kim S, Lim B T H, Kim J 2019 Tenants' motivations to lease Green Office Buildings: an exploratory study of Sydney Central Business District *International Journal of Structural and Civil Engineering Research* **8**(1) 59–62
[10] Hair J F, Black W C, Babin B J and Anderson R E 2010 *Multivariate data analysis* 7th edn. (Upper Saddle River: Prentice Hall) pp 1–734
[11] Comrey A L and Lee H B 1992 *A first course in factor analysis* 2nd edn. (Hillsdale: Lawrence Erlbaum Associates) pp 1–443
[12] Howard M C 2015 A Review of Exploratory Factor Analysis Decisions and Overview of Current Practices. What We Are Doing and How Can We Improve? *International Journal of Human-Computer Interaction* **32**(1) 51–62
[13] Nunnally J C 1978 *Psychometric theory* 2nd edn. (New York: McGraw-Hill) pp 1–640
[14] Office of Environment and Heritage 2018 NABERS: Office Buildings Retrieved on September 30, 2018 from https://www.nabers.gov.au/ratings/spaces-we-rate/office-buildings
[15] Kim S E and Jung C S 2015 The Effects of Status Symbols in the Office on Employee Attitudes in a Human Service Agency *Human Service Organizations: Management, Leadership & Governance* **39**(4) 306–322
[16] Loosemore M and Lim B T H 2016 Linking corporate social responsibility and organizational performance in the construction industry *Construction Management and Economics* **35**(3) 90–105
[17] Dyllick T and Hockerts K 2002 Beyond the business case for corporate sustainability *Business Strategy and Environment* **11**(2) 130–141
[18] World Commission on Environment and Development 1987 *Our Common Future* UN Documents (Oxford: Oxford University Press) pp 1–247
[19] Hair J F, Hult G T M, Ringle C and Sarstedt M 2017 A primer on partial least squares structural equation modeling (PLS-SEM) (Los Angeles: Sage Publications) pp 1–384
[20] Carlson K D and Herdman AO 2010 Understanding the Impact of Convergent Validity on Research Results *Organizational Research Methods* **15**(1) 17–32
[21] Bertea P and Zait A 2011 Methods for testing discriminant validity *Management & Marketing* **IX**(2), 217–224

IOP Conf. Series: Materials Science and Engineering **601** (2019) 012023 doi:10.1088/1757-899X/601/1/012023

Utilization of Fly Ash in Construction

Norhaiza Ghazali[1*], Khairunisa Muthusamy[1] and Saffuan Wan Ahmad[1]

[1]Faculty of Civil Engineering & Earth Resources, Universiti Malaysia Pahang, Lebuhraya Tun Razak, 26300 Gambang, Kuantan, Pahang
*Corresponding author: nhaiza84@gmail.com

Abstract. In year 2016, Malaysia produce about 6.8 million tonnes of fly ash from six coal fired electric power station and this figure will increase due high demand of electric production. Fly ash is a waste product from the combustion of pulverized coal in electricity power plants and considered as a contributing factor for air, water and soil pollution that lead to human health problems and various geo-environmental issue. However, fly ash can be converted into valuable and useful product if properly utilized and many researches has been carried out in aim to increase the utilization of fly ash in various sectors especially in construction sector. In line with current economic conditions, all construction works required cost efficient with good quality of work. Therefore, fly ash is suitable to be used as an alternative raw material to enhance or replacing the existing construction material. In produce construction material many different industries and construction because it is oxide-rich material. The paper will be discussing the potential of managing waste of fly ash by utilizing it as construction materials.

1. Introduction

These days, concrete has become main building material with more than 10 billion tons produced annually in construction industry [1][2]. The important characters of concrete compare to other materials are excellent strength, easy to molded in any form, an engineered material that can meet almost any desired specification, adaptable, quite incombustible, afford- able, and easily obtained. Higher demand for construction materials has been rising to sustain the fast-growing global population. As a result, there is continual depletion of raw materials and natural sources and leading to increase the construction material cost. To this effect, engineers are faced with the challenge of resolving this potential sustainability problem [3].

The current global trends are focused on recovery of usable materials from waste as well as utilization of waste as raw materials whenever feasible in construction. Fly ash is a waste product from the combustion of pulverized coal in electricity power plants that faces an increasing production required large area for disposal. In year 2016, Malaysia produce about 6.8 million tonnes of fly ash [4]. The fly ash is deposited either in a dry landfill over a vast area of land which is not possible in urban areas or deposited in an ash pond which also has its shortcomings [5]. Dispose tonnes of waste in without any treatment would cause land pollution, water pollution, air pollution as well as destruction to flora and fauna. [6]. Fly ash has been successfully used in concrete industry since over 50 years but its application is still limited due to lack of understanding of the characteristics of fly ash itself and the properties of concrete containing fly ash [2][7].

2. Fly Ash Generation and Environmental Issue

Fly ash can be defined as a waste residue that is released from coal combustion process in electric power stations [8-15]. Fly ash is the unburned residue that is carried away from the burning zone in the boiler by the flue gases and then collected by either mechanical or electrostatic separators [4][16][17]. Malaysia started the development of coal-fired electric power station in year 1987 and currently there are six coal fired electric power station in Malaysia [18-20]. Those electric power station produce about 6.8 million tonnes of fly ash [4]. According to statistics, fly ash rate of production is clearly far outweighs consumption due to increased amounts of energy being generated by coal-fired power plants and widely available across the globe as shown in Table 1 [17][21].

Table 1. Fly ash production in different countries [22].

Country	Amount of Production (million tons/year)	Country	Amount of Production (million tons/year)
India	112	Malaysia	6.8
China	100	Canada	6
USA	75	France	3
Germany	40	Denmark	2
UK	15	Italy	2
Australia	10	Netherland	2

Fly ash contains silica, alumina, ferric oxide and other oxides material that might turn fly ash into hazardous material. These hazardous material is contributing factor in air, water and soil pollution that lead to human health problems and various geo-environmental issue [23-28]. These bad situations will interrupt the entire ecological cycles if not properly disposed [14][25][26], therefore good waste management practice needed to sustain a healthy environment [20]. Fly ash emissions from coal combustion units show a wide range of composition with present of elements below atomic number 92 and considered as major source of air pollution. The ultrafine particle of fly ash will behave like cumulative poisons after remain for long periods of time when reaches the respiratory region [13][28]. As a result, several physiological disorders and other related health problems such as respiratory problem, cancer, anaemia, hepatic disorder, gastroenteritis and dermatitis will arise [13][23]. Several studies on the present ground showed that wet disposal of this waste causes migration of metal into the soil [29]. The populations located near the fly ash dumping area facing surface water pollution and underground water pollution. However, the surface water pollution is more critical than the underground water pollution. The surface water pollution decreases the fish population and other aquatic organisms due to heavy metal material and organic matter content contained in the water. The surface water contamination also causes skin diseases, diarrhoea and death due to bathing and drinking of water from the contamination river [23].

2.1. Properties of Fly Ash

Physical and chemical characteristics of fly ash is important, as these characteristics influence its subsequent use and disposal [11][30]. Those characteristics of fly ash is depend on the type of coal used, the combustion condition and temperature at which the coal was fired, the collector setup, the air fuel ratio and other factors [11][17][25][31]. Few generalizations were made in particle size distribution, morphology, surface area, hydraulic conductivity or permeability and density of the fly ash [32]. Fly ash consists of predominantly fine particles ranging from grey to tan to reddish brown [33] with an average size of particles is 20μm and the bulk density ranging from $0.54 g/cm^3$ to $0.86 g/cm^3$ [1][11][34][35]. 70% to 90% of the particle were solid glassy spheres, while the balance containing Quartz, Mullite, Hematite and Magnetite and a small portion of unburned Carbon as shown in Figure 1 [11][20][36].

American Society for Testing and Materials (C618 – 12a), classifies fly ash into two classes, Class C and Class F [30][32][37]. Typically fly ash contains significant amount of Silicon Dioxide (SiO_2), Aluminium Oxide (Al_2O_3) and Calcium Oxide (CaO) [38] and the class divides based on its chemical composition and source of origin [16]. Class F fly ash produced by burning anthracite or bituminous coal and only has pozzolanic properties while Class C fly ash produced by burning lignite or sub-bituminous coal and has both pozzolanic and cementitious properties [39]. The main difference between these two classes is the amount of calcium, silica, alumina, and iron content. Class F fly ash, contains less than 10% CaO and minimum SiO_2 +Al_2O^3 +Fe_2O_3 content of 70%, whereas Class C fly ash, contains more than 20% CaO and minimum SiO_2 +Al_2O_3 +Fe_2O_3 content of 50% [16].

Figure 1. Morphology of Fly Ash [20].

3. Utilization of Fly Ash

In last 30 years, many researches has been carry out in aim to increase the utilization of fly ash in various sectors since it is not considered as hazardous waste [13][28]. Figure 2 shows many countries across the globe utilize this residue waste and turn it into useful product as an alternative to another industrial resource, process or application [21][31]. The utilization of fly ash can be categorized into three classes of application, low, medium and high technology application [17].

Fly ash suitable to be used as raw material in many different industries and construction because it is oxide-rich material [28]. Currently, the fly ash was successfully use in improving the construction material and excellent to be used in agriculture sector to enhance the soil properties too [13]. The fly ash also can be utilized in brick manufacturing, ceramics manufacturing, road construction, concrete production and other activities as shown in Figure 3. The most important areas of fly ash utilization is a concrete production [40], either in conventional concrete or geopolymer concrete. Many researches have been done to find use of fly ash in cement, concrete, and other cementitious composites.

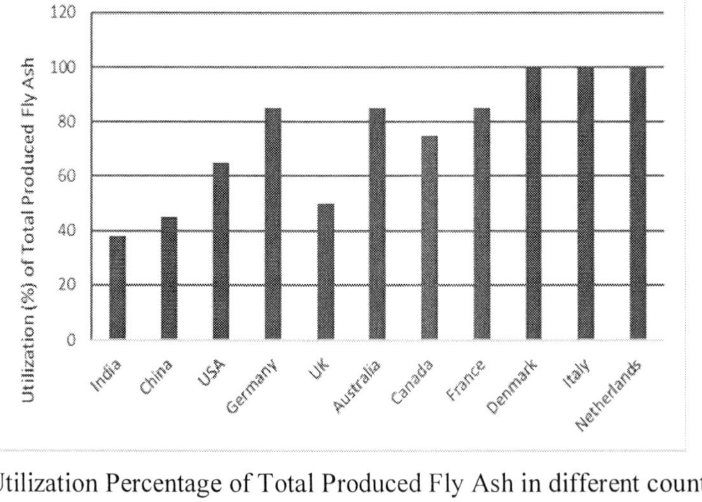

Figure 2. Utilization Percentage of Total Produced Fly Ash in different countries [22].

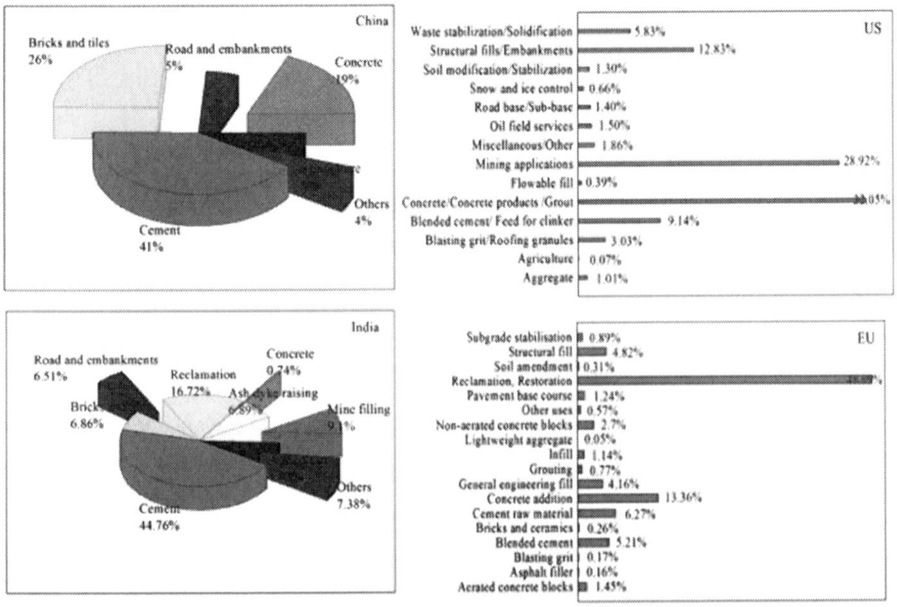

Figure 3. Fly Ash Utilization in China, India, US and EU [11].

3.1. Utilization of Fly Ash in Building Construction

The utilization of fly ash in the construction industry is not a new technology but it is a growing technology in improving the construction quality as well as the environment quality. Adding fly ash in concrete give benefits in term of economical, ecological and technical [2][41-43]. Currently, fly ash used by cement industries as a pozzolanic material for manufacturing of Portland Pozzolana Cement since the SiO_2 and Al_2SiO_3 content is very similar to Portland Cement [27]. In the presence of moisture and at room temperature, it reacts chemically with calcium hydroxide to derive compounds possessing cementitious properties [20][44]. Technically, the use of fly ash together with Portland cement contributes to the consumption of $Ca(OH)_2$ which is formed during the hydration of cement and leads to the formation of cementitious products [42][45]. The glassy phase of fly ash and calcium hydroxide generated leads to the formation of additional C±S±H gel and results in higher density and strength [43]. The compressive strength increased at later ages compared to the early ages [45]. This green technology enhances the durability and service life of concrete structures [17][46].

Fly ash can be used as a partially cement replacement by cement weight or as an admixture to the concrete mix [32]. The utilization fly ash as a component of blended cement can save a significant amount of energy and cost in cement manufacturing [17][36]. The cement replacement using fly ash can be up to 75% of cement weight, however there are strict standards governing its use such as the American Society for Testing Materials ASTM C 618 and European Standard EN 450-1[35]. The new cement alternative in the field of building and construction materials is geopolymer concrete with fly ash [47][48]. This concrete mixture produced a concrete with high compressive strength, low creep, good acid resistance and low shrinkage [20].

Normally fly ash was used in blended cement to product insitu concrete mix however fly ash also can be used in high strength precast and prestressed concrete. Having slow strength development at early age, the use of fly ash in high strength precast and prestressed concrete has been limited. Somehow, current studies have validated that superplasticized fly ash concrete with low water-to-cement ratio can be proportioned to meet the very early age strength as well as other requirements for precast or prestressed concrete products [17].

Significant research about an artificial lightweight aggregate has been carried out using fly ash to replace normal aggregates [49][50]. Fly ash can be used in manufacturing of light-weight aggregates by using sintered (fired) and unfired (cold bonded) processing methods [51-53]. This aggregates are light due to the presence of air voids [24] and used in structural lightweight concrete to reduce the dead load of a concrete structure and leads to the economic savings in terms of structural design [17][36][38][54].

3.2. Utilization of Fly Ash in Road and Embankment Construction

Utilization fly ash in road and embankment construction has many advantages compared with the conventional method. Saves top soil which otherwise is conventionally used, avoids creation of low lying areas (by excavation of soil to be used for construction of embankments) [54]. Fly Ash may be used in road construction for filling purposes, stabilizing and constructing sub base or base [13][17][20][28][55].

Embankment can be constructed using fly ash as fill material. However, the graters concern using fly ash as a fill is the potential of leachate and an associated contamination to the ground water. There are several techniques such as a site grading, compaction, and surface water control, that can be used to prevent leachate from happen [32]. The well-compacted fly ash shows good shear strength comparable to normal soils used in earth-fill operation. Fly ash also has good permeability and even moisture-density curve therefore, it's become desirable material use for embankment construction, backfilling and land development works[36].

Fly ash alone or in combination with lime, is used to stabilize the subgrade in order to reduce plasticity, enhance strength, and improve workability of weak soils [17]. Soil mixed with fly ash and lime will increases California Bearing Ratio (CBR), increased (84.6%) on addition of only fly ash to soil [13][55]. For construction of base and subbase courses for pavements, fly ash is used either with lime or portland cement, and aggregate [17]. Concrete containing large amounts of fly ash can be proportioned to meet strength and durability requirements for road paving work. The 50% fly ash Class C mixture provides an excellent alternative for paving even though this mixture gained strength more slowly at the beginning [56].

3.3. Utilization of Fly Ash in Masonry Work and Other Application

Fly ash can be utilized in manufacture of the fired, unfired, and steam cured bricks and become top 3 among the sector of utilization [11]. Fly ash bricks have a number of advantages over the conventional burnt clay bricks [24][27][35][57][58]. 40% to 70% fly ash can be used in brick production and the fly ash brick is technically acceptable, economically viable and environment friendly [13]. Blocks and paving stones can also be manufactured by adding appropriate amount of coarse aggregate to the mixture [17].

Fly ash pipes are more watertight and more resistant to weak acids and sulphates attack relative to ordinary cement concrete pipes. Fly ash also used in manufacture high-flexural strength ceramics. These include railroad ties, electric line insulators, fence posts and others suitable application [17].

4. Conclusion

Fly ash is waste residue that is released from coal combustion process in electric power station and it can cause air, water and soil pollution if not treated will properly method and procedure. Luckily, this waste residue can be turn into valuable material when properly utilized in various sectors. Generally, fly ash will to be utilized in construction industries to enhance the existing construction material because of its pozzolanic property. The utilization of fly ash gives benefits in term of economical, ecological and technical aspects. However, further study needs to be done on recovery of products from fly ashes with a wide range of characteristics.

5. References

[1] Aprianti S E 2017 A huge number of artificial waste material can be supplementary cementitious material (SCM) for concrete production – A review part II *J. Clean. Prod.* **142** 4178–4194

[2] Hsu S, Chi M, and Huang R 2018 Effect of fineness and replacement ratio of ground fly ash on properties of blended cement mortar, *Constr. Build. Mater.* **176** 250–258
[3] Ibrahim HA and Abdul Razak H 2016 Effect of palm oil clinker incorporation on properties of pervious concrete *Constr. Build. Mater.* **115** 70–77
[4] Rafieizonooz M, Mirza J, Salim MR, Hussin MW and Khankhaje E 2016 Investigation of coal bottom ash and fly ash in concrete as replacement for sand and cement Constr. Build. Mater. **116** 15–24
[5] Abubakar AU, Baharudin KS 2012 Potential Use of Malaysian Thermal Power Plants Coal Bottom Ash in Construction *Int. J. Sustain Constr. Eng. Technol.* **3** 2180–3242
[6] Muhammad Nazrin Akmal AZ, Muthusamy K, Mat Yahaya F, Mohd Hanafi H and Nur Azzimah Z 2017 Utilization of fly ash as partial sand replacement in oil palm shell lightweight aggregate concrete, *IOP Conf. Ser. Mater. Sci. Eng.* **271** 012003
[7] Chindaprasirt P, Jaturapitakkul C and Sinsiri T 2007 Effect of fly ash fineness on microstructure of blended cement paste, *Constr. Build. Mater.* **21** 1534–1541
[8] Tkaczewska E 2014 Effect of the superplasticizer type on the properties of the fly ash blended cement, *Constr. Build. Mater.* **70** 388–393
[9] Shamshad A, Fulekar MH and Bhawana P 2012 Impact of Coal Based Thermal Power Plant on Environment and its Mitigation Measure, *Int. Res. J. Environ. Sci.* **1** 60–64
[10] Yin K, Ahamed A and Lisak G 2018 Environmental perspectives of recycling various combustion ashes in cement production – A review *Waste Manag.* **78** 401–416
[11] Yao ZT, Ji XS, Sarker PK, Tang JH, Ge LQ and Xia MS 2015 A comprehensive review on the applications of coal fly ash *Earth-Science Rev.* **141** 105–121.
[12] Ahmaruzzaman M 2010 A review on the utilization of fly ash *Prog. Energy Combust. Sci.* **36** 327–363
[13] Dwivedi A and Jain MK 2014 Fly ash – waste management and overview: A Review. *Recent Res. Sci. Technol.* **6** 30–35
[14] Nadesan MS and Dinakar P 2017 Structural concrete using sintered flyash lightweight aggregate: A review *Constr. Build. Mater.* **154** 928–944
[15] Ahmad MH and Mohd S 2007 Mechanical Properties Of Palm Oil Clinker Concrete *1st Engineering Conference on Energy & Environment (Sarawak)* pp 172-176
[16] Thomas MDA 2007 *Optimizing the Use of Fly Ash in Concrete* Retrieved on September 24, 2018 from https://www.cement.org/docs/default-source/fc_concrete_technology/is548-optimizing-the-use-of-fly-ash-concrete.pdf
[17] Naik TR and Singh SS 1993 *Fly Ash Generation and Utilization - An Overview* Retrieved on September 26, 2018 from http://www4test.uwm.edu/cbu/Papers/1993%20CBU%20Reports/REP-190.pdf
[18] Mokhtar MM, Taib RM and Hassim MH 2014 Measurement of PCDD/Fs emissions from a coal–fired power plant in Malaysia and establishment of emission factors *Atmos. Pollut. Res.* **5** 388–397
[19] Wong JJ, Abdullah MO, Baini R and Tan YH 2017 Performance monitoring: A study on ISO 14001 certified power plant in Malaysia *J. Clean. Prod.* **147** 165–174
[20] Nordin N, Abdullah MMAB, Tahir MFM, Sandu AV and Hussin K 2016 Utilization of Fly Ash Waste As Construction Material *Int. J. Conserv. Sci.* **7** 161–166
[21] Adegoke KA, Oyewole RO, Lasisi BM and Bello OS 2017 Abatement of organic pollutants using fly ash based adsorbents *Water Sci. Technol.* **76** 2580–2592
[22] Technology Information Forecasting and Assessment Council (TIFAC) 1990 *Technologies for Disposal of Thermal Power Station Flyash* Retrieved from https://tifac.org.in/index.php/8-publication/91-technologies-for-disposal-of-thermal-power-station-flyash
[23] Ghosh KG, Mukherjee K and Saha S 2015 Fly Ash of Thermal Power Plants: Review of the Problems and Management Options With Special Reference To the Bakreshwar Thermal Power Plant, Eastern India *Int. J. Geol. Earth Environ. Sci.* **5** 2277–2081

[24] Nadesan MS and Dinakar P 2017 Mix design and properties of fly ash waste lightweight aggregates in structural lightweight concrete, *Case Studies in Const. Mater.* **7** 336–347
[25] Pandey SK 2014 Coal Fly Ash : Some Aspects of Characterization and Environmental Impacts *J. Environ. Sci. Comput. Sci. Eng. Technol.* **3** 921–937
[26] Carlson CL and Adriano DC 1993 Environmental Impacts of Coal Combustion Residues *J. Environ. Qual.* **22** 227–247
[27] Bahu BK and Jayaram S 2000 Environmental Impacts of Fly Ash *Botswana J. Technol.* 9–13.
[28] Tiwari MK 2016 Fly Ash Utilization A Brief Review in Indian Context Fly Ash Utilization : A Brief Review in Indian Context *Int. Res. J. Eng. Technol.* **3** 949–956
[29] Manas Ranjan Senapati 2011 Fly ash from thermal power plants-waste management and overview *Curr. Sci.* **100** 1791–1794
[30] Elsayed A and Swan C 2007 Compression Behavior of Synthetic Lightweight Aggregates *World of Coal Ash Conferences (Northern Kentucky)* Vol. 5 (USA: http://www.flyash.info/2007/143-elsayed.pdf)
[31] Scheetz BE and Earle R 1998 Utilization of fly ash *Curr. Opin. Solid State Mater. Sci.* **3** 510–520
[32] Openshaw SC 1992 *Utilization of Coal Fly Ash* Master Dissertation (Florida: University of Florida)
[33] William R, Thiery RG, Schuller RM and Subway JJ 1981 *Coal fly ash: a review of the literature and proposed classification system with emphasis on environmental impacts* (Illinois State Geological Survey) pp1–78
[34] Mehta PK 1985 Influence of fly ash characteristics on the strength of portland-fly ash mixtures *Cem. Concr. Res.* **15** 669–674
[35] Blissett RS and Rowson NA 2012 A review of the multi-component utilisation of coal fly ash, *Fuel* **97** 1–23
[36] Sett R 2017 Flyash : Characteristics , Problems and Possible Utilization *Advances in Applied Science Research* **8** 32-50
[37] ASTM *Standard Specification for coal fly ash and raw or calcined natural pozzolan for use in concrete* (United State: ASTM) pp 1-5
[38] Mo KH, Ling TC, Alengaram UJ, Yap SP and Yuen CW 2017 Overview of supplementary cementitious materials usage in lightweight aggregate concrete *Constr. Build. Mater.* **139** 403–418
[39] Chandra S and Bendapudi K 2015 Contribution of Fly ash to the properties of Mortar and Concrete Contribution of Fly ash to the properties of Mortar and Concrete *Int. J. Earth Sci. Eng.* **4** 1017-1023
[40] Malhotra VM 1990 Durability of Concrete Incorporating High-Volume of Low-Calcium (ASTM Class F) Fly Ash *Cement and Concrete Composites* **12** 271–277
[41] Al-Kutti W, Nasir M, Megat Johari MA, Islam ABMS, Manda AA and Blaisi NI 2018 An overview and experimental study on hybrid binders containing date palm ash, fly ash, OPC and activator composites *Constr. Build. Mater.* **159** 567–577
[42] Hemalatha T, Ramaswamy A 2017 A review on fly ash characteristics – Towards promoting high volume utilization in developing sustainable concrete *J. Clean. Prod.* **147** 546–559
[43] Liu B, Xie Y, Zhou S and Yuan Q 2000 Influence of ultrafine fly ash composite on the fluidity and compressive strength of concrete *Cem. Concr. Res.* **30** 1489–1493
[44] Fauzi A, Nuruddin MF, Malkawi AB and Abdullah MMAB 2016 Study of Fly Ash Characterization as a Cementitious Material Procedia Eng. **148** 487–493
[45] Kocak Y and Nas S 2014 The effect of using fly ash on the strength and hydration characteristics of blended cements *Constr. Build. Mater.* **73** 25–32
[46] Bhanumathidas N and P. Kumar Mehta 2001 Concrete Mixtures Made with Ternary Blended Cements Containing Fly Ash and Rice-Husk Ash, *Am. Concr. Inst.* **199** 379–392
[47] Palomo A, Grutzeck MW and Blanco MT 1992 Alkali-activated fly ashes: A cement for the future *Cem. Concr. Res.* **29** 1323–1329

[48] Temuujin J, Van Riessen A and MacKenzie KJD 2010 Preparation and characterisation of fly ash based geopolymer mortars *Constr. Build. Mater.* **24** 1906–1910
[49] Vakhshouri B and Nejadi S 2018 Size effect and age factor in mechanical properties of BST Light Weight Concrete *Constr. Build. Mater.* **177** 63–71
[50] Mousa A, Mahgoub M and Hussein M 2018 Lightweight concrete in America: Presence and challenges *Sustain. Prod. Consum.* **15** 131–143
[51] Courts GD 1991 The aggregate of the future is here today *Proceedings of the Ninth International Ash Use Symposium (Florida) Vol. 23* (United State: https://inis.iaea.org/search/searchsing - lerecord.aspx?recordsFor=SingleRecord&RN=23052884)
[52] Hey P D and Dunstan E R D 1991 Lightweight aggregate production and use in Florida *Proceedings of the Ninth International Ash Use Symposium (Florida) Vol. 23* (United State: https://inis.iaea.org/search/searchsinglerecord.aspx?recordsFor=SingleRecord&RN=23052884)
[53] Liew KM, Sojobi AO and Zhang LW 2017 Green concrete: Prospects and challenge *Constr. Build. Mater.* **156** 1063–1095
[54] National Ready Mixed Concrete Association 2003 *CIP 36 - Structural Lightweight Concrete* Retrived on September 24, 2018 from https://www.nrmca.org/aboutconcrete/cips/36p.pdf
[55] Mukherjee SP and Vesmawala G 2013 Exploring Fly Ash Utilization in Construction of Highways in India *IOSR J. Mech. Civ. Eng.* **8** 23–32
[56] Naik T, Ramme B and Tews J 1994 Use of High Volumes of Class C and Class F Fly Ash in Concrete *Cement, Concrete and Aggregates* **16** 12-20
[57] Bicer A 2018 Effect of fly ash particle size on thermal and mechanical properties of fly ash-cement composites *Therm. Sci. Eng. Prog.* **8** 78–82
[58] Ding J, Ma S, Shen S, Xie Z, Zheng S and Zhang Y 2017 Research and industrialization progress of recovering alumina from fly ash: A concise review *Waste Manag.* **60** 375–387

IOP Conf. Series: Materials Science and Engineering **601** (2019) 012024 doi:10.1088/1757-899X/601/1/012024

Establishing the Level of BIM implementation – A Case Study in Melaka, Malaysia

Yasser Yahya Al-Ashmori[1*], Idris Bin Othman[1], Hisham Bin Mohamad[1], Yani Rahmawati[1] and Madzlan Napiah[1]

[1*] Department of Civil & Environmental Engineering, Universiti Teknologi Petronas, Seri Iskandar, 32610 Bota, Perak
Corresponding author: Yasser.alashmori@gmail.com

Abstract. The implementation of Building Information Modelling (BIM) in Malaysian construction industry is considered very challenging. Although there is significant support from the government to enhance BIM adoption, the implementation of BIM is still in a very low level. Therefore, this paper is evaluating the level of BIM use among Malaysian construction companies. The research is based on a survey and observation conducted in the state of Melaka in which 90 Malaysian companies were surveyed. Based on the evaluation of 90 construction companies, 63 questionnaires were returned and only 46 were valid for further analysis which covered the main objectives of this research. Result revealed that there is a significant drawback of BIM implementation among construction companies due to the lack of awareness of application process and technology concept. The main factor responsible for the lack of awareness was found to be the delay of organizations to use BIM on their project practices. Hence, it is recommended that organizations should be practicing the BIM process within their company. Also, organizations should have the initiative to develop their roadmap for implementation and development of individual transformation plan that gradually overcomes challenges.

1. Introduction

Building Information Modelling (BIM) is described as new technology to model building and infrastructure projects and associated data in a digital manner to produce, communicate, analyse, integrate, and collaborate project processes throughout the construction project life cycle [1–3]. BIM is also called n-D Modelling or Virtual Prototyping Technology [4]. It enhances inter-organizational communication, collaboration, and enables interoperability [5–8]. Moreover, BIM implementation enhances productivity and efficiency during design stage, scheduling, quantity estimating, procurement, construction execution, and project commissioning operation [4,9–12]. In other words, BIM represents the building in 3D geometric objects which is integrated with non-geometric attributes (cost and time) that can be used and maintained throughout the building lifecycle.

The new technology of BIM was introduced to the Malaysian construction industry in 2006 during the launching of the Construction Industry Master Plan (CIMP). The idea of BIM implementation was first introduced by the Public Works Department (PWD) in 2007 [13,14], in order to overcome the weaknesses of the current construction industry [12]. Furthermore, in the launching of the Construction Industry Transformation Program (CITP) in 2015, construction industry tended to eliminate the limited adoption of information technology [15]. As a result, the government represented by CIDB had made a step forward to adopt and promote BIM to the construction industry. The latest development on enhancing the concept of BIM was conducted in 2018. Figure 1 illustrates the milestones of BIM initiation taken by CIDB to promote BIM implementation in Malaysia from 2014 to Sept. 2018.

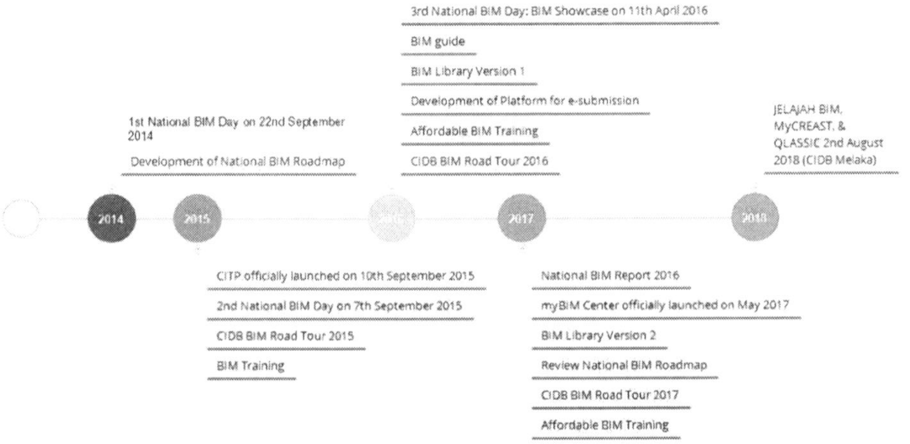

Figure 1. The milestones of BIM initiation by CIDB to promote BIM implementation.

Project development in the Malaysian construction industry is still organized by traditional processes that caused lack of development and global competitiveness [16]. The outcome of the adopted traditional processes results in delay, cost overrun, poor quality, low performance and low productivity. According to Memon et al. [17], BIM implementation is meeting the requirements during the design phase compared to poor implementation during the construction phase. The usage of BIM in the design organisations is more than that implemented in the construction companies [17]. This differences in the implementation causes fragmental application of BIM and keeps the industry in low level of BIM implementation. There are many reasons for this fragmented process such as the lack of understanding the process of BIM implementation, lack of skill in preparation of plans in BIM and the ability to effectively utilise it with stakeholders, resistance to changing current ways of working, lack of collaboration and coordination among various disciplines, limited availability of usage guidelines, etc. [15].

To the best of the author's knowledge, no efforts have been undertaken to explore the shortcomings of BIM implementation in building and infrastructure projects. In addition, there is a lack in the establishment of the current level of BIM application in Malaysia. Therefore, an analysis of the awareness level of professionals in construction industry, usage of technology in the Malaysian organization, and interference of employer with BIM technology is the focus of this study. This research also investigates the level of BIM implementation and the impact of BIM usage within the organizations and compares the level of awareness. Descriptive analysis and correlation were performed to achieve the objectives of this research. The research outcome promotes BIM utilization among construction companies in Malaysia.

2. Methodology

The questionnaire was developed based on the findings from the variable elements identified in the literature review [18–22], and designed to explore the participant background, awareness of BIM technology implementation process, usage of BIM, and to extract the organization challenges and success factors for BIM implementation. The questionnaire survey adopted in this research was distributed to the audience in JELAJAH BIM, MyCREAST, & QLASSIC 2018, Melaka. In this research, participants were from different backgrounds and different organizations including client/developers, contractors, consultants, participants from CIDB and MYBIM centre. Additional information was recorded during the Q&A session to extract the most frequent questions asked by participants regarding BIM technology and implementation. This helped to develop an empirical knowledge to the level of BIM implementation and factors influencing companies to consider BIM in their projects.

The questionnaires were distributed to 90 participants from different construction classification and agency interested in BIM technology. Only 63 questionnaires were returned and after conducting the first screening, 46 returned questionnaires were valid for further analysis and discussion as they fulfil the requirements of this research. Meanwhile, the first part of the questionnaire was established to classify respondents according to their qualification, years of experience, and position as shown in Figure 2. The qualification of respondent measured to be 50%, 26%, and 22% for bachelor degree, master degree, and diploma, respectively. The figure also shows that 33% of the respondents have working experience of less than 5 years, which is the highest percentage. 30% of the respondents have working experience of approximately 5 to 10 years, while 11% of the respondents have 10-15 years working experience, which is the lowest. Furthermore, it can be observed from the collected data that 17% of the respondents are executive management, 17% are junior management, 22% are the senior management and 44% are others. The others included engineers, lecturers, and administration staff, which is the highest percentage among the respondents.

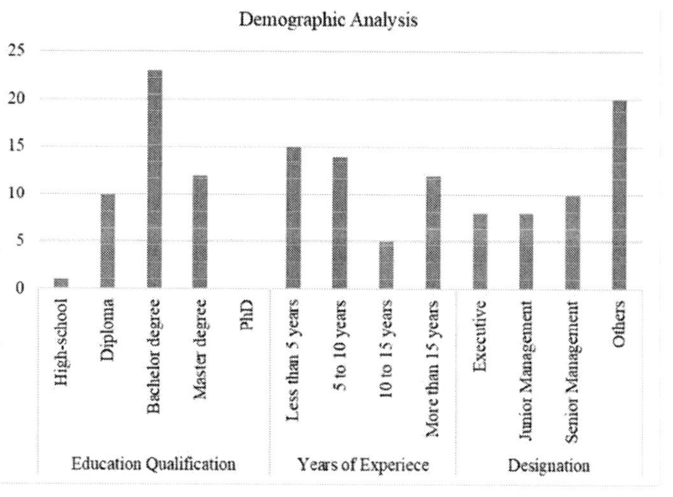

Figure 2. The classification of respondents.

Figure 3 shows that the respondents' sector was classified as public and private. It is clearly shown that 52% of the companies practicing BIM are from the private sector, which was a slightly higher rate than the participants from the public sector (48%).

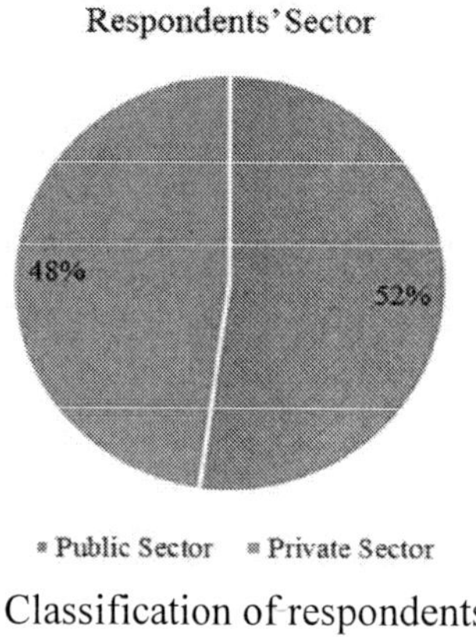

Figure 3. Classification of respondents' sector.

The variation of respondents according to their nature of business was satisfactory except for MEP engineering where it shows zero participants in this research. Figure 4 represents the number of respondents for business practices from each organization in relation to the respondents' classification. It is clearly seen that the results were similar for client / developer from both public and private sectors. The number of contractors was the highest compared to other business classes. When comparing the public and private sectors, the number of contractors from private sector was significantly higher. Furthermore, the majority of civil and structural engineer respondents in this research were from public sector, which is similar to those from other nature of business such as management and education. All respondents from Architecture were from private sector. Overall, variety of participants in such activity provided a significant sign of readiness of these organizations to transfer into BIM. The higher number of contractors participated in this seminar showed their willingness to understand BIM and its benefits that could enhance BIM implementation among Malaysian construction companies.

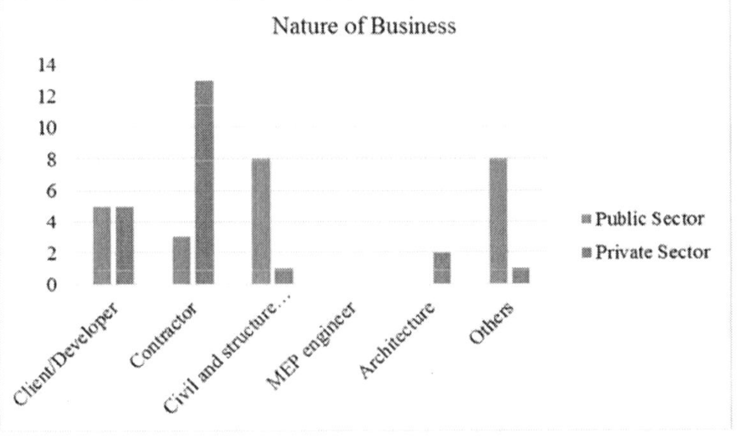

Figure 4. Respondent nature of business.

Professionals representing the participating organization work in different projects as illustrated in Figure 5. 26% and 23% work in Housing and commercial projects, respectively, while organizations working in infrastructure and high-rise projects accounted for 17% and 14%, respectively. The availability of different organizations with different project types interested in BIM technology shows that the construction industry is motivated to make changes and adopt the new technology. Similarly, according to Bui, Merschbrock, & Munkvold [23], BIM is seen as a catalyst for innovation and productivity in the construction industry. Thus, while organizations acknowledge its benefit, the application of BIM incorporates a range of risks [4–6], and required change in management as it reshapes the Architecture-Engineering-Construction (AEC) industry [4,24]. Therefore, organizations are reluctant to adopt the technology unless they know how, where and when to use it.

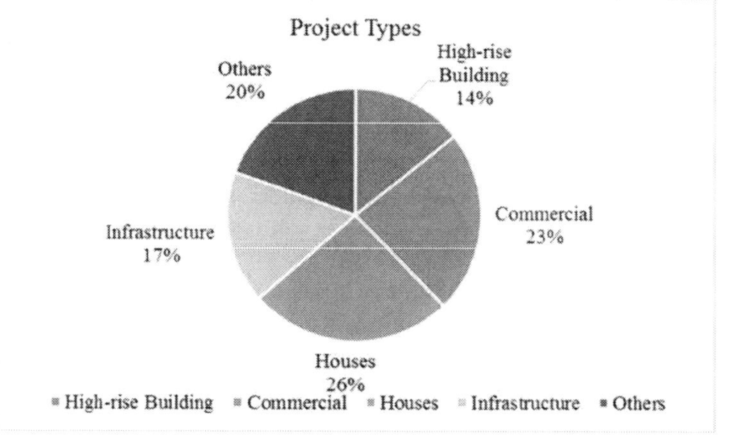

Figure 5. Types of projects.

3. Result and Discussion

This questionnaire analyses BIM diffusion among organizations in Malaysian construction industry with respect to BIM implementation and usage of BIM within their projects. The results of the descriptive analysis are illustrated and discussed in this section.

3.1. BIM Implementation in Malaysian Construction Industry

BIM implementation in Malaysia is still very low and construction players are still facing major challenges in implementing BIM [17,25–27]. The analysis of the conducted survey revealed very low rate of BIM implementation. It can be seen in Figure 6 that only 9 organizations implemented BIM within their projects. Most of the investigated companies reported that they are not using BIM due to various reasons such as poor knowledge about BIM. It is clearly seen that some respondents from four companies did not have a clear idea as to whether their companies are using BIM or not. Figure 6 illustrates the number of participants representing organizations using BIM in Malaysia.

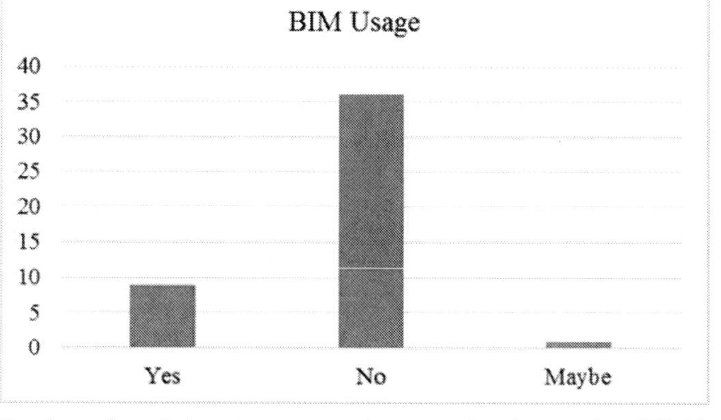

Figure 6. Number of participants representing organizations using BIM in Malaysia.

Furthermore, the survey examines the period of BIM use among those organizations. It is clearly shown that only three organizations have been using BIM for 3 to 4 years. This represents only 30% of the total organizations using BIM. Also, it is obviously seen from the Figure 7 that six organizations have been using BIM for 1 to 2 years. Figure 7 depicts the period of BIM implementation among organizations.

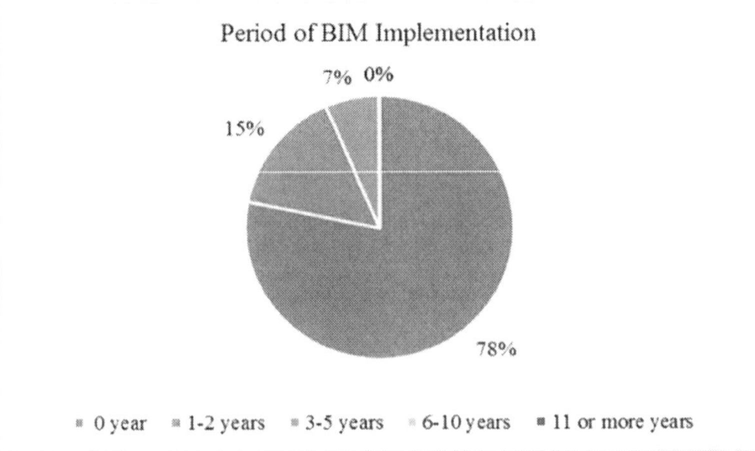

Figure 7. Period of BIM implementation among organizations.

Construction organizations that implement BIM do not always use BIM for all projects [27]. Therefore, Figure 8 represents the frequency of time that Professionals in the organization interfere with the BIM during the development of the building and infrastructure project. The frequent of interference is termed as never, rarely, sometimes, often, and always. The results indicated a very clear difference between those who never use BIM and vice versa. Approximately 72% of those who responded said that they never use BIM. Around 13% of the professionals representing the organizations used BIM in some of the projects, while 11% rarely use BIM in their projects. Low rate of continuous BIM usage among the construction projects was indicated as 4%. This rate gives a drawback indicator of the BIM implementation in Malaysian construction industry.

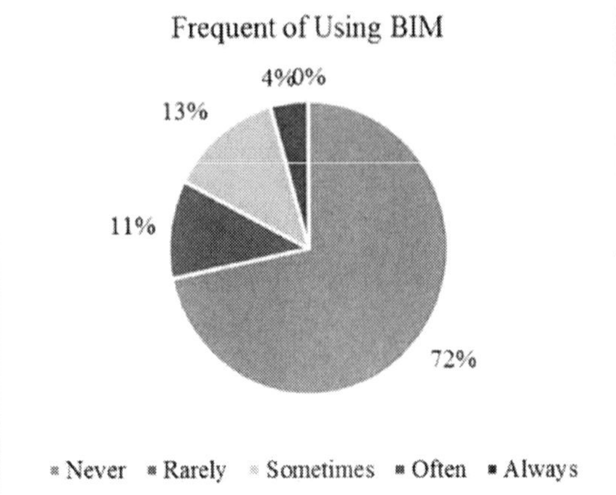

Figure 8. Frequency of time professionals interfere with the BIM.

It can be concluded from the above result and discussion of the respondents that the practice of BIM technology within the organization has a significant impact on improving awareness. Therefore, it is recommended that the organizations should start practicing BIM processes within their projects. Organizations also should take the initiative to develop their roadmap to adopt BIM and develop individual transformation plan and gradually overcome challenges.

3.2. Awareness of BIM Implementation

The awareness of BIM processes implementation was investigated in this section. It can be seen in Figure 9 that the awareness of BIM implementation is still very low [27]. Organizations acknowledge the benefits of BIM technology as a reflection of their interest to attend the promotion programs organized by CIDB, Malaysia. Result shows that 46% of the respondents are not aware of the BIM process implementation, whereas 39% are somewhat aware. However, there is still a significant unawareness of the usage of BIM technology. This was observed through the discussion that was opened for the audience, where all the questions raised were on BIM implementation and how the information is transformed through different organization or parties in the same project.

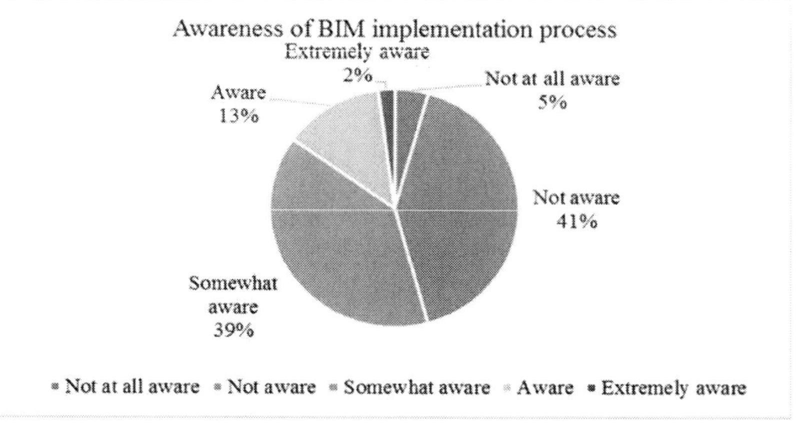

Figure 9. Awareness of BIM implementation process.

4. Conclusion

This paper investigates BIM implementation level and explores the usage of BIM in Malaysia. The investigation was carried out using a questionnaire survey and personal observation. Results revealed that BIM implementation in Malaysia is still in the low level and organizations still lack awareness on BIM processes throughout the project phases. While significant interest has been expressed by the participants to learn the technology, the companies are still not using or promoting the usage of BIM in its projects, with the result that professionals are facing difficulties to practice the technology. This research recommended organizations to initiate the adoption of BIM and encourage professionals to develop their skills on the BIM implementation. Attending training or seminars will enhance the theoretical knowledge of participant, while practicing the technology will increase skills and knowledge. In addition, organizations should adopt best practices from experts and develop individual transformation plan to gradually overcome challenges.

5. References

[1] Construction Industry Development Board (CIDB) 2012 *Promotion of BIM in Malaysia* Retrieved on September 19, 2018 from www.cidb.gov.my/index.php/en/focus-areas/teknologi-dan-inovasi

[2] Ilhan B and Yaman H 2013 Meta-analysis of building information modeling literature in construction, *Int. J. Eng. Innov. Technol.* **3** 373–379

[3] Eastman C, Teicholz P, Sacks R and Liston K 2011 *BIM Handbook: A Guide to Building Information Modeling for Owners, Managers, Designers, Engineers and Contractors* (USA: John Wiley & Sons)

[4] Azhar S, Khalfan M and Maqsood T 2012 Building Information Modeling (BIM): Now and beyond, *Australas. J. Constr. Econ. Build.* **12** 15–28

[5] Alreshidi E, Mourshed M and Rezgui Y 2017 Factors for effective BIM governance, *J. of Build. Eng.* **10** 89-101

[6] Ghaffarianhoseini A, Tookey J, Ghaffarianhoseini A, Naismith N, Azhar S, Efimova O and Raahemifar K 2017 Building Information Modelling (BIM) uptake: Clear benefits,

understanding its implementation, risks and challenges, *Renewable and Sustainable Energy Reviews* **75** 1046-1053
[7] Miettinen R and Paavola S 2014 Beyond the BIM utopia: Approaches to the development and implementation of building information modelling, *Automation in Construction* **43** 84-91
[8] Oraee M, Hosseini M R, Papadonikolaki E, Palliyaguru R and Arashpour M 2017 Collaboration in BIM-based construction networks: A bibliometric-qualitative literature review, *Int. J. of Project Mgmt* **35** 1288-1301
[9] Bentley 2017 *How AIA Partner Bentley's BIM Applications Designed the World-Class Architecture of One Blackfriars* Retrieved on September 10, 2018 from https://www.aia.org/articles/164986-design-innovation-for-large-scale-projects
[10] H. Sone, 2017 "Productivity Improvement with Construction BIM," *Concrete Journal*, 55, 824–829
[11] Migilinska D, Popov V, Juocevicius V and Ustinovichius L 2013 The benefits, obstacles and problems of practical BIM implementation, *Procedia Eng.* **57** 767-774
[12] Harona A T, Marshall-Pontingb A J, Zakariac Z, Nasrun M, Nawid M, Hamide Z A and Kamarf K A M 2015 An industrial report on the Malaysian building information modelling (BIM) taskforce: Issues and recommendations, *Malays. Constr. Res. J. (MCRJ)* **17** 21-36
[13] Latiffi A A, Mohd S, Kasim N and Fathi M S 2013 Building information modeling (BIM) application in Malaysian construction industry, *Int. J. of Constr. Eng. and Mgmt* **2** 1-6
[14] Latiffi A A, Mohd S and Brahim J 2015 Application of Building Information Modeling (BIM) in the Malaysian construction industry: A story of the first government project, *Appl. Mech. Mater.* **773** 943-948
[15] Construction Industry Transformation Programme 2016-2020 "*Productivity*",. Retrieved on September 29, 2018 from http://www.citp.my/strategic-thrusts/productivity..
[16] Abdul Rahim, Syukran and Abdul Nifa, Faizatul Akmar and Mohd Nawi, Mohd Nasrun (2015) Integrated project delivery (IPD) in Malaysian construction industry dealing with waste issue. *Advances in Environmental Biology*, **9** (4). pp. 35-38
[17] Memon A H, Rahman I A, Memon I and Azman N I A 2014 BIM in Malaysian construction industry: Status, advantages, barriers and strategies to enhance the implementation level *Res. J. Appl. Sci. Eng. Technol.* **8** 606–614
[18] Tsai M H, Mom M and Hsieh S H 2014 Developing critical success factors for the assessment of BIM technology adoption: Part I. Methodology and survey, *J. of the Chinese Institute of Engineers* **37** 845-858
[19] Alias Z, Zawawi E M A, Yusof K and Abra A 2014 Determining critical success factors of project management practice : A conceptual framework, *Procedia - Soc. Behav. Sci.* **153** 61–69
[20] Ozorhon B and Karahan U 2016 Critical success factors of building information modeling implementation, *Journal of Management in Engineering* **33** 04016054
[21] Chin K S, Chan B L and Lam P K 2008 Identifying and prioritizing critical success factors for coopetition strategy, *Industrial Management & Data Systems* **108** 437-454
[22] Yong Y C and Mustaffa N E 2013 Critical success factors for Malaysian construction projects: An empirical assessment, *Constr. Manag. Econ.* **31** 959–978
[23] Bui N, Merschbrock C and Munkvold B E 2016 A Review of Building Information Modelling for construction in developing countries, *Procedia Eng.* **164** 487–494
[24] Succar B 2009 Building information modelling framework : A research and delivery foundation for industry stakeholders, *Autom. Constr.* **18** 357–375
[25] Zahrizan Z, Nasly M A, Ahmad Tarmizi H, Amanda M P and Zuhairi A 2014 Exploring the barriers and driving factors in implementing building information modelling (BIM) in the Malaysian construction industry: A preliminary study, *The Institution of Engineers Malaysia* **75** 1-10

[26] Zakaria Z, Mohamed Ali N, Tarmizi Haron A, Marshall-Ponting A J and Abd Hamid Z 2013 Exploring the adoption of Building Information Modelling (BIM) in the Malaysian construction industry: A qualitative approach, *Int. J. of Research in Eng. and Technology* **2** 384-395

[27] Memon A H, Rahman I A, Memon I and Azman N I A 2014 BIM in Malaysian construction industry: Status, advantages, barriers and strategies to enhance the implementation level, *Research Journal of Applied Sciences, Engineering and Technology* **8** 606-614

Acknowledgment

The authors would like to express their utmost gratitude to Universiti Teknologi PETRONAS (UTP) (for providing the Graduate Assistance (GA) scheme), CIDB-Melaka (for give permission conduction the data collection during the JELAJAH BIM, MyCREAST, & QLASSIC 2018), and the form of the grant Cost Centre: 0153AB-D47 for supporting this research. The authors remain indebted to them for their generosity.

IOP Conf. Series: Materials Science and Engineering **601** (2019) 012025 doi:10.1088/1757-899X/601/1/012025

Flexure Behaviour of Foamed Concrete Incorporating Banana Skin Powder and Palm Oil Fuel Ash Strengthened with Carbon Fibre Reinforced Plate

Noridah Mohamad[1*], Muhammad Afif Iman[1], Abdul Aziz Abdul Samad[1], Md Azree Othuman Mydin[2], Suriani Jusoh[3], Anis Sofia[4], Khalidah Aziz[1] and Bryan Lee[1]

[1] Faculty of Civil and Environment Engineering, Universiti Tun Hussein Onn Malaysia, 86400, Parit Raja, Batu Pahat, Johor, Malaysia
[2] School of Housing, Building and Planning, Universiti Sains Malaysia, 11800, Penang, Malaysia
[3] School of Ocean Engineering, Universiti Malaysia Terengganu, 21300 Kuala Nerus, Terengganu, Malaysia
4 School of Materials Engineering, Universiti Malaysia Perlis, 02600 Arau, Perlis, Malaysia *Corresponding author: noridah@uthm.edu.my

Abstract. This paper investigated the chemical properties of banana skin powder (BSP) and palm oil fuel ash (POFA), and mechanical properties of lightweight foamed concrete (LFC) incorporating BSP and POFA strengthened with carbon fibre reinforce polymer (CFRP) plate. The BSP and POFA are added in the LFC mixture at various percentages as cement and sand replacement, respectively. LFC cubes incorporating BSP and POFA, LFC-BSP-POFA, were cast and tested under compression to determine its compressive strength. LFC-BSP-POFA prisms strengthened with 100 mm and 150 mm length of CFRP plates glued on its bottom mid-span surface were cast and tested under four point bending load to determine its flexure behaviour. From the chemical test, it was found that BSP and POFA each contain pozzolan materials, which are 55.98% and 51.83% silicon dioxide, and 2.71% and 2.32% aluminium oxide, respectively. From the mechanical property tests, compressive and flexural strength increased when percentage of BSP and POFA incorporated in the LFC increased. Meanwhile, CFRP plate managed to strengthen the LFC prism further where higher ultimate load recorded when longer CFRP plate was used.

1. Introduction

Foamed concrete has emerged as an attractive innovated concrete which is lightweight, sustainable, and feasible to be developed as structural IBS system [1-3]. However, foamed concrete, just like conventional concrete, is a brittle material; hence, it needs to be refitted with various kind of additive materials, either as cement and sand replacement, or as filler, to enhance its physical, chemical, and mechanical properties. By-product materials from agricultural wastes, such as palm oil fuel ash (POFA), rice husk ash (RHA), coconut fiber (CF) and banana skin powder (BSP), have been investigated on its potential to replace cement and sand in concrete mixture [4-7]. Various researches were conducted where concrete was incorporated with POFA as cement replacement [8-11]. From these researches, it was found that with proper amount of superplasticiser added, the concrete's compressive strength manage to increase up to 40% POFA as cement replacement. A research was conducted on concrete with RHA at 0%, 10% and 20% as cement replacement [12]. It was observed that 10% RHA as cement replacement recorded higher compressive strength by 4%. Similar trend was observed when higher percentage of RHA showed a reduction of strength by 6% [13]. Meanwhile, it was also reported that incorporating banana skins in foamed concrete, due to its fibrous properties increased the bonding in the cement mixture; thus, enhanced its compressive strength and impact resistance [10]. Mohamad et al. [14] conducted an experimental study on the incorporation of BSP and POFA in the foamed concrete mixture. It was found that the mechanical properties which include compressive strength, tensile strength, and modulus of elasticity of this foamed concrete were higher compared to control foamed concrete.

Alternatively, the usage of fibres in strengthening the concrete has also been explored. The type of fibers that could be incorporated in various types of concrete mixture include both artificial and natural fiber. In a report by Awang et al. [15], a density of 1000 kg/m^3 foamed concrete prism were incorporated with polypropylene and kenaf fiber and tested under flexure. It is shown that addition of fibers had contributed to the higher peak flexural strength achieved by the specimens before failure. While specimens without fibers experienced immediate failure and broken into two parts after the peak flexural strength was reached. This statement is in agreement with the work done by Elsaid et al. [16] where it was found that when concrete cracks, the fibers prevents the micro-cracks from spreading and breaking into two parts. Ahmad & Awwad [17] conducted a research on added polypropylene fibers in a concrete mixture observed that these fibers had created bridging across the concrete matrix which help to control shrinkage cracking during plastic stage and matrix micro cracks that occur as the concrete is loaded. It is also found that fibers can control the crack propagation and prevent the occurrence of large cracks widths.

Meanwhile, the use of Fibre Reinforced Polymer (FRP) materials for the repair and strengthening of reinforced concrete structures and for other types of construction has become common practice. As a material, FRP offers outstanding combination of physical and mechanical properties such as its high tensile strength, light weight, high stiffness properties, high fatigue strength and excellent durable properties [18, 19]. It was reported that the application of the FRP composites has been proven to increase the service life contributing to the structures longevity [20-22].

It has been proven that certain additive materials could enhance the strength and properties of foamed concrete. This makes foamed concrete possible to be used as structural element, provided it is strengthened with reinforcement. This research investigated the flexural behaviour of lightweight foamed concrete (LFC) prisms incorporated with banana skin powder (BSP) and palm oil fuel ash (POFA), strengthened with carbon fiber reinforce polymar (CFRP) which was glued at its centre. The LFC developed in this study is considered as sustainable and environmental friendly material because no natural coarse aggregate used, BSP and POFA as partial cement and sand replacement.

2. Materials and Method

The materials for casting LFC include cement, sand, water, foam, BSP and POFA. The type of cement used is Ordinary Portland Cement (OPC) which complied with BS 12:1996 or BS EN197: Part 1: 2000 [23]. Fine sands used in LFC mixture complied with BS 882:1992, with sizes up to 5mm [24].

2.1. Materials Preparation

The materials for casting LFC include cement, sand, water, foam, BSP and POFA. The type of cement used is Ordinary Portland Cement (OPC) which complied with BS 12:1996 or BS EN197: Part 1: 2000 [23]. Fine sands used in LFC mixture complied with BS 882:1992, with sizes up to 5mm [24]. For this research, the banana skin were produced into a powder form where it was dried in the oven at temperature 105 C ± 5 C for 24 hours. The dried banana skins were then grinded and sieved under 150 μm sieve. POFA was obtained from Ban Dung Palm Oil Industries Sdn. Bhd. at Parit Sulong, Batu Pahat, Johor. Figure 1 and Figure 2 show the BSP and POFA, both in powder form.

IOP Conf. Series: Materials Science and Engineering **601** (2019) 012025 doi:10.1088/1757-899X/601/1/012025

Foam agent used to produce foam was from a syntethic type which was mixed with water at the ratio of 1 to 40. The foam was generated from this mixture by using foam generator connected to air compressor where the concentrated foaming agent was added with water. The foam generated was a form of stable bubbles which were added to the LFC-BSP-POFA mixture in stages to control its density.

Figure 1. Banana skin powder (BSP).

Figure 2. Palm oil fuel ash (POFA).

2.2. LFC-BSP-POFA Mixture

The suitable lightweight foamed concrete mixture incorporating BSP and POFA (LFC-BSP-POFA) was determined from several trial mixing processes. The mixture proportion of LFC-BSP-POFA was investigated by adding various percentages of BSP and POFA as cement and sand replacement in all the trial mixtures as shown in Table 1. The percentages of BSP added as cement replacement are 0% to 2%. These values were chosen based on the investigation conducted by Mohamad, 2018[14] which reported that both compressive and tensile strength increased by more than 100% at 2% of added BSP. However, more crack was observed in concrete with BSP prisms compared to plain concrete prisms. Meanwhile, the percentage of POFA used are 0% and 15%. These percentages were chosen based on the observation by Islam et al. [25] and Tangchirapat et al. [10]. Islam et al.[25] investigated the effect of 0%, 10% and 20% POFA added as cement replacement and reported that 10% added POFA increased the concrete's strength up to 11%. Tangchirapat et al. [10] on the other hand, had proven that 20% of POFA added resulted with 24% of strength increased in concrete. From these trial mixtures, the actual mixture to cast the LFC-BSP-POFA prism was determined, which is the mixture that produced the targeted density of 1800 kg/m^3. The mixture of LFC-BSP-POFA chosen is M1 as shown in Table 2. Control prism was cast form LFC with 0% BSP and 0% POFA.

Table 1. Trial mixtures for LFC-BSP-POFA.

Mix	Mixes	Amount in kilograms (kg)				
		BSP	**POFA**	**Cement**	**Sand**	**Water**
Control	0%BSP 0%POFA	0	0	12.03	24.05	6.62
	0%BSP 15%POFA	0	3.6	12.03	20.45	6.62
M1	0.2%BSP 0%POFA	0.02406	0	12.01	24.05	6.62
	0.2%BSP 15%POFA	0.02406	3.6	12.01	20.45	6.62
M2	0.4%BSP 0%POFA	0.04812	0	11.98	24.05	6.62
	0.4%BSP 15%POFA	0.04812	3.6	11.98	20.45	6.62
M3	0.6%BSP 0%POFA	0.07218	0	11.96	24.05	6.62
	0.6%BSP 15%POFA	0.07218	3.6	11.96	20.45	6.62
M4	0.8%BSP 0%POFA	0.09624	0	11.93	24.05	6.62
	0.8%BSP 15%POFA	0.09624	3.6	11.93	20.45	6.62
M5	1%BSP 0%POFA	0.1203	0	11.91	24.05	6.62
	1%BSP 15%POFA	0.1203	3.6	11.91	20.45	6.62
Total Material Required		0.7218	21.6	143.64	267	79.44

Table 2. Mixture of control LFC and LFC-BSP-POFA prism.

Mix	Mixes	Amount in kilograms (kg)				
		BSP	**POFA**	**Cement**	**Sand**	**Water**
Control	0%BSP 0%POFA	0	0	7.8	15.6	4.29
	0%BSP 15%POFA	0	2.34	7.8	13.26	4.29
M1	1.0 %BSP 0%POFA	0.078	0	7.72	15.6	4.29
	1.0 %BSP 15%POFA	0.078	2.34	7.72	13.26	4.29
Total Material Required		0.156	4.68	31.04	57.72	17.16

2.3. Casting of Specimens

The specimens cast for LFC-BSP-POFA include cube and prisms tested under compression and flexure to determine its compressive strength and flexure behaviour. LFC-BSP-POFA cubes of 100 am x 100 mm x 100 mm and prism with the size of 500mm x 100mm x 100mm in length, width and height were cast. The preparation to cast the LFC-BSP-POFA cubes are as shown in Figure 3.

Figure 3. Cubes mould (left), wet mix LFC-BSP-POFA in the mould (center), the LFC-BSP-POFA once demoulded (right).

For LFC-BSP-POFA prism, CFRP was placed at the bottom of the prism with aim to control the crack propagation. Epoxy paste was first spread onto the prism bottom's surface. CFRP with 100 mm and 150 mm length were glued to the surface of different prisms and left to harden. The preparation for these prisms strengthened by CFRP are shown in Figure 4.

(a) (b)

Figure 4. (a) Epoxy paste spread onto the prism surface (b) CFRP glued onto the paste at the bottom surface of the prism.

3. Experimental Programme

The experimental programme conducted were chemical test, compressive strength test and flexural test to determine the chemical properties of BSP and POFA, and mechanical properties of LFC-BSP-POFA. The mechanical properties include its compressive strength, flexural strength and crack pattern. The tests for obtaining the chemical properties of BSP and POFA was X-ray Fluorescence (XRF) test. To determine the mechanical properties of LFC-BSP-POFA, compressive strength test and flexural test were conducted.

3.1. XRF Test

XRF test were conducted on BSP and POFA according to ASTM C618-17a [26]. The material were sieved between 50μm to 75μm (63μm) sieve which is to remove larger particles to avoid the failure of XRF. Hydraulic machine was used to compress the sample to become the pallet coin shaped as shown in Figure 5. The pallet sample is then analyzed under XRF machine.

Figure 5. Pallet sample of POFA (left) and BSP (right).

3.2. Compressive Strength Test

Compressive strength test was conducted on LFC-BSP-POFA cube samples of 100 mm x 100 mm x 100 mm according to BS EN 12390-3:2009 [27]. The test was conducted on cube samples (100 mm x 100 mm x 100 mm) of LFC-BSP-POFA at 7 and 28 days. The cube samples were placed in the center of the plate of the testing machine and the load was applied on the cross section are of the cube samples.

3.3. Four-Point Bending Test

Four-point bending test on LFC-BSP-POFA prisms was conducted by using Universal Testing Machine according to ASTM D6272 [28]. Prisms with the dimension of 500 mm x 100 mm x 100 mm was placed on this machine which was supported by pin at one end and rolled at the other end as shown in Figure 6. Two axial loads from the top were applied on the prism at distance from the end supports as shown in Figure 6.

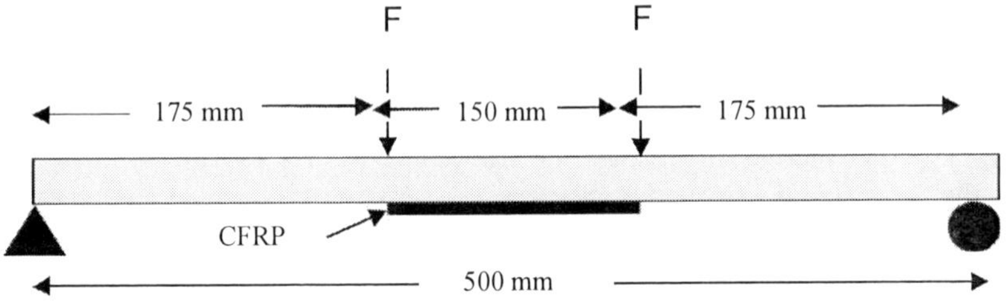

Figure 6. Four load bending test for prism 500mm length.

4. Results

The results recorded from the experiment include chemical composition of BSP and POFA, and mechanical properties of LFC-BSP-POFA mixture. The mechanical properties of LFC-BSP-POFA obtained include its compressive and flexural strength.

4.1. Chemical Properties of BSP and POFA

The chemical composition of BSP and POFA were obtained from the XRF test, as shown in Table 3. From the table, it is seen that BSP and POFA contain high silicon dioxide, which are 55.98% and 51.83% in BSP and POFA, respectively. Both materials contain small amount of aluminium oxide which are 2.71% and 2.32%, for BSP and POFA, respectively. Therefore, both materials could be considered as pozzolanic. A pozzolan is defined as siliceous and aluminous materials which contain little cementitious value, which in finely divided form, will chemically react with calcium hydroxide at ordinary temperature to form compound possessing cementitious properties [29].

Table 3. Chemical composition of BSP and POFA.

Chemical Composition	BSP (%)	POFA (%)
Calcium Oxide, CaO	8.95	8.10
Silicon Dioxide, SiO_2	55.98	51.83
Carbon, C	-	0.17
Aluminium Oxide, Al_2O_3	2.71	2.32
Sulphur Trioxide, SO_3	0.10	2.23
Ferric Oxide, Fe_2O_3	1.36	7.60
Magnesium Oxide, MgO	1.08	3.13
Potassium Oxide, K_2O	28.75	13.72
Phosphorus Pentoxide, P_2O_5	-	4.30

4.2. Compressive Strength

From the data gathered, it is obvious that the compressive strength was found to increase with the increase of BSP and POFA. This is due to high content of SiO2 and CaO in BSP and POFA. Some studies have reported the possibility of a successful replacement of artificial powder to cement required molar ratio of 0.5 [30-32]. However, in Table 3, the calculated molar ratio is less than 0.5 for both BSP and POFA. Interestingly, the existence of these two chemical compositions in the mixture had caused pozzolanic reaction which is shown in the high compressive strength achieved even though molar ratio used was less than 0.5. This could be due to the distribution of finer particle size of BSP and POFA used. Pozzolanic reaction which perform better with finer particles [33].

In pozzolanic reaction, the final products were not eliminated from the cement hydrations; hence, built up their own contribution to the strength and other mechanical properties of the hardened cement paste and concrete mixture [34]. This is clearly indicated in the chemical reaction that took place in the pozzolanic reaction as presented in Equation (1). When calcium hydroxide in cement, $Ca(OH)_2$, reacts with silicon dioxide (SiO_2), calcium silicate hydrate, $3(CaO)_2(SiO_2)3(H_2O)$, or known as C-S-H gel is produced. This gel increased the bonding between particles of various materials in the LFC, resulted with higher compressive strength achieved.

$$3[Ca(OH)_2] + 2[SiO_2] \rightarrow [3(CaO)2(SiO_2)3(H_2O)] \quad (1)$$

From Figure 7, for every percentage of BSP added (0% to 1.0%), the compressive strength increased about 3% to 10% for both LFC with 0% and 15% POFA. Meanwhile, for mixture of LFC-BSP-POFA incorporated with 15% POFA, the compressive strength achieved at every percentage of BSP added was about 5% higher than the compressive strength achieved in mixture of LFC-BSP-POFA with 0% POFA. This shows that effect of incorporating BSP on the compressive strength of LFC is greater than the effect of incorporating POFA. This may be due to higher content of pozzolanic materials in BSP compared to POFA. From Table 3, it is seen that BSP contain silicon dioxide of

55.98% compared to 51.83% in POFA. Both materials contain small amount of aluminium oxide which are 2.71% and 2.32%, for BSP and POFA, respectively.

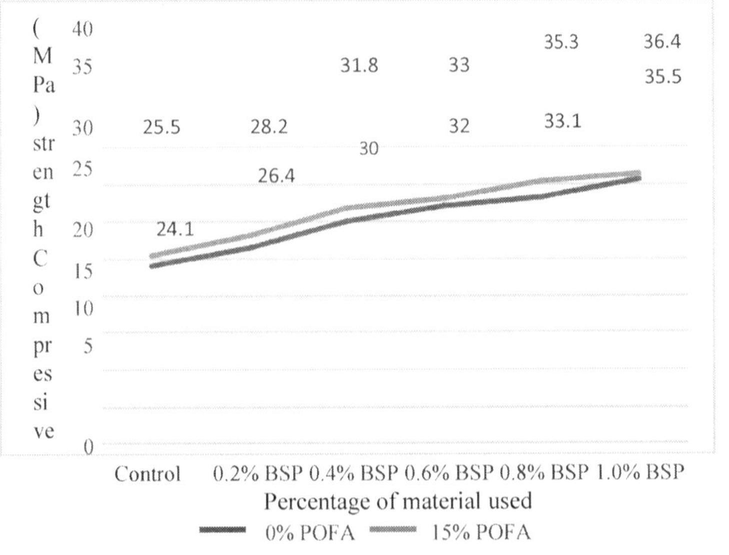

Figure 7. Compressive strength of LFC versus percentage of BSP and POFA at 28 days.

4.3. Ultimate Flexure Load

Under four point bending load test, the ultimate load of LFC-BSP-POFA prisms and LFC-BSP-POFA prisms strengthened with CFRP were recorded and presented in Table 4 and Figure 8. It is noticed that the flexural strength of LFC-BSP-POFA were likely to increase with the increasing portion of cement replacement with BSP. It is noticed to increase even more with the existence of 15% POFA as sand replacement. From Table 4, it is shown that LFC prisms incorporating both BSP and POFA achieved highest ultimate flexure load of 7.82 kN. It is also shown that LFC-BSP-POFA strengthened with CFRP plate achieved higher ultimate flexure strength compared to LFC-BSP-POFA prisms without CFRP glued to it. From the table, it is also noticed that the longer the CFRP plate, the higher the ultimate flexure load achieved. LFC-BSP-POFA with 100 mm and 150 mm length CFRP plate glued at its bottom surface achieve ultimate load of 8.46 kN and 8.74 kN, respectively.

Table 4. Ultimate load of LFC-BSP-POFA w/o CFRP, 100 mm and 150 mm length CFRP.

Foamed concrete incorporating with banana skin powder and palm oil fuel ash	Ultimate flexure load (kN)		
	Without CFRP	CFRP (100mm)	CFRP (150mm)
0.0% BSP 0% POFA	5.06	6.45	6.90
0.0% BSP 15% POFA	6.10	7.69	8.10
1% BSP 0% POFA	6.86	8.20	8.74
1% BSP 15% POFA	7.82	8.46	8.74

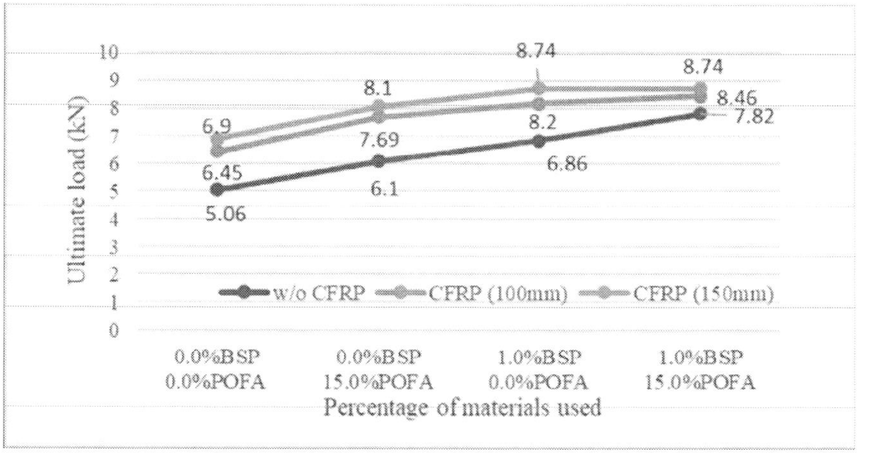

Figure 8. Ultimate flexure load versus the percentage of BSP and POFA incorporated for LFC-BSP-POFA prisms with and without CFRP

4.4. Crack Pattern

From the flexural strength test, the prisms with CFRP and without CFRP illustrated the difference in a crack pattern which affected from the properties of CFRP. For LFC-BSP-POFA prisms without CFRP strengthening, failure in the prisms was accompanied by rupture after reaching its ultimate load as shown in Figure 9(a) and Figure 9(b). From the figures, it is noticed that the crack line occurred in the control prism and prism with 1% BSP inclined at an angle of 4°. The crack was noticed to initiate from the middle span of the bottom surface. For the prisms with added POFA, the crack line developed line of crack with slightly higher degree and propagated to the top surface as shown in Figure 10 (a) and Figure 10(b).

Figure 9. Crack pattern of LFC-BSP-POFA without CFRP strengthening with (a) (0% BSP 0% POFA), and (b) (1% BSP 0% POFA).

Figure 10. Crack pattern of LFC-BSP-POFA without CFRP strengthening with (a) (0% BSP 15% POFA), and (b) (1% BSP 15% POFA).

The additional of CFRP in the mid-span of bottom surface was intended to arrest the debonding cracks that were observed to propagate in the middle of the prisms. CFRP wrapped along the entire beam length were used to anchor the longitudinal laminates since it develops the maximum strength of the composite in tension [35]. Hence, it is agreed with this finding that the longer the CFRP used, the higher the flexural load achieved. For LFC-BSP-POFA strengthened with CFRP, in all prisms the crack was noticed to initiate at a point outside the CFRP region at its bottom surface and propagate to the upper surface and splitted into two parts at failure load as shown in Figure 11 and Figure 12. LFC-BSP-POFA prisms with 100 mm CFRP cracked instantly with line of crack developed at an angle less than 8°. Meanwhile, in LFC-BSP-POFA prisms with 150 mm CFRP developed line of crack at an angle about 12°.

Figure 11. Four point bending load test on prism strengthened with CFRP.

Figure 12. Crack pattern for LFC-BSP-POFA prisms with (a) 100 mm, and (b)150 mm CFRP.

5. Conclusion

This paper conclude that the percentage of the cement and sand replacement incorporated BSP and POFA increased, the higher the compressive strength and flexural strength. From the chemical test, it was found that BSP and POFA each contain pozzolan materials, which are 55.98% and 51.83% silicon dioxide, and 2.71% and 2.32% aluminum oxide, respectively. From the mechanical property tests, compressive strength increased when percentage of BSP and POFA incorporated in the LFC increased. Higher ultimate load under flexure also recorded for LFC prisms with higher percentage of BSP and POFA. Meanwhile, CFRP plate managed to strengthen the LFC prism further where higher ultimate load recorded when longer CFRP plate was used. Unlike LFC without CFRP, LFC strengthened with CFRP experienced crack outside the CFRP region, initiated from the bottom surface and propagated to the upper surface until the prism splitted into two.

6. References

[1] Tonyan T D and Gibson L J 1992 Structure and Mechanics of Cement Foams *Journal of Materials Science* 27 6371-6378 Chapman & Hall.

[2] Mohamad N and Hassan N 2013 The structural performance of precast lightweight foam concrete sandwich panel with single and double shear truss connectors subjected to axial load *Advanced Materials Research* **634-638** 2746-2751

[3] Hadipramana J, Samad A A A, Zaidi A M A, Mohammad N and Wirdawati 2012 Influence of polypropylene fiber in strength of foamed concrete *Advanced Materials Research* **488-489** 253-257.

[4] Mydin M O, Mohamad N, Samad A A A, Johari I and Munaaim M C 2018 Durability performance of foamed concrete strengthened with chemical treated (NaOH) coconut fiber *In AIP Conference Proceedings* vol 2016 (New York: AIP Publishing) pp 020109

[5] Mat Jusoh S, Ghazali C M R, Mat Amin K A, Mohamad N and Jarkoni N K 2018 Innovative uses of recycle waste materials as an artificial concrete reef for estuarine ecosystem *IOP Conference Series: Materials Science and Engineering* vol 374 (UK: IOP Publishing) pp 012088

[6] Tambichikk M A, Mohamad N, Samad A A A, Bosro M Z M and Iman M A 2018 Utilization of construction and agricultural waste in Malaysia for development of green concrete: A review *IOP Conference Series: Earth and Environmental Science* vol 140 (UK: IOP Publishing) pp 012134

[7] Mohamad N, Iman M A, Mydin M A O, Rosli J A and Noorwirdawati A 2018 Mechanical properties and flexure behaviour of lightweight foamed concrete incorporating coir fiber *IOP Conference Series: Earth and Environmental Science* vol 140 (UK: IOP Publishing) pp 012140

[8] Altwair N M, Megat Johari M A and Saiyid Hashim S F 2012 Flexural performance of green engineered cementitious composites containing high volume of palm oil fuel ash *Constr. and Build. Mater.* **37** 518-525

[9] Awal A S M A and Abubakar S I 2011 Properties of concrete containing high volume palm oil fuel ash: A short - term investigation *Malaysian J. of Civil Eng.* **23** 54-66

[10] Tangchirapat W, Jaturapitakkul C and Chindaprasirt P 2009 Use of palm oil fuel ash as a supplementary cementitious material for producing high-strength concrete *Constr. and Build. Mater.* **23** 2641-2646

[11] Rodríguez de Sensale G 2006 Strength development of concrete with rice-husk ash *Cement and Concrete Composites* **28** 158-160

[12] Thambichik M A, Samad A A A, Mohamad N, Mohd Bosro M Z and Iman M A 2018 Effect of combining POFA and RHA as partial cement replacement to the compressive strength of concrete *Int. J. of Intergrated Eng.* **10(8)** pp 61-67

[13] Madandoust R, Ranjbar M M, Moghadam H A and Mousavi S Y 2011 Mechanical properties and durability assessment of rice husk ash concrete *Biosystems Eng.* **110** 144-152

[14] Mohamad N, Samad A A A, Lakhiar M T, Sofia A and Efendi S A 2018 Effects of incorporating banana skin powder (BSP) and palm oil fuel ash (POFA) on mechanical properties of lightweight foamed concrete *Int. J. of Integrated Eng.* **10(9)** 169-176

[15] Awang H, Ahmad M H and Al-Mulali M Z 2015 Influence of kenaf and polypropylene fibres on mechanical and durability properties of fibre reinforced lightweight foamed concrete *Eng. Sci.and Tech.* **10** 496-508

[16] Elsaid A, Dawood M, Seracino R and Bobko C 2011 Mechanical properties of kenaf fibre reinforced concrete *Constr. and Build. Mater.* **25** 1991-2001

[17] Ahmad O A and Awwad M 2015 The effects of polypropylene fibers additions on compressive and tensile strengths of concrete *Int. J. of Civil and Environmental Eng.* **37** 1365-1372

[18] Khalifa A and Nanni A 2000 Improving shear capacity of existing RC T-section beams using CFRP composites *Cement and Concrete Composites* **22** 165-174

[19] Khalifa A and Nanni A 2002 Rehabilitation of rectangular simply supported RC beams with shear deficiencies using CFRP composites *Constr. and Build. Mater.* **16** 135-146

[20] Alferjani M B S, Abdul Samad A A, Elrawaff B S and Mohamad N 2015 Experimental and theoretical investigation shear strengthening of RC precracked continuous T-beams using CFRP strips *Int. J. of Eng., Transactions B: Applications* **28** 671-676
[21] Ali N, Abdul Samad A A, Mohamad N and Jayaprakash J 2013 Shear behaviour of precracked continuous beams repaired using externally bonded CFRP strips *Procedia Eng.* **53** 129-144
[22] Shahidan S, Sharifah Salwa M Z, Keong C K and Jayaprakash J 2016 Repaired of fire damaged concrete-filled double skin steel tubular (CFDST) columns with fiber reinforced polymer (FRP) *ARPN J. Eng. of Applied. Sci* **11** 3718-3725
[23] British Standard EN197: Part 1: 2000 *Cement Composition, Specifications and Conformity Criteria for Common Cement* (London: British Standard)
[24] British Standard 882:1992 *Specification for Aggregates From Natural Sources for Concrete* (London: British Standard)
[25] Islam M M U, Mo K H, Alengaram U J and Jumaat M Z 2016 Durability properties of sustainable concrete containing high volume palm oil waste materials *J. of Cleaner Production* **137** 167-177
[26] ASTM C618-17a 2017 *Standard Specification for Coal Fly Ash and Raw or Calcined Natural Pozzolan or Use in Concrete* (USA: ASTM International)
[27] BS EN 12390-3 2009 *Testing Hardened Concrete: Compressive Strength of Test Specimens* (London: British Standard)
[28] ASTM D6272 2016 *Standard Test Methods for Flexural Properties of Unreinforced and Reinforced Plastics and Electrical Insulating Materials* (USA: ASTM International)
[29] Dodson V H 1990 *Pozzolans and The Pozzolanic Reaction in Concrete Admixtures* (Boston MA: Springer) pp 159-201
[30] Thomas R J and Peethamparan S 2015 Alkali-activated concrete: Engineering properties and stress–strain behaviour *Constr. Build. Mater.* **93** 49–56
[31] Bilodeau A and Malhotra V M 2000 High-volume fly ash system: Concrete solution for sustainable development *Mater. J.* **97** 41-48
[32] Malhotra V 2002 High-performance high-volume fly ash concrete, *Concr. Int.* **24** 30–34
[33] Kroehong W, Sinsiri T and Jaturapitakkul C 2011 Effect of palm oil fuel ash fineness on packing effect and pozzolanic reaction of blended cement paste, *Procedia Eng.* **14** 361-369
[34] Eldagal O E A 2008 *Study on The Behaviour of High Strength Palm Oil Fuel Ash (POFA) Concrete* PhD Thesis (Johor: Universiti Teknologi Malaysia)
[35] GangaRao H V S and Vijay P V 1998 Bending behavior of concrete beams wrapped with carbon fibre *J. of Struct. Eng.* **124** 3-10

Acknowledgement
The authors would like to thank Universiti Tun Hussein Onn Malaysia for its financial support (GPPS Vot U788)

Innovative and sustainable green concrete–A potential review on utilization of agricultural waste

Noridah Mohamad[1*], Muhammad Tahir Lakhiar[1], Abdul Aziz Abdul Samad[1], Md Azree Othuman Mydin[2], Ashfaque Ahmed Jhatial[1], S. Anis Sofia[3], Wan Inn Goh[1] and Noorwirdawati Ali[1]

[1] Department of Structural and Material Engineering, Facility of Civil and Environmenatal Engineering, UTHM, Johor, Malaysia, 84600, MALAYSIA.
[2]School of Housing, Building and Planning, Universiti Sains Malaysia, 11800, Penang, Malaysia
[3] School of Materials, Universiti Malaysia Perlis
*Corresponding author: noridah@uthm.edu.my

Abstract. Concrete is the most versatile product used as building material in world's construction industry, and it generates carbon to the atmosphere. In manufacturing of cement, 8% to 10% of global carbon dioxide (CO_2) gas is generated. The maximum utilization of cement contributes to global warming and climate change. Therefore, many scholars have conducted a research to develop green and sustainable concrete incorporating different waste materials. By utilizing waste materials as cement replacement, the CO_2 gas emissions will decrease. This paper reviews the potential and innovative utilization of agricultural waste as partially cement replacing material to make green concrete. Agricultural waste possesses the pozzolanic materials, when finely grinded, generate pozzolanic reaction, in which silica responds with calcium oxide to create calcium silicates which is responsible for higher strength attained in the concrete. It is concluded from previous studies the agricultural waste which consist the high silica contain generated the pozzolanic reaction in concrete mix which contributes the enhancement in strength. The RHA, SCBA POFA are the pozzolanic material, 10%-20% utilization of these waste is optimum and enhance the 20%-30% strength of concrete. The Banana Skin Powder (BSP) also possesses the high silica contain which will generate the pozzolanic reaction in concrete mix and will improve the strength of concrete such as the other agricultural waste improves.

1. Introduction

Concrete is the most versatile product used as building material in the world's construction industry. It possesses high compression strength, resistance to water and fire, lower maintenance cost and higher service life respect to other building material [1]. These are among many advantages of using concrete as building material. However, concrete production has caused interference with the conservation of natural resources. This is because, concrete is prepared by mixing cement, coarse aggregate, fine aggregate, water and admixtures [2]. Cement is a water-based binder utilized to bind other constituent of concrete. In construction industry, cement is broadly utilized as binding material. This makes it the most demanding material in concrete production. It is recorded that the world cement production produces 8% to 10% of anthropogenic global carbon dioxide (CO_2) gas emissions into the atmosphere as shown in Figure 1, which creates the weathering and climatic changes. With the rapid growth of cement industry, it is expected that CO_2 emission will continue to increase [3]. Therefore, to protect the environment from global warming, the utilization of cement must be minimized.

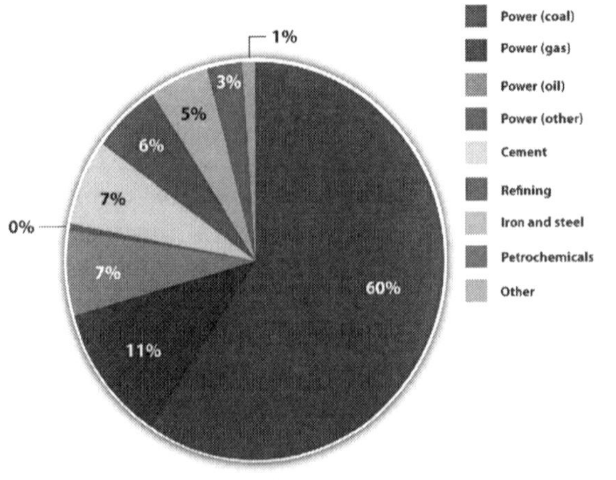

Figure 1. Carbon dioxide (CO_2) emission in atmosphere [3].

Meanwhile, municipal solid waste is the waste produced from residential area, commercial business and institutions. Municipal waste includes paper, food, fruits, glass and other daily life things as shown in Figure 2. The municipal waste contains the 55% organic waste (agricultural waste), which will be significantly increased more due to increasing in urbanization, growing economy and changing in livings standards [3]. In present 1.3 billion tons per annum global municipal solid waste is generated because of increase in population and improve in standard of living, approximately 1.2 kg/capita/day municipal waste is generated daily, but it rises to 1.42 kg/capita/day in 2020 [4]. One of the attractive solution to overcome this problem is to recycle these wastes by adding them as new materials in concrete admixtures. The proportions of the particles from these wastes, and the proportion to be added are determined from the chemical properties possessed in the waste.

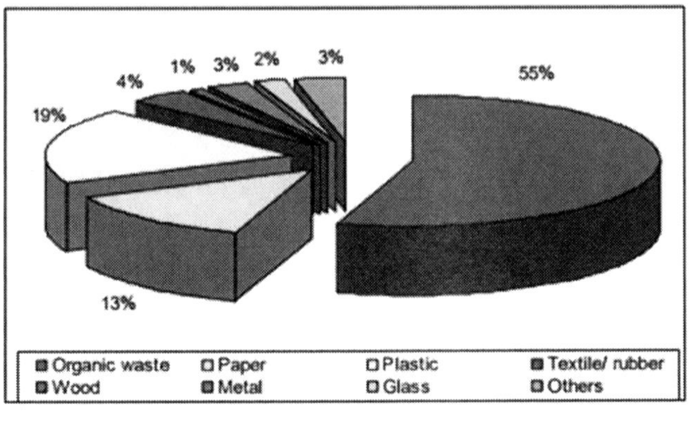

Figure 2. Typical municipal solid waste composition in Malaysia [7].

Green concrete does not represent the colour green, but it shows the surrounding environment. Concrete produced from concrete wastes which are eco-friendly can be named as green concrete. It may be defined as green concrete since it saves the natural resources and decreased the environmental pollution in the context of CO_2 emission. Green concrete is a concrete in which waste material is utilized, and at least one of its ingredients or its manufacturing process does not affect the environment. Manufacturing procedures, life cycle sustainability impacts and quantity of cement replaced are the key factors which are used to categorize weather a concrete is green or not [5]. The main purpose of producing green concrete is to minimize the CO_2 gas emissions, to utilize the natural resources and the utilization of waste materials in concrete, which causes environmental pollution.

This paper reviews the potential usage of agricultural wastes in concrete to form a green innovative sustainable concrete by utilizing agricultural waste which may solve the environmental issues and reduce the CO_2 gas emissions and save the natural resources which are widely utilized in the manufacturing of cement.

2. Agricultural waste

Agricultural wastes are the byproduct of manufacturing and processing of raw agricultural products like fruits, vegetables and crops. Agricultural wastes include the animal wastes (manure, animal carcasses), food manufacturing waste (only 20% of maize is preserved and 80% is waste), crop wastes (corn stalks, sugarcane bagasse, drops and culls from fruits and vegetables, pruning's), and hazardous and toxic agricultural wastes (pesticides, insecticides and herbicides) as shown in Figure 3 [6-8].

Malaysia is counted as among the top agricultural countries in the world. Rice, sugar cane, palm oil, fruits, coconut and rubber are the major agricultural product of Malaysia. Almost 40% of the total land in Malaysia is covered by agriculture land. The rice and sugarcane product are also cultivated in a large scale because their demand is high. However, development of these products creates a lot of agricultural waste which cause the environmental pollution [8].

This paper will discuss the materail and mechanical properties of various agro-wastes (rice husk ash, sugarcane bagasse ash, palm oil fuel ash and banana skin powder) which possess pozzolanic properties; namely, silica calcium gel, which would enhance the strength of concrete up to certain limit. The discussion on RHA, SCBA, and POFA are based on reviews of previous work while discussion on properties of BSP is based on authors' experimental work. The discussion revolves around their physical, fresh and hardened properties, and mechanical properties. From the review, the content of pozzolanic materials in each type of agro-waste will be determined. From this result, the fresh and hardened state properties and mechanical properties of each agro-waste could also be determined because these properties are mainly influenced by the content of the pozzolanic materials.

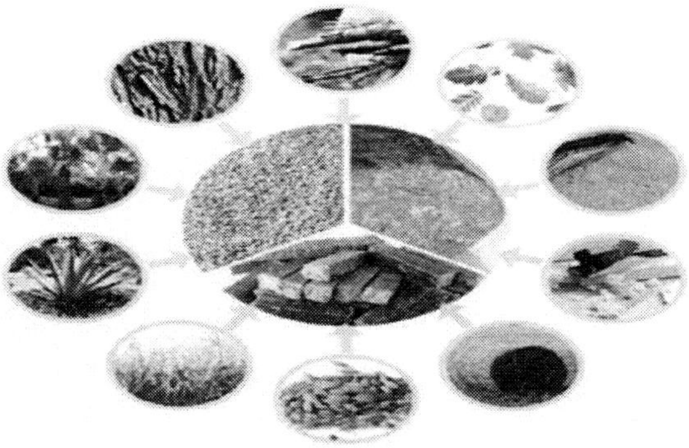

Figure 3. Various types of agricultural wastes [7].

2.1. Rice husk ash (RHA)

Rice milling industry has generated approximately 600 million tons of rice paddy annually [8] and is predicted to increase significantly in the future. Almost 20% rice husk is generated from every paddy ton in rice mills, which is utilized as fuel in the boiler to generate electricity for the mill. Approximately 25% of the rice husk is converted into solid waste known as Rice Husk Ash (RHA) as present in Figure 4 [9].

Figure 4. Rice husk and rice husk ash [9].

The chemical compositions of RHA, Sugar Cane Bagasse Ash (SCBA), Palm oil fuel ash (POFA) and Banana Skin Powder (BSP) are shown in Table 1. The compositions of OPC, RHA, SCBA, and POFA were obtained from previous researches [10-12, 31, 37] whilst the compositions of BSP were conducted in the analytical laboratory faculty of civil and environmental engineering, UTHM, Malaysia.

Table 1. Setting Word's margins.

Chemicals	OPC (%) [10]	R.H.A (%) [11-12]	S.C.B.A (%) [31]	POFA (%) [37]	B.S.P (%)
CaO	65.96 %	1.03	11.8	6.4	8.95
SiO_2	20.99 %	93.46	62.43	65.3	55.98
Al_2O_3	6.19 %	0.58	4.38	2.6	2.71
MgO	0.22 %	0.51	2.51	3.1	1.08
SO_3	---	0.6	1.48	0.5	0.10
K_2O	0.60 %	1.82	3.53	5.7	28.75
Fe_2O_3	3.86 %	0.52	6.98	2.0	1.356
LOI	1.73	7.76	4.73	10.1	1.121

From Table 1, RHA possesses a great amount of silica oxide, aluminum oxide (Al_2O_3) and ferric oxide (Fe_2O_3). According to Table 2. ASTM C613 specifications the RHA is Class N pozzolanic material [13-15].

Table 2. ASTM C618 Chemical Requirements.

Class			
	N	F	C
Silicon dioxide (SiO2) plus aluminium oxide (Al2O3) plus iron oxide (Fe2O3), min, %	>7 0	<70	50
Sulphur trioxide (SO3), max, %	4	5	5
Loss on ignition, max, %	10	6	6

Pozzolanic characteristics allows it to be used as a cementitious material. When it is finely grinded, it consumes calcium hydroxide [Ca(OH$_2$)] at ordinary temperatures in the presence of water. The silica and calcium hydroxide reaction generate the calcium silicate gel participating to strength gain in the concrete mixture. However, OPC has certain limit to calcium hydroxide that RHA can utilize, thus providing higher strength up to certain limit. The pozzolanic reaction depend upon the calcium present in the OPC up to certain limit it improves the strength then it reacts negative or reduces the strength of concrete.

When RHA is added to the concrete mixture, the pozzolanic gel (Calcium Silicate gel) will be formed which will improve the strength performance of concrete [16]. The summary of previous studies on physical and mechanical properties of RHA shown in Table 3. It concludes the outcomes while utilizing the RHA as cement replacement concrete. It was predicted by many researchers that the RHA has great potential to execute as a Pozzolanic material, which will have considered as conventional or high strength concrete as cement replacement. Through the utilization of RHA can decrease the environmental pollution (Disposal waste and CO_2 emission) and produces the sustainable construction material. The following paragraphs will discuss the previous studies on physical and mechanical properties of RHA.

Table 3. Summary on previous research on RHA.

Ref No.	Design & % used of RHA	Findings
[15]	w/b ratio: 0.53 Curing: 1, 7, 14 and 28 days RHA: 5%,10%, 15% & 20%	10% RHA gave compressive strength rises up to 30% at 0.53 w/b ratio was used.
[16]	w/b ratio: 0.45 and 0.50 Curing: 28 days RHA: 5%,10%, 15% 20%, 25%,30%,35% & 40%	20% RHA exhibited compressive and split tensile strength increased up to 20% at 0.5 w/b ratio was utilized.
[17]	w/b ratio: 0.34 Curing: 7, 14, 28, 56 & 91 days RHA: 10%, 20%, 30% & 40%	10% RHA indicated that compressive, splitting tensile and flexural strength improved 20%.
[18]	w/b ratio: 0.32, 0.40 and 0.5 Curing: 7, 14, 28, 56 & 91 days RHA: 10% and 20%	20% RHA improved the 10% compressive and split tensile strength when 0.32 w/b ratio was adopted.
[19]	w/b ratio: 0.5 Curing: 7, & 28 days RHA: 0%, 5%, 10%, 15%, 20% and 25%.	5% RHA improved the compressive, flexural and split tensile strength 4%, 5% and 18% respectively.
[20]	w/b ratio: 0.23, 0.35 and 0.47 Curing: 1, 3, 7, 14, 28, 56 & 91 days RHA: 10%, 20% & 30% size of RHA: 12 μm	20% RHA showed equivalent compressive strength when 0.35 w/b ratio was used.
[21]	M40 and M50 grade concrete Curing period: 7, 28 & 56 days RHA: 5%, 10% & 15%	10% RHA was optimum, where 5% compressive strength increased while increase in RHA content, the workability of concrete decreases.
[22]	Curing period: 7, 14 & 28 days. 600 μm passing RHA: 5%, 10%, 15%, 20% & 25%	Compressive Strength reduced with increase in RHA. Optimum was determined to be 0-20% for better performance.
[23]	Curing period: 7, 14 & 28 days Target concrete grade: M20 63 μm passing RHA: 5%, 15% and 25%	5% RHA was optimum, showed lower compressive strength but achieved the target strength of M20 grade. Workability decreased with increase in RHA content.
[24]	Curing period: 7, 14 & 28 days Target concrete grade: M15 RHA: 5%, 10%, 15%, 20% & 25% RHA as sand replacement Volume and Weight Batching	It was observed that compressive strength 6% decreased with increase in RHA content. Volume-based batching showed slightly higher strength compared to weight-based batching.

From the discussion of previous research, it can be concluded that the optimum percentage of RHA that could be added in the concrete admixture is about 10 % which will increase the compressive, tensile and flexural strength of concrete up to its maximum at 30 MPa and 20 MPA, respectively.

2.2. Sugarcane bagasse ash (SCBA)

Sugarcane is counted in major crops grown in over 110 countries and the production is almost 1500 million tons per year in worldwide. SCBA is 10 million tons annually, generated by the sugar mills. The sugarcane waste possesses 26% of bagasse and 0.62% of residual ash as shown in Figure 5 [25-27].

Figure 5. Sugar cane bagasse and sugar cane bagasse ash [9].

Bagasse ash possessing high amounts of un-burnt matter, silicon, aluminum, iron and calcium oxides [27]. The chemical composition of SCBA is shown in Table 1. The chemical composition contains the high silica contain and other pozzolanic material. According to ASTM C618 shown in Table 2. The SCBA is Class N pozzolanic material because the sum of silica oxide, aluminum oxide (Al_2O_3) and ferric oxide (Fe_2O_3) greater seventy.

SCBA possesses the pozzolanic material which will develop the pozzolanic reaction in manufacturing of concrete and it will contribute to rise the strength in concrete. Various researchers conducted the experimental work to analyses the physical and mechanical performance of concrete consisting SCBA as cement replacement. The summary of previous study on SCBA as cement replacement is presented in Table 4.

The summary of previous studies on material, fresh and hardened properties of SCBA are shown in Table 3. It concludes the results while utilizing the SCBA as cement replacement concrete. It was prophesied by various scholars that the SCBA has great tendency to perform as a Pozzolanic material, which will have considered as conventional or high strength concrete as cement replacement. Through the consumption of SCBA can decline the environmental pollution (Disposal waste and CO_2 emission) and develops the sustainable construction material. The following paragraphs will elaborate the previous studies on material and mechanical properties of SCBA.

From the discussion of previous research, it can be concluded that the optimum percentage of RHA that could be added in the concrete admixture is about 10 % which will increase the compressive, tensile and flexural strength of concrete up to its maximum at 30 MPa and 20 MPA, respectively.

Table 4. The summary of previous study on SCBA.

Ref No.	Design & % used & Fineness of SCBA	Findings
[28]	w/b ratio: 0.45 Curing: 7 and 28 days Target mean strength: 30 Mpa SCBA: 0%,5%,10%,15% &20%.	5% SCBA was optimum and the compressive strength was increased up to 28% compared to control sample.
[29]	w/b ratio: 0.48 Curing: 7 & 28 days SCBA: 0%, 5%, 10%, 15%, 20% and 25%	Slump flow was maximum at 25% replacement. The compressive strength was 27.5% increases at 5% replacement where as it decreases as percentage of SCBA rises.
[30]	w/b ratio: 0.44 & 0.63 Curing: 7, 28, 56 & 90 days SCBA: 0%, 10%, 15%, 20%, 25% and 30%. Size of SCBA: 300 μm	10% SCBA gave 15% better compressive, splitting tensile and flexural strength performance compared to other mix proportions.
[31]	w/b ratio: 0.49 Curing: Compressive and split tensile strength: 7, 14 & 28 days. Flexural strength: 7 & 28 days Target mean strength: 25 Mpa SCBA: 0%, 5%, 10%, 15% & 20% . Size of SCBA: 63 μm	The compressive, split tensile and flexural strength increase linearly up to 10% SCBA replacement. In compressive 20% strength increases where as in tensile and flexural the 10% and 15 % strength rises respectively.
[32]	w/b ratio: 0.5 Curing: 7, 14, 28 & 60 days SCBA: 0%, 5%, 10%, 15%, 20% and 25%. Size of SCBA: 150 μm	The optimum percentage was 15% SCBA replacement where the compressive and tensile strength increases 19% and 12.5 % respectively compared to control sample.
[33]	w/b ratio: 0.40 & 0.45. Curing: 7, 28, 56 & 90days SCBA: 0%, 5%, 10%, 15% and 20%	The 5% SCBA replacement gave the better slump flow compare to other proportions. The 15% replacement performance was better than other proportions. The compressive strength was increases up to 15% and the split tensile strength performance was 5% better than conventional concrete.
[34]	M40 and M50 grade concrete w/b: 0.5 Curing period: 7, 28 & 90 days Target mean strength: 30 Mpa SCBA: 10%,20%,30%, and 40%	20% SCBA improves the 9% compressive strength compared to conventional concrete, whereas tensile and flexural was 5% improved compared to control sample.

2.3. Palm oil fuel ash (POFA)

Palm oil fuel ash is produced from burning of palm oil husk and palm shell in the palm oil mill as shown in Figure 5. The chemical composition of POFA is shown in Table 1, where it is shown that POFA contains the pozzolanic materials. According to ASTM C618 specifications shown in Table 2. POFA is concluded that Class F pozzolanic material because it contains silica oxide (SiO_2), aluminum oxide (Al_2O_3) and ferric oxide (Fe_2O_3) less then seventy. The summary of previous studies on material and mechanical properties of POFA presented in Table 5.

Figure 6. Palm oil husk, palm shell and palm oil fuel ash [35].

The various study will deliberate the previous studies on material, fresh and hardened properties of POFA.

Table 5. Summary of previous research on POFA.

Ref No.	Design & % used & Fineness of POFA	Findings
[35]	w/b ratio: 0.40 Curing: 28 days POFA: 10%, 20% & 30%.	20% POFA gave 15% more strength compared to conventional concrete.
[36]	w/b ratio: 0.38 Curing: 7 & 28 days POFA: 10%, 15% and 20%	The 15% POFA replacement possessed the better workability results compare to other samples. The compressive, split tensile and flexural performance enhanced 10% at 20% POFA.
[37]	w/b ratio: 0.36 Curing: 7, 14, 28, 56 & 90 days. POFA: 0, 5, 10, 15, 20, 25 and 30%. Finesses = 45 μm	The durability and slump of SCC was examined, AT 15% POFA replacement the slump was higher than the other proportions whereas in water absorption test the 30% replacement absorb less water compared to other mixes.
[38]	w/b ratio: 0.36 Curing: 1, 7, 28, 56 & 90 days SCBA: 0%, 30%, 50% & 70% Finesses = 45 μm	The workability increases with increment of POFA percentage as cement replacement. Compressive strength and split tensile strength was 20 % maximum at 30% POFA compared to all replacement.

The results showed that the strength (compressive, split tensile and flexural strength) increases significantly by incorporating the POFA in concrete up to 20% cement replacement because up to 20% replacement the pozzolanic reaction generated in cement paste contributes to strength after that the calcium and silica did not generate the pozzolanic activity. The pozzolanic reaction is shown below.

$$Cao + H_2O \longrightarrow Ca(OH)_2 \quad (1)$$

$$Ca(OH)_2 + SiO_2 \longrightarrow C\text{-}S\text{-}H \text{ (Calcium Silicate Hydrate gel)} \quad (2)$$

After 20 % replacement of POFA no extra calcium left which will react with silica to generate the pozzolanic gel. So, alone silica creates negative impact on the strength of concrete.

2.4. Banana Skin Powder (BSP)

In Malaysia banana plantation is second most planted fruit, it covers the 26,000-hectare area and produced 530,000 metric ton banana per year [39- 41]. Banana waste is almost 4 ton produced annually and creates the environmental pollution [42]. Banana fruit is most used among all the fruits due to that the banana skin waste generates in huge amount which create the environmental problem such as odor smell and generate leachate which will affect the atmosphere and ground water respectively. The banana skin and banana skin powder are shown in Figure 7.

Figure 7. Banana skin and banana skin powder.

Many researchers did experimental work to utilize the banana as reinforcement and banana leaves as cement replacement. Zhu et al. [43] investigated the flexural performance of concrete utilizing banana fiber at different percentages as filler, between 8% and 10%. The results showed that 2 to 3% split tensile and flexural strength increased due to banana fiber reinforcement. Lukman [44] analyzed the material properties of banana trunk ash and mechanical properties of concrete utilized as cementing material. The banana trunk ash was utilized 0% to 3% with an increment of 1%. The outcomes demonstrated that banana trunk ash contains the pozzolanic material and it reduces the 30% compressive strength compared to conventional concrete.

In literature review no one utilized the banana skin powder as cement replacement so in this research the chemical composition of banana skin powder was examined the outcomes are shown in Table 1, it is seen that the banana skin powder contains the pozzolanic material in ample amount. According to ASTM C618 the banana skin powder is Class F pozzolanic material.

Due to this pozzolanic material, it can be utilized as cement replacement material which will definitely enhance the mechanical performance of concrete, made the economical, green and sustainable concrete and reduces the agricultural waste.

3. Conclusion

From the above previous research, the durability, workability and strength (compressive, split tensile and flexural strength) was investigated the strength performance of concrete containing different types of agricultural waste. The RHA gave the 20% better strength performance from 10%-20% cement replacement. The SCBA improves the strength up to 30% at 5%-15% replacement where as POFA enhances the strength 20% at 10%-20% partially replacement with POFA.

The workability performance of concrete containing the agricultural waste was better than control sample. The workability increases significantly as agricultural waste replacement percentage increased. It is concluded from the literature that RHA, SCBA and POFA possessed the pozzolanic materials due to that pozzolanic activity develop which enhance the strength of concrete. From literature it is concluded that 20% RHA, 5%-10% SCBA and 10%-20% POFA replacement with OPC are optimum.

It is concluded from this review that banana skin powder has the pozzolanic material as other pozzolanic material. Therefore, banana skin powder can be utilized as partially cement replacing material. For further extensive research is required to understand the performance of banana skin powder as partially cement replacing material on strength behavior of concrete. The use of banana skin powder as partially cement replacement in concrete leads to develop the green and sustainable concrete and resolve the environmental pollution issues produced by waste.

4. References

[1] Khaw A H 2010 Performance of Lightweight Foamed Concrete Using Laterite as Sand Replacement *Journal of Chemical Information and Modelling* **53**(9) 1689–1699

[2] Zongjin L I 2014 *Advanced Concrete Technology* (1st ed.) (New Jersey: John Wiley and Sons Inc.)

[3] Suhendro B 2014 Toward Green Concrete for Better Sustainable Environment *Procedia Engineering* **95** 305 – 320

[4] Tahir M, Hussain T, Behaylu A and Tilahun A 2015 Scenario of Present and Future of Solid Waste Generation in Metro Cities *Journal of Environment and Earth Science* **5** 164-169

[5] Zhu M, Fan X, Alberto R and Qichang H 2009 Municipal Solid Waste Management in Pudong New Area, China *Waste Management* **29** 1227–1233

[6] Obi F, Ugwuishiwu O B and Nwakaire J N 2016 Agricultural Waste Concept, Generation, Utilization and Management *Nigerian Journal of Technology* **35**(4) 957 – 964

[7] Agamuthu P 2009 Challenges and Opportunities in Agrowaste Management: *An Asian perspective. Inaugural meeting of First Regional 3R Forum in Asia* 11 -12 Nov., Tokyo, Japan

[8] Brown and Root 1997 Environmental Review of National Solid Waste Management Plan *Environmental Consultancy Group* Interim report submitted to the Government of Mauritius

[9] Baikerikar A 2014 A Review on Green Concrete *J. Emerging Technol. Innovative Res.* **1**(6) 472–474

[10] Obla K H 2009 What is Green Concrete? *The Indian Concrete Journal* **4** 26-28

[11] Krishna N K, Sandeep S K, Mini M 2016 Study on Concrete with Partial Replacement of Cement by Rice Husk Ash, *IOP Conf. Series: Materials Science and Engineering,* **149** 1 – 11

[12] Rambabu P V, Varma V C and Ramarao G V 2015 Experimental Study on Rice Husk Ash for Optimum Level of Replacement of Cement in Concrete *International Journal of Engineering Research & Technology (IJERT)* **4**(10) 354 – 359

[13] Habeeb G A and Mahmud H B 2010 Study on Properties of Rice Husk Ash And Its Use as Cement Replacement Material *Materials Research* **13**(2) 185 – 190

[14] Sharmin A, Alengaram U J, Jumaat M Z, Yusuf M O, Kabir M S M A and Bashar I I 2017 Influence of Source Materials and The Role of Oxide Composition on The Performance of Ternary Blended Sustainable Geopolymer Mortar *Construction and Building Materials* **144** 608 – 623

[15] Kartini K 2011 Rice Husk Ash – Pozzolanic Material for Sustainability *International Journal of Applied Science and Technology* **1**(6) 23-29

[16] Karim M R, Hossain M M, Khan M N N, Zain M F M, Jamil M and Lai F C 2014 On the Utilization of Pozzolanic Wastes as an Alternative Resource of Cement *Materials,* **7** 7809-7827

[17] Gupta N, Kumar P, Mishra A and Srivastava R P 2015 Characterization and Application of Rice Husk Ash as Pozzolanic Material in Concrete *IOSR Journal of Mechanical and Civil Engineering* **12**(3) 28 – 32

[18] Tangchirapat W, Tangpagasit J, Waew-kum S and Jaturapitakkul C A 2003 New Pozzolanic Material from Palm Oil Fuel Ash *KMUTT Research and Development Journal* **26** 459-473

[19] Sathawanea, Vikrant S and Vairagadeb 2013 Effect of Rice Husk Ash and Fly Ash on Concrete by 30% Cement Replacement *Chemical, Civil and Mechanical Engineering Track of 3rd Nirma University International Conference Combine* **51** 35-44

[20] Ghassan A B and Hilmi B M 2010 Study on Properties of Rice Husk Ash and Its Use as Cement Replacement Material *Research and Materials* **13**(2) 16-20

[21] Deepa G, Sivaraman N and Job T 2013 Mechanical Properties of Rice Husk Ash (RHA) - High strength Concrete *American Journal of Engineering Research (AJER)* **3** 14-19

[22] Dong D C, Pham D H and Nguyen N L 2008 Effect of Rice Husk Ash on Properties of High Strength Concrete *The 3rd ACF International Conference* **44** 442-449
[23] Dabai M U, Muhammad C, Bagudo B U and Musa A A 2009 Studies on The Effect of Rice Husk Ash as Cement Admixture *Nigerian Journal of Basic and Applied Science* **17**(2) 252-256
[24] Rauf M A and Arsalan M 2017 Experimental Study on Properties of Concrete with Partial Replacement of Cement by Sugarcane Bagasse Ash using Manufactured Sand *International Research Journal of Engineering and Technology* **4**(5) 1238-1244
[25] Gupta R B, Chandraul K and Singh M K 2017 Experimental Study on Strength of Concrete Using Sugar Cane Bagasse Ash *International Research Journal of Engineering and Technology* **4**(6) 2392-2395
[26] Yashwanth M K, Avinash G B and Raghavendra A 2017 An Experimental Study on Alternative Cementitious Materials: Bagasse Ash as Partial Replacement for Cement in Structural Lightweight Concrete *The Indian Concrete Journal* **6** 51-58
[27] Mukundrao G S, Changdeo D S and Vikas J S 2017 Experimental Study of Sugarcane Bagasse Ash Blend & Its Application in M-30 Grade of Concrete for Moderate Exposure Conditions *International Research Journal of Engineering and Technology* **4**(4) 1156-1159
[28] Tasnim S, Rahman M E and Ahmad R 2018 Mechanical Performance of Modified Cement Paste Made With Microfine POFA in Ammonium Nitrate Environment *Construction and Building Materials* **162** 534-542
[29] Ofuyatan T, Olutoge F and Olowofoyeku A 2015 Durability Properties of Palm Oil Fuel Ash Self Compacting Concrete *Engineering, Technology & Applied Science Research* **5**(1) 753-756
[30] Piyush K and Anil P S 2015 Studies on Partial Replacement of Cement by Bagasse Ash in Concrete *International Journal of Civil and Structural Engineering Research* **3** 308-310
[31] Srinivasan R and Sathiya K 2010 Experimental Study on Bagasse Ash in Concrete *International Journal for Service Learning in Engineering* **5**(2) 60-66
[32] Dhengare S W, Raut S P, Bandwal N V and Anand K 2015 Investigation into Utilization of Sugarcane Bagasse Ash as Supplementary Cementitious Material in Concrete *International Journal of Emerging Engineering Research and Technology* **3**(4) 109-116
[33] Kumar T S, Meeravali K and Balaji V G D T 2014 Partial Replacement of Cement in Concrete with Sugar Cane Bagasse Ash-Behaviour in HCl Solution *Intenational Journal of Advance Research In Science and Engineering* **3**(7) 368-373
[34] Ahmadi R and Zainudin N 2016 Micro Fine Sized Palm Oil Fuel Ash Produced Using a Wind Tunnel Production *Hindawi Publishing Corporation Advances in Materials Science and Engineering* **6** 22-28
[35] Awal A and Shehu I A 2011 Properties Of Concrete Containing High Volume Palm Oil Fuel Ash: A Short-Term Investigation *Malaysian Journal of Civil Engineering* **23**(2) 54-66
[36] Irfan A N, Devinder S and Munish K 2008 Experimental Investigation on The Use of Bagasse Ash in The Construction of Low Volume Traffic Roads *International Research Journal of Engineering and Technology* **04**(09) 1535-1538
[37] Aiswarya V S, Beyoola W and Harsha V N 2017 Palm Oil Fuel Ash as Partial Replacement of Cement in Concrete *International Journal of Engineering Research & Technology* **6**(3) 544-546
[38] Subramani T and Anbuchezian A 2017 Experimental Study of Palm Oil Fuel Ash as Cement Replacement of Concrete *International Journal of Application or Innovation in Engineering & Management* **6**(3) 1-5
[39] Mekhilef S, Saidur R, Safari A and Mustaffa W S B 2011 Biomass Energy in Malaysia: Current State and Prospects *Renewable and Sustainable Energy Reviews* **15**(7) 3360-3370
[40] Goh C S, Tan K T, Lee K T and Bhatia S 2010 Bio-ethanol from Lignocellulose: Status, Perspectives and Challenges in Malaysia *Resource Technology* **101**(13) 4834-4841

[41] Sellin N N, Souza O S, Medeiros H W and Afuso R K S 2013 Characterization of Banana Culture Wastes as A Potential Energy Source *Engineering and Technology International Journal of Energy and Power Engineering* **8**(8) 815-819
[42] Kanning R and Kleber C 2014 Banana Leaves Ashes as Pozzolan for Concrete and Mortar of Portland Cement *Construction and Building Materials* **54** 460–465
[43] Zhu W H and Tobias B C 1994 Air-cured Banana-Fibre-Reinforced Cement Composites *Cement and Concrete Composite* **16**(01) 3-8
[44] Lukman M H 2014 The Use of Banana Trunks Ash as Cement Replacement in Mortar *Universiti Malaysia Pahang* Master Thesis

Strategizing contractor firms to deliver green construction projects: conceptual framework

Ayisha Powmya[1*] , Nazirah Zainul Abidin[1] and Nurul Sakina Mokhtar Azizi[1]
[1]School of Housing, Building and Planning, University Sains Malaysia (USM), 11800, Pulau Pinang, Malaysia
*Corresponding author: ampowmya@gmail.com

Abstract. Delivering green construction (GC) projects is not an easy task for inefficient or inexperienced contractor due to added requirements. This paper conceptualizes firm strategies in delivering green projects by focusing on enhancing firm resources. Journal articles with keywords such as 'green construction', 'green project', 'sustainable construction', 'resources and capabilities', 'contractor strategies', and so forth in the title, abstract or keywords of the publications were reviewed. Articles published from 2000 to 2018 in Scopus database were selected. The current literature on firm resources mainly focused on four assets (personal, financial, technological and organizational) and four capabilities (technical, innovative, managerial competency and collaborative). These assets and capabilities are necessary for contractor firms to survive in the industry. Capitalizing on these resources, this paper proposes seven strategies: human resource strategy, innovation strategy, financial strategy, smart collaboration strategy, environmental pro-activeness strategy, management enhancement strategy and information and communication strategy, for contractor to effectively deliver green project. Using Resource Based View (RBV) and Resource Mobilisation Theory (RMT), a conceptual framework is produced connecting firm resources and Resource Mobilisation Strategies (RMS). This paper shall benefit contractors and academician as it extended the knowledge on firms' resources and recommend ways to strategize green delivery.

1. Introduction

Contractors implement responsible construction practices to support environmental sustainability at construction site. However, GC project construction may appear to be an uneasy task for inexperienced and incompetent contractor in that type of projects due to the use of green technology, additional design requirements and rigorous practices at site. The level of contractor firm success depends on their resources and processes, in addition to management practices used during implementation of technologies [1]. Contractor firm's internal strength such as competence, knowledge to do the job and financial arrangements are very important to ensure successful project delivery [2]. Strategies are the patterns of decisions that organisations plan to execute for achieving stated and desired objectives [3]. Strategy provides direction and focus for the organistaion. An organizational capability is its ability to perform a task using its resources to achieve its goals [4]. This indicates that resources will influence strategies of firms. While commitment of efforts and resources from contractors is generally influenced by pressure exerted by environmental regulations and stakeholder demands [5], resources account as key factors affecting the adoption of GC practices in projects. Contractor firms are likely to have greater need for resources and RMSs in order to be able to deliver GC projects.

Although there have been attempts to identify enablers of green project construction, the existing works have only managed to provide fragmented insights into practical strategies of contractor firms to implement green project construction using firm resources as basis. The aim of this paper is to systematically review related researches on contractor firm resources and strategies to provide an overall summary of current body of knowledge and to present a conceptual framework for strategizing contractor firm resources to deliver GC projects using RBV and RMT. The conceptual framework is produced from extensive literature review obtained from scopus database. The timeframe of 2000 to 2018 was chosen for the sample, since the year 2000 shows growth in research work on green projects. The framework relates contractor firm resources to RMSs in order to ensure successful construction of green projects.

2. Contractor and green project involvement

Construction contractors who are responsible for practical realization of designs play critical role in achieving environmental sustainability of construction projects [5–7]. Although all parties involved in construction project should cooperate, being the implementers of construction projects, it is imperative for the construction firms to be knowledgeable and well equipped to deliver GC projects. For a contractor to be able to adopt green practices in its organization, factor such as managerial concern is an important driver [5]. As such, improvement in environmental performance of construction projects is due to the resources and strategies of contractor firm.

3. Research approach

The objectives of this paper are to provide information on contractor firm resources and strategies to deliver GC projects. Scopus search engine was selected as most of the research papers in construction, management and engineering are archived in Scopus. Literatures from 2000 to 2018 were selected using keywords such as 'green construction', 'green project', 'sustainable construction', 'resources and capabilities', 'contractor strategies', and so forth in the title, abstract or keywords of the publications. The literatures were explored to identify resources of contractor firms delivering GC projects and the RMSs to be adopted by contractor firms engaged in green project construction. Finally the findings of the review were summarized. The resources of contractor firms which are basis of RMSs are presented in the next section.

4. Contractor firm resources

Contractors' responses to the government imposed requirements and project stakeholders' pressure differ according to the capability of firm in tackling varied issues. Firms use resources as input factors to conceive of and implement their strategies [8, 9]. Efficient use of resources will lead to better competitive ability, diversification of good and services and growth potential.

For contractor firms to deliver GC projects, they need to ensure the employees are knowledgeable and able to cope with the technological sophistication and complexity of green projects [5, 10]. Financial assets are assets relating to company's finance capital base and turnover and profitability which are indicators of the financial strength of a contractor firm [11]. Realization of strategic plans requires investment-in facilities, knowledge, personnel, and so on. Financial capital base is of primary importance for contractors as they are responsible to pay suppliers [12]. Hence, cash flow forecast and budget allocation is essential for green project delivery. Technological assets of firm such as information technology resources and plants and equipments which are necessary for physical realization of projects [9]. In the implementation of the GC projects, GC technology is the foundation [13] as green projects are more sophisticated than conventional projects [14]. Appropriate IT systems are also necessary to deliver and communicate the information from green consultants to the project team members [15]. Organizational policies, its procedures for planning and control, and its environmental management system constitute the organizational assets. Environmental protection being one of the major project success criteria requires contractor firms' environmental protection plan [11, 16]. Therefore, presence of Environmental Management System (EMS) has been identified as helpful for GC project [17]. Thus, organizational assets which relate to organizational infrastructure such as sustainability policies, procedures and Environmental Management System (EMS) are required for GC project success. High technical capability of contractor firms is critical to achieve environmental objectives of GC projects [4]. A firm with strong technical capabilities can certainly deliver a GC project at a lower cost because it understands GC project needs, relevant technologies to such project and able to identify suitable green material suppliers [18]. Innovative capability is a firm's ability to develop new products and processes and to achieve superior technological and/or management performance [19]. Research & Development facilitates transition to GC by continuous development of new products and processes [20]. Contractor firms should have strong innovation capability to undertake a GC project successfully [21]. Managerial competency is important for contractor as they are responsible to plan, manage and oversee all construction activities. Since GC projects are embedded with innovative practices, contractor requires managerial competency to operate within the acceptable standards of GC [21]. Ingenious collaboration skills promote goal alignment and helps to reduce counterproductive behavior, more frequent and open sharing of information, the exchange of expertise and resources [22]. The Table 1 shows the resources required by contractor firms, which are identified from literatures.

Table 1. Resources of contractor firms delivering green development projects.

	[20]	[22]	[11]	[10]	[23]	[15]	[9]	[21]	[14]	[5]	[24]	[17]	Frequency
1			√	√			√	√		√			5
2			√				√						2
3			√			√	√		√				4
4			√									√	2
5			√				√						2
6	√						√	√					3
7							√	√			√		3
8	√	√			√			√					4

1=Personnel assets; 2=Financial assets; 3=Technological assets; 4=Organizational assets; 5=Technical capability; 6=Innovative capability; 7=Managerial competency; 8=Collaborative capability

5. Resource Based View and Resource Mobilization Theory

RBV, in the field of strategic management emphasizes that every firm possesses a unique resource bundle that influences its strategic choices [25]. The researchers recognized that the RBV is the most appropriate theoretical background on which the resources of contractor firm engaged in GC project delivery can be explored. However, RBV is unable to explain what routines are applied by contractor firms to mobilise the firm resources. Effective mobilization of resources leads to success of a social movement [26]. RMT is the dominant philosophy in the study of social movement which emphasizes the importance of resource mobilization in order to advance a social movement's motives [27]. Hence, the researchers observed that RMT need to be associated with RBV. This is illustrated in Figure 1.

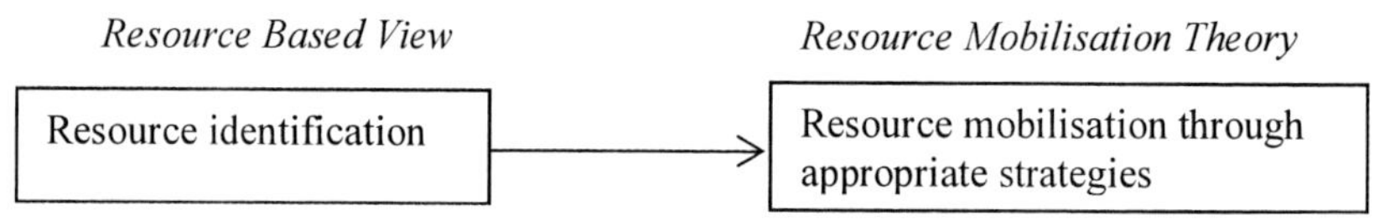

Figure 1. Theoretical model.

6. Resource Mobilisation Strategy of contractor firms

Contractor firm strategies in this study are planned actions of firms to improve resources which facilitate green project construction [28]. Resource mobilisation is activities carried out to prepare and organise resources. RMS may trigger firm's actions towards quality improvement, better GC technologies application and reduction of any negative environmental impacts during construction processes [29]. This is supported by [10] who indicated through empirical test that innovation strategies are to be triggered within firms to obtain complete effect of any GC policy on firms' business performance. Resources that are required for contractor firms engaged in GC project delivery have been explained in section 4. Firms need to devise appropriate strategies to ensure smooth flow and effective utilization of resources during project implementation [12]. This paper proposes seven strategies: human resource strategy, innovation strategy, financial strategy, smart collaboration strategy, environmental pro-activeness strategy, management enhancement strategy and information and communication strategy, for contractor to effectively deliver green project.

Human resource strategy is concerned about managing human aspects of a firm [30]. Experience of technical personnel within the contractor firm boosts it's performance [31]. Periodic training programs conducted have potential effect on environmental performance of construction projects [6]. A well-designed reward system could help to promote employees towards performing environmentally sound practices [15]. Investment in R&D activities can help to improve green practices' innovations which in turn can help to score more points during green certification and have higher "return on investment" [6]. Dedicated budget allocation for promoting GC goals would also influence green project delivery [1]. Construction firms could benefit from conventional contractual elements such as surety bonds and insurance policies as their financial strategies [13] and it is necessary to maintain a strong balance sheet to continually secure surety bonds for ongoing project procurements [30]. Collaboration with international firms engaged in GC that are well experienced and have adopted GC would be prudent for knowledge base of contractor firms [21]. It is necessary for contractors to take initiative to collaborate with government bodies and international/national green building councils for better understanding as well as for their advocacy in implementation [21]. Early involvement of suppliers in GC projects would be advantageous as it enables suppliers to understand the needs of clients and contractors and provide more value to them [14]. Environmental performance of GC projects would be enhanced through effective compliance to environmental regulations [32]. Adopting an EMS showcases a firm's high level of environmental ambitions in the project. This would improve its corporate environmental image which in turn help attracting potential clients [17, 33]. Separate GC department to provide the expert knowledge, resources and support for their LEED projects would be beneficial [1]. Designers' involvement in the construction stage and contractors' involvement in the design stage could act as most effective ways to solve the complicated issue [32]. It is critical to provide uninterrupted flow of information to avoid any miscommunication. The information provided by green consultants should be clearly communicated to project team members at proper timing to update their knowledge to construct green buildings [15]. Sufficient communication channels are required to be provided to communicate effectively. Maintaining GC project database and regularly benchmark their green building performance with their competitors would help improve the capacity of the contractor to understand, develop and implement its project specific green building goals [1]. Thus, the conceptual framework of contractor firm resources and RMSs to deliver GC projects is as shown in Figure 2.

Contractor Firm Resources → **Resource Mobilisation Strategies**

Personnel asset	• Qualified and experienced staff in green project • Skilled project manager
Financial asset	• Finance capital base • Turnover and profitability
Technological asset	• IT resources • Plants and equipments
Organisational asset	• Sustainability poilicies and procedures • Environmental Management system
Technical capability	• Technological knowledge to use available technological assets • Capability to communicate technically in projects
Innovative capability	Capability to create new knowledge or new product
Managerial competency	• Capability to manage technological aspects • Capability to manage financial aspects
Collaborative capability	• Capability to adapt to collaborated firms • Capability to adopt collaborated firms' practices

Human resources strategy	• Recruit new staff with green experiences. • Training of employees. • Reward employees.
Financial enhancement strategy	• Allocate sufficient budget. • Use surety bonds. • Use insurance policy.
Technology innovation strategy	• Engage in Research & Development
Information and communication strategy	• Develop green project database. • Provide sufficient communication channels.
Smart collaboration strategy	• Partnership with material suppliers • Collaborate with international green construction firms. • Collaborate with governmental environmental regulatory bodies.
Environmental pro-activeness strategy	• Comply with sustainability legislations and EMS.
Managerial efficiency strategy	• Establish separate green construction department. • Coordinate with designers during construction stage. • Involve with designers during design stage.

Figure 2. Conceptual framework of Resource Mobilisation Strategy of contractor firms to deliver green projects.

7. Conclusions

Contractor firms' strategies are shaped as a result of their resources. In order to encourage transition towards GC, adequate analysis along with in-depth understanding of contractor firm resources is necessary for selecting appropriate RMS. Contractor firms with resources such as personnel assets, financial assets, technological assets and organizational assets, technical capability, innovative capability, managerial competency and collaborative capability have more tendencies to deliver GC projects successfully. In addition, it is necessary to adopt appropriate strategies to ensure capitalization of firm resources to deliver GC projects. Policies and procedures of contractor firm that are targeted on project delivery need to transform to realize firm resources for GC project delivery. Therefore this study provides a broader view for policy makers when trying to adopt appropriate strategies. This paper presents a conceptual framework to cater to the RBV of strategy development and RMT. The adoption of the framework improves contractors' resources to deliver GC projects. However, further research is required to understand the process of RMS. The model used in this study is not validated by real cases. Further research is required to empirically test and validate the conceptual model developed by this study.

8. References

[1] Ofori-Boadu A, Owusu-Manu D, Edwards D and Holt G 2012 Exploration of Management Practices for LEED Projects : Lessons from Successful Green Building Contractors *Struct. Surv.* **30** 145-162

[2] Tarawneh S 2004 Evaluation of Prequalification Criteria- Client Perspective: Jordan Case Study *J. Appl. Sci.* **4** 354–363

[3] Price A D F 2003 The Strategy Process Within Large Construction Organisations *Eng. Constr. Archit. Manag.* **10** 283-296

[4] O'Regan N and Ghobadian A 2004 The Importance of Capabilities for Strategic Direction and Performance *Manag. Decis*. **42** 292-313

[5] Qi G Y, Shen L Y, Zeng S X and Jorge O J 2010 The Drivers for Contractors' Green Innovation: An Industry Perspective *J. Clean. Prod.* **18** 1358-1365

[6] Li Y Y, Chen P H, Chew D A S and Teo C C 2014 Exploration of Critical Resources and Capabilities of Design Firms for Delivering Green Building Projects: Empirical Studies in Singapore. *Habitat Int.* **41** 229–235

[7] Tan Y, Shen L and Yao H 2011 Sustainable Construction Practice and Contractors' Competitiveness : A Preliminary Study *Habitat Int.* **35** 225-230

[8] Hitt M, Ireland R and Hoskisson R 2009 *Strategic Management: Competitiveness and Globalization: Concepts & Cases* (Mason, OH: South-Western)

[9] Isik Z, Arditi D, Dikmen I and Birgonul M T 2009 Impact of Corporate Strengths/Weaknesses on Project Management Competencies *Int. J. Proj. Manag.* **27** 629–637

[10] Eiadat Y, Kelly A, Roche F and Eyadat H 2008 Green and competitive? An Empirical Test of The Mediating Role of Environmental Innovation Strategy *J. World Bus.* **43** 131–145

[11] Alzahrani J I and Emsley M W 2013 The Impact of Contractors' Attributes on Construction Project Success: A Post Construction Evaluation *Int. J. Proj. Manag.* **31** 313–322

[12] Kuwaiti E A, Ajmal M M and Hussain M 2017 Determining Success Factors in Abu Dhabi Health Care Construction Projects: Customer and Contractor Perspectives *Int. J. Constr. Manag.* **18** 1–16

[13] Yu Z, Lu C and San B 2014 *Application of Green Construction Technology in Construction Projects* In: International Conference on Construction and Real Estate Management (ASCE, Kunming, China) pp 389–397

[14] Mokhlesian S 2014 How Do Contractors Select Suppliers for Greener Construction Projects? *The Case of Three Swedish Companies Sustain.* **6** 4133–4151

[15] Hwang B G and Tan J S 2012 *Sustainable Project Management for Green Construction: Challenges, Impact and Solutions* In: World Construction Conference 2012-Global Challenges in Construction Industry (Colombo, Sri Lanka) p 9

[16] Shen L, Lu W and Yam M C H 2006 Contractor Key Competitiveness Indicators : A China Study *J. Constr. Eng. Manag.* **132** 416–424

[17] Varnäs A, Balfors B and Faith-Ell C 2009 Environmental Consideration in Procurement of Construction Contracts: Current Practice, Problems and Opportunities in Green Procurement in The Swedish Construction Industry *J. Clean. Prod.* **17** 1214–1222

[18] Mayer K J and Salomon R M 2006 Capabilities, Contractual Hazards, and Governance: Integrating Resource-Based and Transaction Cost Perspectives *Acad. Manag. J.* **49** 942–959

[19] Rangone A 1999 A Resource-based Approach to Strategy Analysis in Small-medium Sized Enterprises *Small Bus. Econ.* **12** 233–248

[20] Albino V, Berardi U 2012 Green Buildings and Organizational Changes in Italian Case Studies

Bus. Strateg. Env. **21** 387–400
[21] Mensah S, Ayarkwa J, Nani G 2015 Towards Enabling Construction Organizations' Adaptation to Environmental Sustainable Construction in Developing Countries *Adv. Soc. Sci. Res. J.* **2** 84–100
[22] Allred C R, Fawcett S E Wallin C, Magnan G M 2011 A Dynamic Collaboration Capability as a Source of Competitive Advantage *Decis. Sci.* **42** 129–161
[23] Fergusson H, Langford D A 2006 Strategies for Managing Environmental Issues in Construction Organizations *Eng. Constr. Archit. Manag.* **13** 171–185
[24] Robichaud L B and Anantatmula V S 2011 Greening Project Management Practices for Sustainable Construction *J. Manag. Eng.* **27** 48–57
[25] Chen P H, Ong C F, Hsu S C 2016 The Linkages between Internationalization and Environmental Strategies of Multinational Construction Firms *J. Clean. Prod.* **116** 207–216
[26] Oliver C 1997 Sustainable Competitive Advantage: Combining Institutional and Resource-Based Views *Strateg. Manag. J.* **18** 697–713
[27] McCarthy J D and Zald M N 1997 Resource Mobilization and Social Movements: A Partial Theory *Am. J. Sociol.* **82** 1212–1241
[28] Manley K McFallan S, Kajewski S 2009 Relationship between Construction Firm Strategies and Innovation Outcomes *J. Constr. Eng. Manag.* **135** 764–771
[29] Chen Y, Tang G, Jin J, Li J, Paillé P 2015 Linking Market Orientation and Environmental Performance: The Influence of Environmental Strategy, Employee's Environmental Involvement, and Environmental Product Quality *J. Bus. Ethics.* **127** 479–500
[30] Cheah C Y J and Garvin M J 2004 An Open Framework for Corporate Strategy in Construction *Eng. Constr. Archit. Manag.* **11** 176–188
[31] Aje O I, Odusami K T, Ogunsemi D R 2009 The Impact of Contractors' Management Capability on Cost and Time Performance of Construction Projects in Nigeria *J. Financ. Manag. Prop. Constr.* **14** 171–187
[32] Li Y Y, Chen P H, Chew D A S, Teo C C, Ding R G 2011 Critical Project Management Factors of AEC Firms for Delivering Green Building Projects in Singapore *J. Constr. Eng. Manag.* **137** 1153–1163
[33] Zhang X, Wu Y, Shen L 2015 Embedding "Green" in Project-Based Organizations : The Way Ahead in The Construction Industry? *J. Clean. Prod.* **107** 420–427

Experimental study on the out-of-plane strengthening of unreinforced masonry walls using cement-based fiber composites

Salah Altoubat[1*] , Mohamed Maalej[1], Moussa Leblouba[1] and Abdul Saboor Karzad[1]

[1] Department of Civil and Environmental Engineering, Sustainable Construction Materials and Structural Systems Research Group, University of Sharjah, P.O. Box 27272, Sharjah, UAE
*Corresponding author: saltoubat@sharjah.ac.ae

Abstract. This paper presents results from an ongoing research that is intended to study the behavior of URM walls for out-of-plane action. The effect of fiber reinforced shotcrete (FRS) as a strengthening method for the URM walls is examined. Eight URM walls were built inside a precast reinforced concrete boundary frame and strengthened with cement-based shotcrete mixes reinforced with different types of fibers. The masonry wall is 1.2x 1.2 m, and the masonry units were hollow concrete blocks (10 x 20 x 40 cm). Conventional fiber reinforced concrete (FRC) and engineered cementitious composite (ECC) were used as the strengthening layers sprayed on the surface of masonry walls. Control walls with plain shotcrete and conventional plastering mortar were also examined. Synthetic macro-fibers at dosage rate of 6 kg/m^3 in FRC mix and Polyethylene micro-fibers at a dosage rate of 16 kg/m^3 in ECC mix were considered in this study. The walls were placed in a carefully designed set up and tested under static central load using a hydraulic actuator with a capacity of 500 kN. The parameters recorded during testing were the applied load and out-of-plane displacement of the walls. Cracking pattern and failure modes were also monitored. The reported test results demonstrate that FRC and ECC mixes can significantly increase the out of plane strength and ductility of the URM walls.

1. Introduction

Unreinforced masonry (URM) walls are most vulnerable to damage in the event of an earthquake, wind load, or blast pressure. When a building with URM walls is subjected to lateral forces, the URM walls might experience in-plane/out-of-plane failure modes or a combination of both [1]. Basicaly the URM walls are designed to resist in-plane compression loads without considering the lateral forces and it not considered as part of building seismic load path [2]. Noitng that out-of-plane failure can trigger partial or full collapse of the wall which may endanger integrity of the building [3].

Many existing URM structures are not up to the standards of today's code requirements and are categorized as earthquake-prone buildings or maybe structurally deficient for current use. Studies have reported that the out-of-plane collapse was noted durring earthquake in the buildings where the walls were not designed for lateral loads [4]. Therefore, to enhance their out-of-plane capacity, ductility, energy absorption, and to upgrade their safety requirement, it is necessary to strengthen these structures [5]. Otherwise, even local failure of these masonry elements can trigger partial or full collapse of a structure or pose significant hazard to the occupants and pedestrians [2] [6].

Numerous researches have been conducted in recent years to develop effective methods for retrofitting the URM walls against lateral loads and improve their out-of-plane behavior [1], [6–9]. Those researches have resulted in developing and proposing a number of methods to enhance the out-of-plane response of URM walls. The methods proposed and used for retrofitting the URM walls subjected to lateral foces are: bonding the fiber reinforced polymer (FRP) sheets to the surface of the walls, applying textile reinforced mortar (TRM) on the surface of the walls, installation of unbonded post-tensioning, fiber reinforced cementitious composites with and without additional steel reinforcement, etc. [6] [10].

The proposed methods provides some enhancement to the strengthened structure, however, there are some drawbacks associated with some of those proposed methods. For example, strengthening using FRP sheets is a very simple application and when applied significantly increase the out-of-plane response of URM walls, but due to its high rigidity it eventually fails abruptly and occurs without much warning. It is also reported in the literature that the walls retrofitted with externally bonded FRP system fails prematurely by shear at the end connections of the wall and FRP system [5]. Similarly the application of some other of those strengthening system inheret disadvantages like time-consuming applications [1].

The objective of this research study is to develop a rapid method of strengthening URM walls using FRC & ECC, and to study the out of plane behavior of the strengthened walls. The paper presents and discusses results obtained from static testing of walls strengthened with different cement-based composites. To this end, an experimental program was carried out to test eight URM wall panels strengthened with FRS mixes. The results are used to assess the performance of each wall and compare with that of the control wall.

2. Experimental program

2.1 Test specimens

A total of Eight URM walls were built and tested in this research study. The walls were constructed using masonry infill units inside a precast reinforced concrete boundary frame. Figure 1 shows the dimensions and the reinforcement details of the walls and the boundary frame. The URM walls are built of ordinary hollow concrete blocks acquired locally and have a dimensions of 10 x 20 x 40 cm. The masonry units (hollow blocks) were made of plain concrete that had a compressive strength of 7.5 MPa. The overall dimension of the walls is 1.5x1.55 m including both the boundary frame and the masonry units. The net dimensions of URM walls without the boundary frame is 1.2 x 1.2 m with a thickness of 0.1 m.

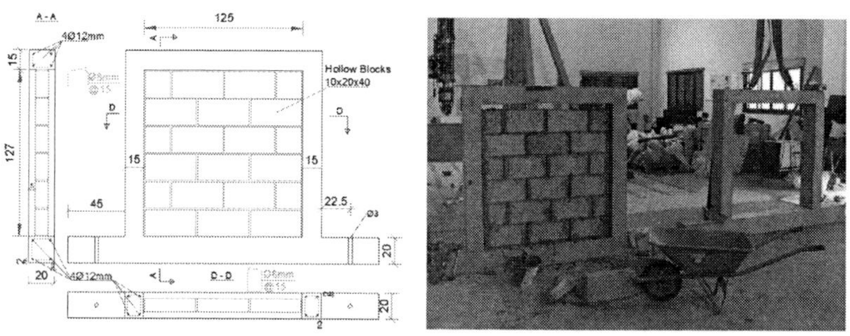

Figure 1. Schematic layout and photo of the test specimens (all the dimensions are in cm).

2.2. Strengthening materials and test setup

Three different types of concrete mixes were sprayed to the surface of the wall using shotcrete technology. First mix was a plain concrete mix (OPC), second mix was Fiber Reinforced Concrete (FRC), and the third mix was Engineerid Cementitious Composite (ECC). The mix proportions of these concrete mixes are presented in Table 1. Each mix was used to strengthen a pair of duplicate URM walls. A total of six walls were sprayed using shotcrete technology with concrete layers from both side with an average thickness of 45mm on each side as shown in Figure 2. In addition, two URM walls were plastered using normal cement mortar with a thickness of 15mm on each side, to present the traditional practice of building walls and assess the effectiveness of proposed strengthening method.

Table 1. Mix details of the concrete mixes used for strengthening the walls.

Mix Proportions	**OPC**	**FRC**	**ECC**
Type 1 Cement (kg)	420	420	1449
Silica Fume (kg)	25	25	145
Aggregates 0-5 mm (kg)	1480	1480	0
Dune Sand (kg)	345	345	0
Water (L)	190	190	478
Superplasticizer (L)	6	12.6	11.5
Macro-Fibers (kg)	0	6	0
Micro-Fibers (kg)	0	0	16

In the fiber reinforced concrete mix, synthetic macro-fibers (Strux^{TM} 90/40) were incorporated to the mix with a dosage of 6 kg/m^3. In the ECC mix a polyethylene micro-fibers commercially called (Spectra^{TM} 900) was used with a dosage rate of 16 kg/m^3. Properties of both the fibers are presented in Table 2. In the FRC and OPC mixes, the water to cement ratio (w/c) was 0.45 and type I Portland cement with a percentage of silica fume was used. The ECC mix had w/c ratio of 0.3 and it is a mixture of cementitious material, water, and fibers only. Adding fibers in the concrete mix reduce the workability of concrete. Therefore, to improve the workability without compromising the compressive strength, a locally available superplasticiser called ADVA Flow 480 was added to the concrete mixes. Table 3 shows the 28-day average compressive strength of the mixes measured according to ASTM C39 and flexural strength measured according to ASTM C78.

Table 2. Materials properties of nonmetallic fibers used in the shotcrete mixes.

Type	**Length** (mm)	**Diameter** (mm)	**Modulus of Elasticity** (GPa)	**Tensile Strength** (MPa)
Strux^{TM} 90/40	40	1.4 x 0.105 (rec)	9.5	620
Spectra^{TM} 900	12	0.039	66	2610

Table 3. Average compressive strength and flexural strength of the mixes

Mixes	**Avg Compressive Strength** (MPa)			**Avg Flexural Strength** (MPa)
	One day	**Seven days**	**28 days**	**28 days**
OPC	46	60	68	4.7
FRC	31	46	60	5.6
ECC	38	45	53	4.6

The walls are named with specific labels showing the type of shotcrete mix used for strengthening as reported in Table 4. For example, RF stands for the reference wall (wall with plastering mortar), PLS stands for the walls strengthened with plain shotcrete (OPC), FRS stands for fiber reinforced shotcrete, and ECCS which stands for engineered cementitious composites shotcrete, specify the walls sprayed with a pure cement-based mix reinforced with polyethylene micro-fibers.

Table 4. Strengthening Matrix.

Labels	Strengthening Mixes	Qty	Wall Dimensions			Shotcrete or Mortar Plaster Thickness (mm)	Total Thickness (mm)
			Height (mm)	Width (mm)	Thickness (mm)		
RF	Reference	2	1200	1200	100	15 (mortar plaster)	130
PLS	Control	2	1200	1200	100	45	190
FRS	FRC (6 kg/m^3)	2	1200	1200	100	45	190
ECCS	ECC(16 kg/m^3)	2	1200	1200	100	45	190

Figure 2. Photos of the walls before and after shotcrete application.

A test set up was designed to test the walls for out-of-plane action as shown in Figure 3. The load was applied at the center point of the walls in a displacement control mode with the rate of 0.01 mm/s. The applied load and wall deflection at different locations were measured. The load was measured using the load cell attached to the actuator head and the displacement was recorded using LVDTs deployed at different locations on the walls. Cracking patterns and failure mods were also monitored and documented.

Figure 3. Test setup with a wall during testing.

3. Experimental results

3.1. Load-deflection relationship

Figure 4 shows the load-deflection diagram of all the tested walls plotted together for ease in comparing the behavior of the walls. Each curve in Figure 4 presents the average test results obtained from testing two duplicate walls. However, and since the compressive strength of the FRS mix and ECCS mix was lower than the PLS mix, a direct comparison of the results will not reveal the real effect of macro-fibers and micro-fibers in enhancing the out-of-plane behavior. Therefore, the results of the FRS and ECCS walls were normalized using the ratio of the square root of their compressive strength relative to the plain concrete and then plotted in Figure 4.

It can be revealed from Figure 4 that the reference walls (RF) showed a linear behavior until the occurrence of the first crack at a load level equal to 20 kN and turns into a nonlinear behavior until it reached a failure load of around 45 kN and a central deflection of almost 25mm at peak load. Thereafter, the load carrying capacity drops and associated with punching failure at a load level of around 40 kN followed by a sudden drop in load deflection diagram. The load deflection curve in Figure 4 correspondent to the plain shotcrete wall (PLS), shows that the wall is behaving linearly until the load level of around 40 kN. With further increase in the applied load, the wall started exhibiting nonlinear behavior until the appearance of the first crack at load level around 70 kN after which it drops slightly. After the first drop, the wall started again resisting the applied load and eventually failed at load level around 130 kN with the central deflection of 30 mm at peak load. The loading was continued a bit more after failure until the wall resistance was degraded noticeably & dropped from peak to 100 kN, then the test was stopped.

Figure 4 also shows the load-deflection diagram of the walls sprayed with fiber reinforced shotcrete and it can be inferred from the figure that the FRS wall was behaving linearly until the onset of the first crack at a load level around 80 kN. After which a slight drop was noticed in the load resisting curve. However, after the first drop, the wall started resisting the applied load again and kept increasing. The FRS wall started failing at a peak load of 150 kN and central deflection of almost 30 mm. Similar to the PLS walls, the loading was stopped for the FRS walls after the applied load dropped to around 100 kN. The load-defllection relashionship for the ECCS walls plotted in Figure 4 shows that the load was constantly increasing without any significant drop even though the ECCS wall exhibited multiple distributed crack at different load levels. It can be noticed that the ECCS walls were continuously resisting the applied load until it reached its maximum capacity at a load level of around 170 kN with the corresponding central deflection of around 30mm. Thereafter the load resistance trend started to degrade until it dropped gradualy from peak load to a load level equal to 140 kN and then the test was stopped manually.

From comparison of the load-deflection relationship of all the tested walls, it can be realized that the walls strengthened with ECC mix followed by the wall sprayed with FRC mix, exhibited a significant increase in the load carrying capacity and ductility of the walls relative to the control and reference walls. Strengthening the walls with ECC shotcrete increased the load resisting capacity by upto almost 4 times than the RF walls. Similarily, the walls strengthened with FRC layers resisted around 3 times higher applied load compared to the RF walls. Comparison of the load deflection of the PLS walls with that of the FRS walls indicates that the addition of macro-fibers and micro-fibers further improved the out of plane behavior of the URM walls.

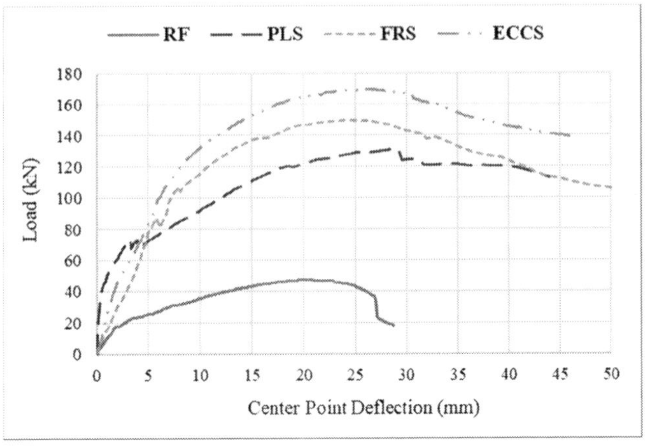

Figure 4. Normalized average load deflection diagram of the tested walls.

3.2. Failure modes

Failure modes of the tested walls with a sketch of cracking pattern and the load levels correspond to the initiation of the cracks are shown in Figure 5. As can be seen in the photos, all of the walls exhibited multiple cracking behaviors. Fiugre 5a shows that the reference (RF) walls developed a vertical crack at a load level around 16 kN, followed by a horizontal crack at 20 kN. With further increase in the

applied load, some diagonal cracks were also developed along with the expansion of the previous cracks before punching at around 40 kN. It can be seen in the photo of Figure 5a that a big chunk of the wall is in the spalling position due to the punching failure.
It can be seen in Figure 5b that the PLS walls exhibited major horizontal and vertical cracks at a load level of around 70 kN, followed by some smaller diagonal cracks before failure at load level around 130 kN. Similar cracking behavior was noticed in the FRS walls but with more distributed cracks (see Figure 5c), indicating more energy absorption and improved toughness of the FRS walls relative to the PLS walls. In the FRS walls, the cracks initiated at load level around 80 kN which can also be confirmed with a slight drop in Load deflection curve shown in Figure 4. Eventually the FRS walls failed due to the propagation of multiple cracks and after reaching a peak load of 150 kN.
The ECCS walls exhibited a diferent cracking pattern compared to the other walls. It can be seen in Figure 5d that a highly distributed cracks are developed on the surface of ECCS walls indicating a very high energy absorption behavior. Moreover, it can be seen that in contrary to the other walls, the ECCS walls exhibited a balanced cracking behavior (two similar major diagonal cracks) indicating a highly uniform distributed load resisting behavior. The ECCS walls resisted a maximum load of 170 kN after which the resistance of the wall started declining and the diagonal cracks were widely expanded resulting in dividing the wall to almost four equal triangles. In all the walls the crack width reduced after unloading, particularly this behavior was more pronounced in the walls with FRS and ECCS mixes suggesting existance of some residual strength.

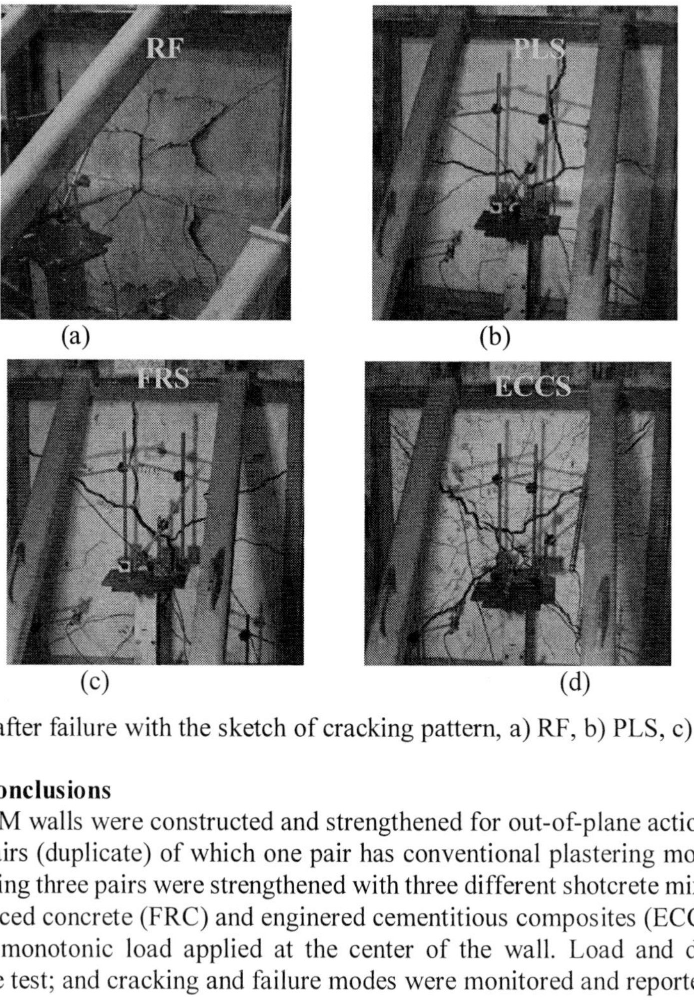

Figure 5. Walls after failure with the sketch of cracking pattern, a) RF, b) PLS, c) FRS, d) ECCS.

4. Summary and conclusions

A total of Eight URM walls were constructed and strengthened for out-of-plane action. The walls were divided into four pairs (duplicate) of which one pair has conventional plastering mortar applied to the surface. The remaining three pairs were strengthened with three different shotcrete mixes: plain concrete (OPC), fiber reinfroced concrete (FRC) and enginered cementitious composites (ECC). The walls were tested under static monotonic load applied at the center of the wall. Load and displacement were measured during the test; and cracking and failure modes were monitored and reported.

The test results revealed that the out-of-plain resistance, toughness, and ductility of the walls can be significantly enhanced with application of fiber reinfroced shotcrete layers. The use of ECC shotcrete layer increased the out of plane resistance of the walls by up to 4 times relative to the referecne wall with conventional plaster. Similarly, the use of FRC shotcrete increase the out of plane strength by up to 3 times relative to the reference wall. In summary, The results of the load-deflection and cracking behavior demonstrated that the use of synthetic macro-fibers and polyethelyne micro-fibers in the shotcrete mix can be an effective strengthening method to improve the out-of-plane response of URM walls.

5. References

[1] Maalej M, Lin V W J, Nguyen M P, Quek S T 2010 Engineered cementitious composites for effective strengthening of unreinforced masonry walls, *Eng. Struct.* **32**(8) 2432–2439
[2] Stephan C 2006 *Out-of-plane Seismic Performance Of Unrinforced Clay Brick Masonry Walls* MSc Thesis (The University Of British Colombia)
[3] Vaculik J 2012 *Unreinforced Masonry Walls Subjected To Out-of-plane Seismic Action* PhD Thesis (The University Of Adelaide)
[4] Rosa M, Francesca V, Garbin E, Panizza M 2014 Out-of-plane Behaviour of Infill Masonry Panels Strengthened with Composite Materials, *Mat. & Struc.* **47** 2131-2145
[5] Hamoush B S A, Mcginley M W, Mlakar P, Scott D, Murray K 2001 Out-of-plane Strengthening of Masonry Wall with Reinforced Composites, *J. of Comp for Const.* **5** 139-145
[6] Dizhur D, Griffith M, Ingham J 2014 Out-of-plane Strengthening of Unreinforced Masonry Walls Using Near Surface Mounted Fibre Reinforced Polymer Strips, *Eng. Struct.* **59** 330-343
[7] Willis C, Seracino R, Griffith M C 2010 Out-of-plane Strength of Brick Masonry Retrofitted with Horizontal NSM CFRP Strips, *Eng. Struct.* **32**(2)
[8] Triantafillou T C and Karlos Æ K 2007 Textile-reinforced mortar (TRM) versus FRP as Strengthening Material of URM Walls : In-Plane Cyclic Loading, *Mat. & Struc.* **40**(10) 1081-1097
[9] Galati N, Tumialan G, Nanni A 2006 Strengthening with FRP Bars of URM Walls Subject To Out-of-plane Loads *Constr. Build. Mater.* **20** 101-110
[10] Amiraslanzadeh R, Ikemoto T, Miyajima M 2012 A Comparative Study on Seismic Retrofitting Methods for Unreinforced Masonry Brick Walls *15th World Conf. Earthq. Eng.*, 2–10

6. Acknowledgments

The authors would like to acknowledge the funding provided by the sustainable construction materials and structural systems (SCMASS) research group. The research team would also like to appreciate the in kind support provided by the industrial partners including the GCP Applied Technologies, Emirates Stone Co., Gulf Rock Tunneling, and Siji Ready Mix Plant.

How fair is construction business in Australia? Perception of Consultant Quantity Surveyors

Benson Teck-Heng Lim[1*], **Rachel Miu**[1]

[1] Faculty of Built Environment, University of New South Wales Sydney, Kensington, NSW 2052, Australia
*Corresponding author: b.lim@unsw.edu.au

Abstract. Creating and managing a fair work environment is a crucial undertaking for management to increasing talent retention and improving organisational efficiency. Hitherto, little or no empirical study has done to investigate how construction employees perceive fairness at work and if their perceived fairness is positively associated with their emotion and pro-social behaviours. In addressing this gap, an online questionnaire survey was undertaken across 50 consultant quantity surveyors in the Australian construction industry to examine their perception of fairness at work and if their perceived fairness would associate with their emotion and behaviours. Overall, the findings suggest that there is a higher level of interpersonal fairness than distributive, procedural and information fairness in the Australian construction workplace. Furthermore, it is found that employees' perceived interpersonal, informational and procedural fairness have strong positive correlations with their level of confidence, satisfaction, trust, harmony, enthusiasm and stress within the company, and that, in presence of a trusting, harmonious and active working relationship employees are very likely to adopt righteous, sportsmanship and courteous behaviours. In summary, the findings inform company management the types of perceived fairness that are associated with employees' emotion and behaviours, and thus enable them to develop more targeted human resource management strategies to improve their organisational performance.

1. Introduction

Studies have shown that employees' behaviours will vary according to how they have been treated or how they perceive justice occurring in the workplace in terms of procedures (procedural fairness), rewards (distributive fairness), information exchange (informational fairness) and interpersonal treatment (interpersonal fairness) [1]. In the presence of fairness, employees are more likely to feel positive and engage in organisational citizenship behaviours (OCBs), and thereafter contribute to organisational effectiveness [2]. OCBs are employees' discretionary behaviours that help promoting the effective operation of an organisation [3].

Over the past few decades, there is an increasing number of reported cases about unfair practices in the global construction industry (see [4] for example). Some of examples unfair practices include: gender and racial discrimination, corruption, sham contracting, abuses of safety and human rights, discrimination and unfair business treatment and practices. Adding to the ongoing discourse of reducing unfairness in construction, Kadefors' [5] analysis of inter-organizational fairness across construction stakeholders has shown that an intuitive cost-based standard of fair pricing shapes interaction in Swedish construction projects, and that it is important for clients to design appropriate procurement strategies to developing mutual trust with contractors to gain their cooperative behaviour. More recently, Loosemore and Lim [6] explored the dimensions of fairness across project stakeholders in the Australia's construction industry and found that inter-organizational fairness exists in three forms (i.e. distributive and procedural fairness, informational fairness and interpersonal fairness) and that project participants experienced different level of injustice across projects of different nature and value. Most recently, Lim and Loosemore [7] studied the effect of perceived fairness of project participants on their OCBs and found that interpersonal fairness is a key ingredient in bringing about positive OCBs in construction projects and that project performance can be enhanced if project managers treat project participants with politeness, respect and dignity.

Hitherto, it appears that a considerable amount of research has documented different types of inter-organisational fairness and how perceived fairness could shape OCBs in construction projects. However, little has been done to investigate employees' perceived fairness of their compensation, relationships with peers and supervisors, and the decision-making processes within organisations, and how their perceived fairness, emotion and OCBs are interrelated. Although Loosemore and Lim's [8] analysis of intra-organisational fairness had offered insights into how employees perceived fairness at work, they did not consider the relationship between employees' perceived fairness, and their emotion and OCBs. This is an important gap in construction business management to explore since by better understanding the relationships between different dimensions of perceived intra-organisational justice and their association with employees' emotion and intention to engage in OBCs, more targeted human resource management strategies can be developed to improve organisational efficiency. Furthermore, hitherto, little or no emphasis has been placed on the fairness perception of quantity surveyors (QS). Thus, the aim of this research is to explore the perceived fairness of consultant QS and how their perceived fairness relates with their emotion and intention to engage in OCBs.

2. Literature Review

A considerable amount of mainstream organisational research has documented how employees perceive fairness differently and could engage in different types of mood and behaviours, and further sought if there are associations between perceived fairness, emotion and OCBs. Numerous theoretical positions have been taken to explain this phenomenon. For example, Organ [3] conceptualized the relationship between people's perceived fairness and OCBs by using the theories of social exchange and equity. From the social exchange perspective, the author highlighted that people define working relationships as forms of social and economic exchange and that if people judge the relationships as more social than economic in nature, then the quality of those relationships will influence their perception on other job-related matters, emotion and OCBs. From the economic aspect of social exchange theory, it is conceptualised that people judge fairness by assessing the association between the reward they received and their inputs. Upon perceiving inequitable rewards, people tend to feel sad and unmotivated and thus are more likely to display a lower level of OCBs.

It is widely accepted that there exists four main types of fairness in any organisation. Procedural fairness (PF) refers to employees' perceived fairness about the policies, processes and procedures through which decisions were made [9]. Distributive fairness (DF) refers to employees' perceived fairness about the rewards they received given their contributions [10]. Interpersonal (IPF) and informational fairness (ITF) refer to the degree to which employees had been treated with politeness, respect and dignity [11], and had been given adequate information and consulted during the enactment of any decisions [1], respectively. Watson et al.'s [12] PANAS model, which has been widely used by organisational psychologists, shows that people's emotion could be classified into positive affectivity (PA) and negative affectivity (NA). The former reflects the extent to which employees feel enthusiastic, active, alert and motivated, whereas the latter refers to subjective distress and unpleasurable engagement of employees that include ranges of aversive mood states such as anger, contempt, disgust, guilt, fear, and nervousness. Organ's [3] framework of OCBs show that people can engage in five dimensions of pro-social behaviours: conscientiousness (i.e. behaviours that go beyond those expected by specific role requirements); altruism (i.e. behaviours that help other members); civic virtue (i.e. behaviours that entail political participation); sportsmanship (i.e. behaviours that avoid complaining and/or aggravating unpleasant situations); and courtesy (i.e. behaviours that prevent problems from occurring).

3. Research Method and Data Analysis

An online survey questionnaire, containing four main sections, was developed in this study. In the first section, respondents were required to provide their background information such as age, job title, years of experience as a consultant quantity surveyor, average number of work hours per week, and salary bracket. In the second and third sections, they were asked to rate 38 and 15 statements relating to the perceived, PA, NA and OCBs, based on a seven-point Likert Scale, ranging from 1(Strongly disagree/very low) to 7 (Strongly disagree/very high). Lastly, an open-ended question was included, requesting respondents to highlight if they had encountered any other unfair treatments at work and if they had experienced different forms of moods when undertaking their work. The questionnaire was pretested, validated and amended before an industry-wide survey was undertaken. For the survey, probability sampling was employed whereby members of the Royal Institute of Chartered Surveyors in Australia and employees from a quantity surveying consulting company were invited to participate.

In this study, the relative prevalence indexing (RPI) method was adopted to facilitate the relative comparisons of items relating to employees' perceived fairness. The RPI method was preferred over the arithmetic average method because the former can derive relative indices within the range of 0-1 for each item and therefore enable researchers to undertake relative comparisons of items (following the technique adopted by Loosemore and Lim [8]).

Furthermore, the One-sample Wilcoxon Signed Rank Test was performed to determine if the median values of the sample were significantly different to the test median value of 4 (i.e., the midpoint of the 7-point Likert scale). This helps ascertain if there was significant consensus among the respondents on their perceived fairness. Lastly, the Spearman's rho correlation test was undertaken to examine the relationship between employees' perceived fairness, PA, NA and OCBs. For ease of interpretation, Hinkle et al.'s [13] guideline on the magnitude of association was adopted: 0.00 to 0.30 as negligible correlation; 0.30 to 0.50 as low level of correlation; 0.50 to 0.70 as moderate; 0.70 to 0.90 as high; and 0.90 to 1.00 as very high.

4. Results and Discussion

In this study, a total of 50 valid responses were collected. Of these respondents, about 70% (34) of them were males, 50% were aged from 20 to 30 years old, and about 80% (40) of the respondents held at least a bachelor degree. In terms of their job and role, about 40% of the respondents were cadets and junior QS, 20% were senior QS, and the rest were associate and above. Also, most of the respondents had worked for 45-50 hours a week (32%) and had at least 10 years of relevant work experience (50%) and had a salary of at least $100,000 (57%).

Table 1 summarises our respondents' perceived fairness. The overall results show a moderate level of optimism on how respondents perceived fairness at work. Interestingly, the respondents had shown higher positive reception of interpersonal fairness (IPF), than procedural (PF), distributive (DF) and information fairness (ITF), as characterised by the corresponding RPI values of 0.71, 0.68, 0.67 and 0.65.

Table 1. Perceived fairness at work

Item code	Statements	RPI	Average RPI	Median	One sample Wilcoxon Rank Test
I feel that my overall entitlements at work are fair considering					
DF1	the amount of training and education I have had.	0.69	0.67	5.00	3.695**
DF2	the amount of experience I have.	0.69		5.00	3.398**
DF3	my responsibilities.	0.69		5.00	3.514**
DF4	the number of projects/ jobs/ tasks completed.	0.65		5.00	2.711**
DF5	the effort I've applied to work.	0.66		5.00	2.579*
DF6	the level of stress in my job.	0.66		5.00	2.654**
DF7	what I have contributed to the organisation.	0.69		5.00	3.525**
DF8	my overall performance.	0.65		5.00	2.633**
I feel that:					
PF1	the company upholds my ethical and moral standards when making decisions.	0.72	0.68	5.00	4.125**
PF2	my performance is measured objectively.	0.67		5.00	3.039**
PF3	there is no prejudice in the decision-making process.	0.69		5.00	3.481**
PF4	the company has fair procedures and guidelines.	0.72		5.00	4.330**
PF5	the company makes decisions that align with organisational ethics and values.	0.71		5.00	3.809**
PF6	the company is consistent with disciplinary procedures.	0.66		4.00	2.913**
PF7	the company conducts disciplinary procedures in a timely manner	0.64		4.00	2.414*
PF8	the company fairly implements disciplinary measures.	0.64		4.00	2.383*
PF9	my supervisor has fair procedures and guidelines.	0.71		5.00	4.048**
PF10	my supervisor is consistent with disciplinary procedures.	0.67		4.00	3.306**
PF11	my supervisor conducts disciplinary procedures in a timely manner.	0.68		4.00	3.606**
PF12	my supervisor fairly implements disciplinary measures.	0.67		4.00	3.391**
IPF1	I have been treated equally to others in the same position.	0.69	0.71	5.00	3.463**
IPF2	the company treats staff politely during decision making processes.	0.67		5.00	2.926**
IPF3	my supervisor treats me with dignity and respect.	0.80		6.00	5.397**
IPF4	my supervisor treats me with kindness and consideration.	0.77		5.50	5.065**
IPF5	my supervisor shows concern for my rights	0.69		5.00	3.379**

	as an employee.				
IPF6	my supervisor is completely candid and frank with me.	0.69		5.00	3.269**
IPF7	my supervisor can supress personal biases	0.68		4.50	3.080**
IPF8	my supervisor has made it clear that I am a valuable employee.	0.67		4.50	2.917**
ITF1	all employee concerns are considered in the decision-making processes.	0.60		4.00	0.71
ITF2	I have input in the decision-making process.	0.63		5.00	2.012*
ITF3	the company provides additional information about job decisions when requested.	0.67		5.00	3.112**
ITF4	my performance review is based on complete and accurate information.	0.63		4.50	2.044*
ITF5	the company clearly provides information about disciplinary procedures.	0.68		5.00	3.242**
ITF6	the company has provided opportunities to challenge and/ or reverse decisions.	0.62		4.00	1.61
ITF7	my supervisor makes decisions openly so that everyone always knows what is going on.	0.65		5.00	2.476*
ITF8	my supervisor keeps me informed regardless if the outcome of decisions is good or bad.	0.67	0.65	4.00	2.624**
ITF9	my supervisor considers my viewpoints in important decision-making processes.	0.69		5.00	3.359**
ITF10	my supervisor customises communications to individuals' specific needs.	0.67		4.50	3.064**

The results show that most respondents were particularly appreciative of how their supervisors had treated them with dignity and respect (IPF3) and kindness and consideration (IPF4). This tends to support Loosemore and Lim's [6] conclusion that interpersonal justice is high within the Australian construction industry. However, a handful of respondents had shown high level of grievance on their company, believing that: there were some forms of inequality of treatment among employees (IPF1), and employees were not treated politely during decision making process (IPF2). They further felt that their supervisors were not: completely candid and frank (IPF6) telling them of their values to the company (IPF8); and concerned of their rights (IPF5) and able to suppress personal biases (IPF7). When referring to decision making processes, most respondents felt that the procedures and guidelines adopted by their company and supervisor were fair (PF4 and 9) and ethical (PF1 and 5). However, they felt that their supervisors (PF10 – 12) were much fairer and more consistent and effective than their companies (PF6 –8) when undertaking disciplinary measures and procedures. This further points to Folger and Cropanzano's [14] assertion that the fairness of interpersonal treatment is closely associated with supervisors' sentiment and that employees are more tolerant towards the behaviors of their supervisors than their management. Furthermore, it is notable that respondents had shown higher positive reception of their overall entitlement when reflecting on their education, responsibilities, experience and contribution (DF1-3, and 7) than on the amount of work, stress and efforts that they had undertaken in their role (DF4-6, 8). Lastly, the results indicate that most respondents were positive in the way how their opinions were considered, and information was provided in the decision-making processes (ITF3, 5, and 7-10). However, a handful of the respondents were less impressed with how their performance review was undertaken (ITF4) and their company dealt with employees' feedback and challenges to decisions made (ITF2 and 6).

Adding to the above, the Spearman correlation results indicate that there exists a considerable degree of correlations between respondents' perceived fairness, PA, NA and OCBs. Particularly, it is found that respondents' perceived interpersonal fairness (i.e. IPF2 –8), informational (i.e. ITF1, 3 and 7-10) and procedural fairness (i.e. PF1-5, and 9) have moderate to strong level of correlations with their positive affectivity of 'Confidence' (PA1), 'Satisfaction' (PA2), Trust' (PA3), 'Harmony' (PA4), and 'Enthusiasm'(PA5). Further to this, it is notable that respondents' positive feeling of trust, harmony and enthusiasm strongly correlate with their righteous (OCB2), sportsmanship (OCB4) and courteous behaviors (OCB5). This means that when there exists a strong feeling of trusting, harmonious and active working relationship, employees are very likely to adopt righteous behaviors, striving to reduce aggravation of unpleasant situations and occurrence of problems.

On the other hand, some interesting trends were noted on the low level of correlations between respondents' perceived fairness and negative affectivity, characterised by 'Anger' (NA1), 'Conflict' (NA2), 'Frustration' (NA3), 'Hostility' (NA4), and 'Stress' (NA5). Generally, there was no significant correlation between perceived fairness, NA2 and NA3. The results however show that NA1 negatively correlate with DF5, PF4, PF5, IPF5, IPF8, ITF7 and ITF8. This means that employees are less likely to get annoyed when: they have been rewarded appropriately for their efforts given; their company has put in place fair and ethical decision-making procedures and guidelines; and their supervisor has shown greater concern and appreciation of their rights and contribution to the company. Furthermore, to reduce employees' hostility, our results point to the importance of management to place greater emphasis on the distribution of employees' entitlement (DF2, 5, 6 and 8), and the quality of interpersonal relationship (IPF5 – 8) and informational exchange (ITF6, 7, 9 and 10) between supervisors and employees. This tends to support Skarlicis and Folger's [15] conclusion that supervisors play an important role driving employees' tolerance of unjust treatment. Interestingly, the findings also point to the positive correlations between respondents' stress and perceived interpersonal (IPF2 – 6 and 8) and informational fairness (ITF1, 2, 4, 5 and 7-9). An overall picture that emerges from here is that, for effective employee management of employees, a transparent and controlled communication mechanism needs to be put in place as information overload and candid communication could bring about spill-over effect and unnecessary stress on employees.

5. Conclusion

This exploratory study surveyed 50 consultant quantity surveyors in Australia to examine their perception of fairness at work and determine if their perceived fairness would associate with their emotion and organisational citizenship behaviours (OCB). The overall findings show that respondents were more optimistic of interpersonal fairness than procedural, distributive and information fairness at work, and furthermore, they were generally more content with the behaviours of their supervisors than those of their company. Our results thus tend to support the social exchange theory that people's emotions and behaviours are more likely to be associated with the quality of interpersonal relationships and informational exchange between employees rather than the way how rewards are distributed and decision-making processes are undertaken. This further points to the importance of good communication and relationship management to building employees' loyalty for organisational effectiveness.

It is acknowledged that the findings are indicative and not conclusive mainly due to the exploratory nature of this study and the small sample size of 50 consultant quantity surveyors. However, the findings could add to the existing knowledge of intra-organisational fairness for future studies, whereby a larger sample size, covering different groups of construction workforces, should be collected for generalisation purpose. Further predictive modelling should be conducted to examine the effect of different roles and ethnic backgrounds on perceived fairness, hence their emotion and OCBs. A comparative investigation of perceived intra-organisational fairness across different countries should also be considered in future studies towards identifying exemplar of fair work practices.

6. References

[1] Colquitt J A 2001 On the dimensionality of organizational justice: a construct validation of a measure *Journal of Applied Psychology* **86**(3) 386-400

[2] Podsakoff P M Ahearne B, MacKenzie S B 1997 Organizational citizenship behavior and the quantity and quality of work group performance *Journal of Applied Psychology* **82**(2) 262-270.

[3] Organ D W 1988 *Organizational Citizenship Behavior: The Good Soldier Syndrome* Lexington Books Lexington

[4] Royal Commission into the Building and Construction Industry (RCBCI) 2003 *Final Report, Royal Commission into the Building and Construction Industry* Melbourne Victoria Australia

[5] Kadefors A 2005 Fairness in inter-organizational project relations: norms and strategies, *Construction Management and Economics* **23**(8) 871-878

[6] Loosemore M and Lim B T H 2015 Inter-organisational unfairness in the construction industry, *Construction Management and Economics* **33**(4) 310–326

[7] Lim B T H and Loosemore M 2017 The effect of inter-organizational justice perceptions on organizational citizenship behaviors in construction projects, *International Journal of Project Management* **35**(2) 95–106

[8] Loosemore M and Lim B T H 2016 Intra-organisational injustice in the construction industry. *Engineering, Construction and Architectural Management* **23**(4), 428-447

[9] Thibaut J and Walker I 1975 *Procedural Justice: A Psychological Analysis* Erlbaum Publishing New Jersey

[10] Folger R and Konovsky M A 1989 Effects of procedural and distributive justice on reactions to pay raise decisions *Academy of Management Journal* **32**(1) 115-130

[11] Bies R J and Moag J F 1986 Interactional justice: Communication criteria of fairness In: Lewicki R J, Sheppard B H, Bazerman M H (eds) *Research on Negotiations in Organizations* **1** JAI Press London pp 43–55

[12] Watson D, Clark L A and Tellegen 1988 A Development and validation of brief measures of positive and negative affect: the PANAS scales *Journal of Personality and Social Psychology* 54(6) 1063-1070

[13] Hinkle D E, Wiersma W and Jurs S G 2003 *Applied Statistics for the Behavioral Sciences* 5th ed. Boston: Houghton Mifflin

[14] Folger R and Cropanzano R 1998 *Organizational Justice and Human Resource Management* Sage California

[15] Skarlicki D P and Folger R 1997 Retaliation in the workplace: the roles of distributive, procedural and interactional justice *Journal of Applied Psychology* **82**(3) 434-443

Conceptualising eco-innovation practices in contractor firms – the dynamic capability approaches

Siti Sarah Mat Isa[1*], Nazirah Zainul Abidin[1] and Izyan Yahaya[1]
[1] School of Housing, Building and Planning, Universiti Sains Malaysia, 11800 USM Penang, Malaysia
*Corresponding author: ctsarahmatisa@gmail.com

Abstract. The world has witnessed extreme environmental deterioration over the past few decades due to rapid development in urban areas and unsustainable way of past construction. Recently, eco-innovation (EI) has increasingly been accepted as the solution to major environmental problems. EI involves new or modified processes, techniques, products, systems, and services that can satisfy human needs and provide quality of life whilst significantly decreases environmental impacts. In addition, EI revolutionary path to improve contractor firms' competitiveness and sustaining business endeavour. Thus, the aim of this paper is to develop EI framework practices in contractor firms. Unfortunately, to be eco-innovative, the contractor firms need to strengthen their green knowledge, technology and resources as well as to develop their capabilities and strategies faster than their rivals. They also need to absorb and capitalise any external opportunities for their own advantages. Therefore, this study adopts dynamic capability theory to investigate a firm's ability to integrate, build and reconfigure its competences and resources to accomplish the environmental goal in its business strategy. The outcome of this paper could provide clear directions of EI practices by means of dynamic capability theory in demonstrating how contractor firms can adapt to changing current industry conditions to remain competitive.

1. Introduction

The construction industry in Malaysia is an essential sector in contributing 3% to 7% in Gross Domestic Product (GDP) over the past decade and had played an important function in the country's development by supporting other sectors to expand [1]. Since the 1980s, Malaysia has been driven towards industrialisation [2] and targeted to become a high-income developed nation by 2020 [3]. Despite the rapid advancement in the construction industry, most of the construction projects in Malaysia are still applying the traditional methods which are no longer sustainable. The traditional construction is actually going nowhere without innovation [4]. The construction industry is generally recognised as a major contributor towards environmental degradation, global warming, a massive producer of waste, excessive resource consumption and the building's operation also contribute to greenhouse gases (GHGs) through its lifecycle [3][5]. Thus, the construction industry faces difficulty and an urgent change to a greener mode of overall operations and management with the use of green technology and innovation in the construction project.

Although, the concept of green development along with its advantages has been widely agreed and accepted, however, numerous studies [6-7], revealed that this concept has not been widely applied and the implementation is still below than expectation. Most of the innovation in construction were not aligned with the green obligation whereby the environmental impact is often neglected [6][8]. Hence, new environmental-friendly construction process, as well as improvement in building design and management practices, are needed to attain greater respect for the environment. These objectives can be accomplished by encouraging contractor firms to implement EI. In light of providing green development, EI has emerged as a new paradigm for contractor firms seeking to shift from traditional construction practices that are inefficient, slow, and labour-intensive towards better green practices [3-5]. In addition, EI is increasingly been accepted as one of the best initiatives, considering that its core objectives are directing towards sustainable development and environmental sustainability [9]. Unfortunately, the contractor firms are still unable to blend EI practices which focus on the environmental requirement in their business strategies. Thus, they need to have sufficient capabilities to absorb and transform green knowledge, technology and innovation to improve the construction process. Refining this practice, dynamic capability theory demonstrates how a contractor firm ability to integrate, build and reconfigure its competences and resources to accomplish the environmental goal in its business strategy.

2. Literature review

2.1. The concept of eco-innovation

The term EI can be interchangeably used with green innovation, environmental innovation or sustainable innovation by a different scholar [9-11]. Organisation for Economic Cooperation, and Development (OECD) described EI as *"the creation or implementation of new, or significantly improved, products (goods and services), processes, marketing methods, organisational structures and institutional arrangements which - with or without intent - lead to environmental improvements compared to relevant alternatives"* [12]. Meanwhile, the Eco-Innovation Observatory [13] defined EI as *"the introduction of any new or significantly improved product, process, organisational change or marketing solution that reduces the use of natural resources and decreases the release of harmful substances across the whole life-cycle."* Thus, from the definitions, this innovation differs from general innovations since EI result in producing quality products in both economic and environmental benefits. EI can also move towards more environmental-friendly production processes and services, indirectly able to reduce the GHGs and to more efficient use of various resources. The core element of EI definition is focused on the positive environmental impact towards innovation.

Rennings [14] clarified the term EI addressing explicitly three kinds of changes towards sustainable development named, technological, social and institutional innovation. Meanwhile, Reid & Miedzinski [15] asserted that the EI concept is part of the sustainable development strategy, whereby, EI is the development of products, processes, services and technologies innovation that can make better use of the natural resources and reduce the ecological footprint to the environment. Charter & Clark [16] expressed that EI is a process where sustainability considerations in environmental, social and financial are integrated into firm systems from idea generation through research and development (R&D) and commercialisation which applies to products, services and technologies, as well as new business models. Other industries had improved their environmental performance after enhancing their EI orientation [17-19]. In developed countries, EI is one of the leading strategies to promote resource and energy efficiency and create a low carbon society [12-13][15,19]. In Malaysia, most of the EI practices had been discussed in the chemical sector, the automotive sector, and green technology companies [20-24]. However, in the construction sector, it is barely limited. Slaughter [25] and Fang et al. [26] concluded that successful EI implementation in construction able to contribute in reducing project duration and cost by simplifying construction work, improved quality, increased efficiency and productivity, which ultimately competitiveness and environmental-friendly. In addition, EI is significant in order to reduce the materials consumption and labour usage as well as reducing the carbon footprint on the environment [27-28].

2.2. The implementation of eco-innovation within the contractor firms

EI has been viewed as an effective way for contractor firms to avoid or reduce environmental damages but at the same time still remaining sufficiently profitable [29]. EI practices are crucial to being embedded within the contractor firm as they are responsible for delivering successful completion of a construction project. Thus, contractor plays a mediator role in the interface between the organisations that develop many of the new products and processes such as materials and components suppliers, specialist consultants and sub-contractors, and those which adopt these innovations in their project [25][28]. In addition, as indicated by Xue et al. [28], EI is critical for contractor firms' competitive advantages and at the same time able to modernise the industry. For those reasons, the contractors need to equip themselves not only with innovative knowledge but to combine it with green knowledge and technology. Hence, it is the liability of contractor in connecting the vision of EI between the client, the suppliers, consultants and others project stakeholders in order to eliminate or reduce the impact of construction activities on the natural environment.

2.3. Dynamic capability approach

The innovation literature focusing on EI has shown that increasing investments in EI were influenced by the capabilities of the firms and to firm competitiveness [18]. Firm capabilities have been defined as a firm's collective physical facilities, skills of employees and firm capacity to deploy its assets, tangible or intangible to perform a task or activity to improve performance [30]. Winter [31] viewed dynamic capability as the process of extending, modifying, or creating new capabilities. The key differences between ordinary capabilities and those that are dynamic are that dynamic capabilities are linked with change and more specifically, changing the resource base of a firm [32-33].

The dynamic capability approach is particularly relevant nowadays when global competitive forces are changing landscapes of the construction industry. Thus, firms in this industry must have timely strategies, versatile infrastructures, and an ability to utilise its competences and resources in an innovative way [34]. The effective use of a firm's capabilities defines the firm's overall performance [23]. In practising EI, Xavier et al., [9] claimed that a firm need direction on how to apply their capabilities in a systematic manner in order to achieve environmental goals and maintain continual enhancement in the environmental performance of services, products, and processes. Explicitly, management innovation alone is inadequate for generating success without the dynamic capabilities of a firm to create, extend or modify its resource base [32].

2.4. Eco-innovation components for contractor

A considerable amount of literature comprehends the notion of EI practices is broader than merely reducing environmental impacts through waste minimisation [34]. EI involved utterly new products which a net reduction in environmental impact as well as products that can satisfy human needs and improve the quality of life. Cheng et al. [35] highlighted that EI involved in the development and introduction of new products and process with the use of green technology innovation. Whereas process EI refers to the improvement of existing production processes or the addition of new processes to reduce environmental impact [35].

In the construction industry, Tatum [36] observed that process innovation is an improvement in construction methods that are designed or developed for the accomplishment of usual construction operations or the improvement in the efficiency of the construction process. Suprun & Stewart [37] also viewed the process EI as the introduction of new design and modern method of construction. Meanwhile, organisational EI refers to upgrading the organisation's management processes through a new and eco-method in business practices. Organisational EI generally does not reduce environmental impacts directly but facilitates the implementation of process and product EI [38]. All these EI practices aforementioned are able to change the paradigm of the indigenous construction industry [6, 14][21]. The clarification of each practice needs to clearly explain to allow a creative and thorough transformation of a contractor firm in improving the EI practices that able to improve the firm performance.

3. Conceptual framework

The conceptual framework of this study adopts the concept of dynamic capability theory. According to Teece et al. [30][39], the concept of dynamic capability theory consists, the capacity to sense external opportunities and challenges, seize them and maintain competitiveness through reconfiguring or transforming the business firm's intangible and tangible capabilities. Many studies were done in integrating EI with business strategy for instance in automotive, chemical, electronic sector and many more [17-24][40]. Unfortunately, very few research has done in the construction sector. Whereas, the concept of EI has the potential to be introduced and applied in individual contractor firms to promote sustainable development. In fact, many developed countries have integrated the EI concept in every process of their construction development [12-13][15].

Understanding the current status of a firm is the first crucial steps. Garcés-Ayerbe et al. [40] and Adam [41] stated that, in order to comprehend how resources can be created or altered to address changes, it becomes compulsory to look at the firm's past and how its history and the actions it has implemented may have influenced its current practice. Thus, as this paper looks into a contractor firm, technically, the need to understand what is the EI level of implementation within their firm is indispensable. Nevertheless, most of the contractor firms still unable to blend the suitable EI practices in embedding the environmental requirement in their business strategy due to a misconception of EI term. Therefore, there is a necessity for a clear understanding of EI components and its integration within the construction process to effectively satisfy environmental and economic performance.

The contractor firm has operational capabilities and resources that are involved directly in the EI process by converting inputs into outputs. In the context of contractor firm, activities of sensing involve scanning, learning and interpretation of new green knowledge and technology that is able to improve the process, product and management of construction project towards embracing sustainability. It typically requires investments in R&D works and other related innovation activities [39]. In green development, the contractor firm, sensing basically is about understanding the environmental requirement and their challenges. Meanwhile, seizing refers to the determining and implementing of new opportunities that have been identified as conducive to competitive advantages [39][41] which consists of mobilising resources in order to address an opportunity and capture its value. Lastly, transforming is the activity by which the firms continuously reconfigures its resource base by adjusting internal and/or external resources and operating capabilities as needed to transform the firm's strategies towards embracing the EI practices [39][42]. For examples, one of the challenges contractors may face in venturing to a green development project is lack of knowledgeable and skilled human resources. Because of these lacking, the contractor is able to plan and to seize opportunities on how to improve their human resources by having the training to ensure them to have the adequate environmental knowledge, able to learn and undertake new environmental practices. As a result, the contractor is able to transform their human resources towards embracing green practices.

Therefore, this study adopts a dynamic capability theory to develop a framework for EI practices in contractor firms. EI which has been linked to a business strategy can enhance a firm's capabilities and reduce negative impacts on the environment during the construction process [24]. This framework will take into consideration on the current dependency of the contractor firms, examines the EI components and the capabilities of the contractor firms towards developing the EI strategies through dynamic capability theory, with respect to the activities of sensing, seizing and transforming. This EI framework perhaps would be beneficial to contractor firms to remain competitive and success especially in complex, large-scale management of green development projects based on Malaysian context.

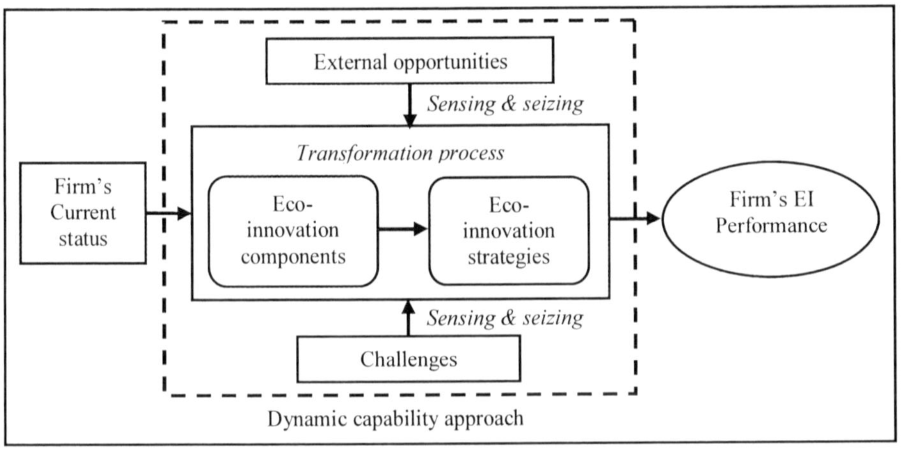

Figure 1. A conceptual framework of eco-innovation practices in contractor firms; adapted and modified from [30, 32, 39, 41,]

4. Conclusion

This study assists contractor firms to understand the concept of EI in improving the current construction practices by enhancing resource efficiency by learning the implications of the EI, which is linked to sustainable firm performance. It is increasingly impossible to deny that the current construction market environment is complex and uncertain coupled with client and customer pressure and government regulation towards embracing environmental sustainability. Thus, under this condition, the survival and development of the contractor firm are critical. The contractor firm must adapt and lead the external environment changes to make itself in an invincible position by means of EI practices through dynamic capability theory to be competitive and survive in local and global competition.

5.0 References

[1] CIDB Malaysia, Country Report *21st Asia Construct Conference* 2016 Tokyo Japan

[2] Kamar K A M and Hamid Z A 2011 Sustainable construction and green building: The case of Malaysia *WIT Trans. Ecol. Environ.* **167** 15–22

[3] Hamid Z A, Ali M C, Anuar K, Kamar M, Zura M and Zain M 2012 Towards a sustainable and green construction in Malaysia *Malaysian Constr. Res. J.* **11**(2) 55-64

[4] Kamal E M, Yusof N and Iranmanesh M 2016 Innovation creation, innovation adoption, and firm characteristics in the construction industry *J. Sci. Technol. Policy Manag.* **7**(1) 43–57.

[5] Shi Q, Zuo J, Huang R, Huang J, and Pullen S 2013 Identifying the critical factors for green construction - An empirical study in China, Habitat Int., **40** 1–8

[6] Mohamad Bohari A A, Skitmore M, Xia B, Teo M, Zhang X, and Adham K N 2015 The path towards greening the Malaysian construction industry *Renew. Sustain. Energy Rev.* **52** 1742–1748

[7] Bamgbade J A, Kamaruddeen A M , Nawi M N M 2017 Towards environmental sustainability adoption in construction firms : Empirical analysis of market orientation & organizational innovativeness impacts *Sustain. Cities Soc.* **32** 486–495

[8] Zhiwei Y, Chen L, and San B 2014 Application of Green Construction Technology in Construction Projects *Smart Construction and Management* 389–397

[9] Xavier A F, Naveiro R M, Aoussat A and Reyes T 2017 Systematic literature review of eco-innovation models: Opportunities and recommendations for future research *J. Clean. Prod.* **149** 1278–1302.

[10] Schiederig T, Tietze F, and Herstatt C 2011 What is Green Innovation? – A quantitative literature review *The XXII ISPIM Conference*

[11] Carrillo-Hermosilla J, Del Río P, and Könnölä T 2010 Diversity of eco-innovations: Reflections from selected case studies *J. Clean. Prod.* **18**(10) 1073–1083.

[12] OECD 2009 *Sustainable Manufacturing Eco-Innovation: Framework, Practices & Measurement* OECD Paris

[13] Eco-Innovation Observatory (EIO) 2013 Europe in transition. *Paving the way to a green economy through eco-innovation* European Commission Brussels

[14] Rennings K 2000 Redefining innovation - Eco-innovation research and the contribution from ecological economics *Ecol. Econ.* **32**(2) 319–332

[15] Reid A. and Miedzinski M 2008 *Eco-Innovation. Final Report for Sectoral Innovation Watch*, ()

[16] Charter M and Clark T 2007 *Sustainable Innovation: Key Conclusions from Sustainable Innovation Conferences 2003-2006 organised by the centre for sustainable design* Uni. College for the Creative Arts

[17] Arnold M G and Hockerts K 2011 The greening Dutchman: Philips' process of green flagging to drive sustainable innovations *Bus. Strategy Environ.* **20** (6) 394-407

[18] Kiefer C P, Carrillo-Hermosilla J, Del Río P, and Callealta Barroso F J 2017 Diversity of eco-innovations: A quantitative approach *J. Clean. Prod.* **166** 1494–1506

[19] Eryigit N and Özcüre G 2015 Eco-Innovation as modern era strategy of companies in developing countries : Comparison between Turkey & European Union *Procedia - Soc. Behav. Sci.* **195** 1216–1225

[20] Keshminder J S 2017 *Eco-Innovation in the Chemical Manufacturing Industry : Status and Its Determinants* PhD Thesis Uni. Malaya

[21] Rashid N Jabar J Yahya S and Shami S 2015 Dynamic eco-innovation practices : A systematic review of state of the art and future direction for eco-innovation study *Asian Soc. Sci.* **11**(1) 8–21

[22] Conding J, Zubir A F M., Hashim S, Lanang N A S and Habidin N F 2013 The investigation of green practices, green innovation and green performance in Malaysian automotive industry *Environ. Manag. Sustain. Dev.* **2**(1)

[23] Fernando Y, Wah W X and. Shaharudin M. S 2016 Does a firm's innovation category matter in practising eco-innovation? Evidence from the lens of Malaysia companies practising green technology *J. Manuf. Technol. Manag.* **27**(2) 208–233

[24] Fernando Y and Wah W X 2017 The impact of eco-innovation drivers on environmental performance : Empirical results from the green technology sector in Malaysia Sustain. *Prod. Consum.* **12** 27–43

[25] Slaughter E, Sarah 2000 Implementation of construction innovations *Bldg Research & Infor* **28**(1) 2-17

[26] Fang C Y 2015, *Mapping Innovation in the Construction Sector: A Study of National Firms in Malaysia* PhD Thesis Uni. Malaya

[27] Miao C, Fang D, Sun L and Luo Q 2017 Natural resources utilization efficiency under the influence of green technological innovation *Resour. Conserv. Rec.* **126**,153–161.

[28] Xue X, Zhang R, Yang R and Dai J 2014 Innovation in construction: A critical review and future research *Int. J. Innov. Sci.* **6**(2) 111–126

[29] Li Y 2014, Environmental innovation practices and performance : moderating effect of resource commitment *J. Clean. Prod.* **66** 450–458

[30] Teece D J, Pisano G and Shuen A 1997 Dynamic capabilities and strategic management *Strateg. Manag. J.* **18**(7) 509–533

[31] Winter SG 2003 Understanding dynamic capabilities *Stra. Manag. J.* **24**(10) 991–995

[32] Lin H F, Su J Q, and Higgins A 2016 How dynamic capabilities affect adoption of management innovations *J. Bus. Res.* **69**(2) 862–876
[33] Abu Bakar J L and Ahmad H 2010 Assessing the relationship between firm resources and product innovation performance: A resource-based view *Bus. Process. Manag. J.* **16**(3) 420–435
[34] Hellström T 2007 Dimensions of environmentally sustainable Innovation: The structure of eco-innovation concepts, Sustain. Dev. **15**(3) 148–159.
[35] Cheng C C J, Yang C L, and Sheu C 2014, The link between eco-innovation and business performance: A Taiwanese industry context *J. Clean. Prod.* **64** 81–90
[36] Tatum C. B. 1990 Organizing to increase innovation in construction firm *J. Constr. Eng. Mng.* **115**(4) 602–617.
[37] Suprun E V and Stewart R A 2015 Construction innovation diffusion in the Russian Federation, Constr. Innov. **15**(3) 278–312.
[38] Murphy J and Gouldson A 2000 Environmental policy & industrial innovation : Integrating environment and economy through ecological modernisation **31** 33–44.
[39] Teece D J 2007 Explicating dynamic capabilities: the nature and microfoundations of (sustainable) enterprise performance *Strategy. Manag. J.* **28** 1319–1350
[40] Garcés-Ayerbe C, Scarpellini S, Valero-Gil J and Rivera-Torres P 2016 Proactive environmental strategy development: from laggard to eco-innovative firms *J. Org. Chang. Manag.* **29**(7), 1118–1134
[41] Adam A 2016 *Managing construction challenges : Viability of a dynamic capabilities approach for the public client* PhD Thesis Chalmers Uni. of Technology
[42] O'Reilly C and Tushman M 2008 Ambidexterity as a dynamic capability: Resolving the innovator's dilemma *Research in organizational behavior* **28** 185-206.

Acknowledgement
The author would like to acknowledge the Ministry of Education Malaysia for providing financial support and Universiti Sains Malaysia for the physical support.

Compressive Strength and Infiltration Characteristic of Pervious Concrete Using Recycled Concrete Aggregate

Nurul Aishah Abd Rahman[1*], Muhammad Aiman Muhamad Shah [2], Noor Safwan Muhamad [1], Ahmad Amzari Yaccob [1] and Libriati Zardasti[2]

[1]Faculty of Civil Engineering, Universiti Teknologi MARA Pahang, 26400, Bandar Jengka, Pahang, Malaysia
[2]School of Civil Engineering, Universiti Teknologi Malaysia, 81310, Johor Bahru, Johor, Malaysia
*Corresponding author: naishah.ar@pahang.uitm.edu.my

Abstract. In construction industry, rapid development will increase the demand versus supply which leads to increasing the production of construction waste. Recycle Concrete Aggregate (RCA) was proposed as the one of the component in the pervious concrete for sustainability in the construction industry to reduce the wastage. The pervious concrete was function as a medium to allow the water penetrate through porosity of the concrete and avoiding water from pooling at the surface. Therefore, this study is conducted to investigate the properties of the pervious concrete by infiltration test and compression strength test using Recycle Concrete Aggregate (RCA) and Natural Aggregate (NA) as the main composition. The cement aggregate ratio of 1:3 and 1:5 were choosing as variable to produce the workability pervious concrete that similar to conventional concrete. The replacement of NA with RCA is likely a viable option for structural use.

1. Introduction

Concrete is a main material used for any structure in the construction industries. The ability of the concrete to resist load varies depending on its application. The reason concrete is widely used in construction is because it is easy to fabricate it to any shape and structure. The composition of concrete is primarily consisting of water, cement and aggregate that are mixed with varies ratio depending on several factors such as strength and cost consideration. Sand or fine aggregate is also added in the mixture as an element to fill the presence gap between larger aggregate. This conventional concrete is different compared to pervious concrete in terms of the mixture content and usage. The pervious concrete functions as a drainage cover or hardstand pavement by allowing the water to seep between the interconnecting void in it to prevent the water from pooling at the surface. [1] states pervious concrete requires less raw materials compared to conventional concrete based on composition content of course aggregate and cement paste with little fine aggregate or without fine aggregate.

Generally, both types of concrete require aggregate as a main element in their mixture but it will deal with the issue of source depletion when it comes to the natural resources. Besides that, the construction industry itself produces a tonnes of concrete volume wastes during concrete casting, renovation, repairing, demolition and maintenance. There are a lot of possibilities of excessive disposal to dumping site due to mishandled wastes during the construction. About 50% from the nature of raw material has been taken consuming 40% of the total energy and 50% of total waste produce for each construction [2]. This problem will lead to inadequate area of landfill and increase the rate of carbon pollutants in the environment. Thus recycling the concrete waste by extracting the aggregate for another use would likely help in minimizing the increasing number of concrete waste and reduce the cost in construction industry. This condition further strengthened by the use of Recycled Concrete Aggregate (RCA) in concrete mixture for reducing the need for opening new aggregates quarries and decreases the amount of construction waste that goes into landfill.

This paper seeks to: (i) compare the compressive strength of pervious concrete using recycled concrete aggregate (RCA) and natural aggregate (NA), (ii) analyse the infiltration rate using recycled concrete aggregate (RCA) and natural

2. Literature Review

Recycled-concrete aggregate (RCA) is one of the ideas for future sustainability in construction industry that can be used for non-structural elements such as drain cover, decoration at park, concrete walkway, pervious concrete and any else than building structure. For United State, RCA was introduced as non-structural basis material [3]. The use of this material in non-structural elements as a filler material or as a base course material of underlying layer for improving soil characteristics has also enabled a reduction in the usage of NA. Some studies on RCA were suggested replacing NA with RCA in structural concrete for non-structural application.

Approximately 10 billion cubic meter of concrete is produced annually and about 60% to 80% of the concrete volume mainly consists of NA [3]. The calculated amount of NA from the annually production, the consumption of the NA is high which cause it depletion and demand at the same time. Recycling of the waste concrete can reduce the uses of NA for non-structural concrete element. A total number of 431 million tons of aggregate was produced in 2008 which was used in the civil engineering field and building industry with percentages of 79% and 21% respectively. About 5 % of 431 million tons is recycling demolition wastes aggregate [4]. Although the production number of RCA is small, but it provides a huge contribution in reducing environmental impact by reducing deterioration of NA. The RCA mechanic properties may be similar to NA of conventional concrete shown by many studies of material made modification. However, RCA has higher absorption of water compared to NA which is caused by old mortar attached to the RCA. In addition, some modifications in the water cement ratio needed in order to have a similar workability of normal concrete.

In United State and Japan, pervious concrete is used for more than 50 years. It is an engineering application to solve engineering problems by using the feature of pervious concrete [5]. Pervious concrete is the concrete that allows air and water to penetrate through porosity of the pervious concrete [6], usually is used to reduce the runoff on top of the surface from precipitation. Pervious concrete has similar constituents as normal concrete. Its total volume consists 15% to 30% of interconnected void network that allows 0.014 to 0.023m3 of water per minute through its open cells for each 0.0929m2 of surface area [7]. These values are larger than the occurrence of the rain and it very effectively can reduce the excessive runoff from heavy rainfall. For special type of pervious concrete, it has large pore proportion range which is between 2 to 8mm and its percentage of porosity is 15% to 30% typically. The water flow easily through the pervious concrete since it is allowed by the present of the large pores interconnected system [8]. Pervious concrete design combination of structural and hydrological for pavement construction to storm water management is the best solution and it manages to receive recognition from the US Environmental Protection Agency recognition. Towards green ecological development, application of RCA is considered as an effective measure in achieving sustainability construction industry development [9].

The major characteristic of pervious concrete that needs to be tested is it strength under compressive load and infiltration rate. Compressive strength of pervious concrete using RCA is weaker than that for the pervious NA concrete [8]. It shows in the Figure 1 that the RCA use to replace the NA in the concrete mixing with percentages of 0% and 30% is almost similar for the 0.25 of water cement ratio.

However, the replacement of 100% RCA in the pervious concrete has resulted in decrease compressive strength about 5N/mm2 [8]. Infiltration characteristic is the main feature of the pervious concrete. The evaluation of the infiltration is on the volume of water per unit time or flow rate of water that can pass through the pervious concrete within the interconnected void. Water penetration through pervious concrete is depending to the air void content [6]. The mix proportion of the pervious concrete will be more challenging in achieving the balance between the acceptable infiltration rates with adequate compressive strength [10].

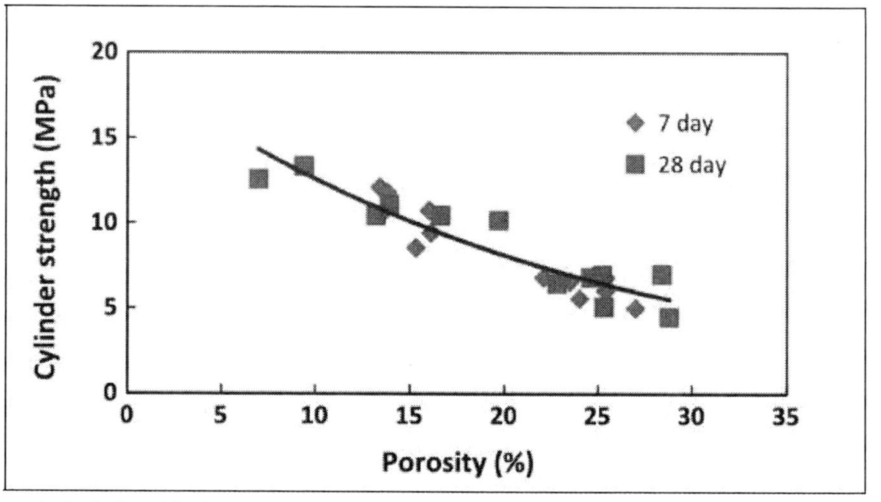

Figure 1. Relationship between strength and porosity for recycled pervious concrete [8]

3. Methodology

Methodology of this research will elaborate on the materials used, the procedure of mixing concrete and the laboratory testing for investigation of compressive strength as and infiltration characteristic as shown on Figure 2. The compressive strength test was conducted using BS EN 12390: Part 3:2009 as a guideline while infiltration test was conducted based on ASTM C 1701 (ASTM 2009) standard.

3.1. Material Selection

In this study, pervious concrete main materials used are Ordinary Portland Cement which follow MS EN 197: Part 1:2007 specifications, coarse aggregate and water. For sampling pervious concrete, fine aggregate or sand was not added to mix design and natural aggregate was used as a coarse aggregate. NA was obtained through supplier whereas RCA was obtained from construction site concrete waste materials.

3.2. Water-Cement Ratio

The water cement ratio is the proportion of the weight of water to the weight of cement that use in concrete mix. The water cement ratio of 0.3 was used in this study to prevent the cement paste settle to the bottom and fill the void. It requires 0.3kg of water for every 1kg of cement used in pervious concrete.

3.3. Concrete Mix Design

Mix design of pervious concrete was prepared into two portions using water, cement and aggregate ratio of 0.3:1:3 and 0.3:1:5. Production of pervious concrete in this study, the system mix proportions for 1 cube and 1 slab:
i. Weight 3kg and 5kg for mix design 0.3:1:3 and 0.3:1:5 respectively of coarse aggregate for both RCA and NA
ii. About 0.3kg of water use for mix design and set as fixed through this study.
iii. Cement required for the mix design is 1kg as fixed.
Concrete density of 2400kg/m^3 is set as fixed value to calculate the mix proportion.

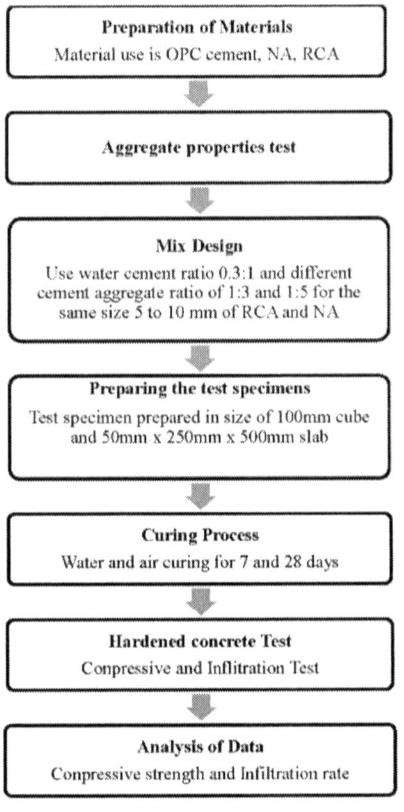

Figure 2. Flow chart of research methodology

3.4. Batching and Mixing

Batching is the process of mix together the water, cement and aggregate done using concrete mixer. The first thing in batching process is the aggregate pour into the mixer and cement little by little. Then water is poured into the mixer while the mixer rotating for 15 minutes until the mortar formed and it depend on the amount of ingredient inserted in the mixer. Lastly, the fresh concrete is casted into the cube and slab mould.

3.5 Casting and Curing

For concrete casting phase, the test specimen prepared in size of 100mm cube for compressive strength test and 50mm x 250mm x 500mm slab mould for infiltration test of pervious concrete. After casting of the concrete, the fresh concrete need to be left for 24 hours before it can be unmould for hydration process taking place. The concrete will be harden and next ready for curing. Curing process in important in order to prevent the moisture loss to allow the concrete gaining it strength. Curing is needed to prevent the moisture of the concrete being hydrated atmosphere that can lead to shrinkage and it will be tested for 7 days and 28 days after casting. Concrete will gain it fully strength at 28 days or curing.

During curing period for hardened concrete, the cube specimen was tested for compressive strength after 7 days and 28 days using Auto Test 3000 compression machine with compression rate 3kN per second. The infiltration test was conducted after 7 days and 28 days using slab specimen with cylindrical pipe-like and slab mould as shown in Figure 3. In this test, the cylindrical pipe-like was placed at two different points on the top of slab before filling with water. The time was taken during pouring the water into the cylindrical. The infiltration rate is equal to the flow rate of the water in cylindrical pipe-like by calculate the volume of the water divided by time taken to all water flow into pervious concrete.

Figure 3. a) Compression Test for cube specimen, b) Infiltration Test for slab specimen

4. Result and Discussion

4.1 Compressive Strength of RCA and NA Pervious Concrete with Different Cement-Aggregate Ratio
The compressive strength result of RCA and NA pervious concrete using mix design 0.3:1 of water cement ratio and aggregate size of 5mm to 10mm with two different cement aggregate ratios which is 1:3 and 1:5 is shown in Table 1.

Table 1. Pervious Concrete Compressive Strength Data

Water Cement Ratio	Agg. Size (mm)	Type Of Agg.	Cement agg. ratio	Curing (Day)	Compressive Stress (N/mm^2)			Average (N/mm^2)
0.3:1	5-10	Natural aggregate (NA)	1 : 3	7	22.16	23.04	23.67	22.960
				28	28.80	18.91	22.24	23.317
			1 : 5	7	12.610	9.447	12.112	11.389
				28	13.630	13.030	12.120	12.927
		Recycled concrete aggregate (RCA)	1 : 3	7	4.960	6.339	5.993	5.764
				28	6.907	8.181	6.561	7.216
			1 : 5	7	4.430	4.569	3.751	4.500
				28	4.953	4.717	5.148	4.939

Figure 4 shows the interpretation result of compressive strength of RCA pervious concrete and NA pervious concrete with given cement aggregate ratio mix proportions. The pervious concrete with different mix ratio gives a different compressive strength. This characteristic with those factor are same with the normal concrete compressive strength.

Obviously the curing factor were normally known for 28days of curing, the concrete achieve in optimum strength compared to 7 days of curing. The pervious concrete compressive strength is very influenced by the type of aggregate. There is so much different in strength between RCA and NA. The compressive strength of the recycled aggregate concrete is weaker than normal concrete. From the result obtain, the average of 1:3 cement to aggregate ratio of NA is 23.317N/mm^2 for RCA is 7.216 N/mm^2. There is a huge different between RCA and NA concrete compressive strength [6].

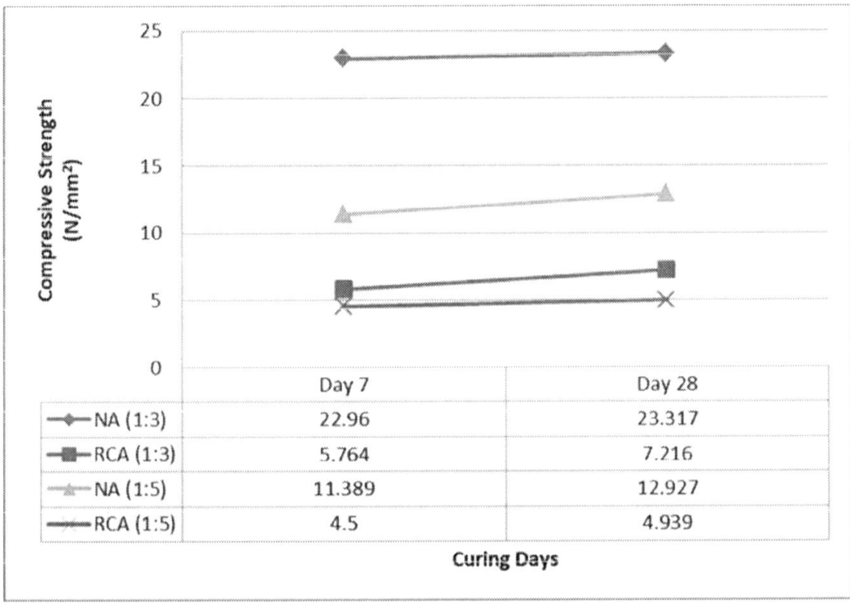

Figure 4. Compressive Strength of Pervious Concrete Using Different Cement Aggregate Ratio

4.2.. Infiltration Rate

The actual outcome of infiltration rate result for the RCA and NA pervious concrete with different cement aggregate ratio is referring to Table 2.

Table 2. Pervious Concrete Infiltration Rate Data for Both RCA and NA

Water Cement Ratio	Agg. Size (mm)	Type Of Agg.	Cement agg. ratio	Curing (Day)	Infiltration Rate (mm/s)
0.3:1	5-10	Natural aggregate (NA)	1 : 3	7	6.53
				28	6.15
			1 : 5	7	6.88
				28	6.79
		Recycled concrete aggregate (RCA)	1 : 3	7	6.73
				28	6.56
			1 : 5	7	7.01
				28	6.92

Figure 5 shows the graph of infiltration rate of RCA pervious concrete and NA pervious concrete with given cement aggregate ratio mix proportions. The infiltration rate of the curing day 28 lowers that curing at day 7 due to the small shrinkage in the pervious concrete. The steady decrease in water penetration coefficient was found after the cement hydration reaction [6].

For NA pervious concrete, the infiltration rate for day 7 was 6.53mm/s and day 28 was 6.15m/s for cement aggregate ratio 1:3. NA pervious concrete, the infiltration rate for day 7 was 6.88mm/s and day 28 was 6.79mm/s of cement aggregate ratio 1:5. For RCA pervious concrete, the infiltration rate for day 7 was 6.73mm/s and day 28 was 6.56mm/s with cement aggregate ratio 1:3. For RCA pervious concrete, the infiltration rate for day 7 was 7.01mm/s and day 28 was 6.92mm/s for cement aggregate ratio 1:5.

The RCA pervious concrete infiltration rate was higher than NA pervious concrete. This was because the pervious concrete sample using NA are having lower water absorption compare to RCA [9]. That mean, RCA water absorption was higher. The data of infiltration rate show that for reading of RCA was higher than NA uses for the pervious concrete for both cement aggregate ratio 1:3 and 1:5.

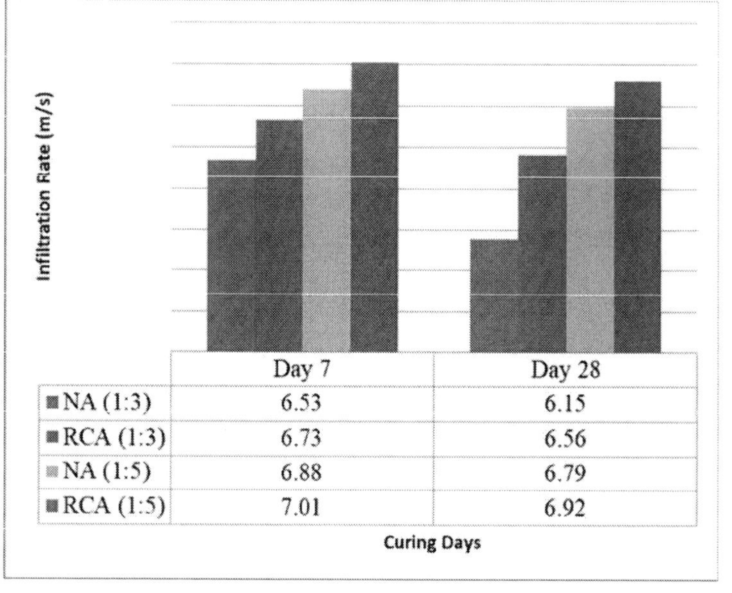

Figure 5. Infiltration Rate of Pervious Concrete Using Different Cement Aggregate Ratio

5. Conclusions

This study proves that recycled-concrete aggregate (RCA) pervious concrete possesses lower compressive strength in comparison to natural aggregate (NA) pervious concrete. Using the right proportion of the cement-aggregate ratio of pervious concrete is necessary in order to determine the optimum strength of the pervious concrete. The RCA pervious concrete with 1:3 cement-aggregate ratios would likely to have higher compressive strength than the RCA pervious concrete with 1:5 cement-aggregate ratios.

Comparing the compressive strength difference between both NA pervious concrete and RCA pervious concrete, it can be concluded that regardless of any cement-aggregate ratio and curing days, the result still shows a lower compressive strength parameters for RCA pervious concrete in comparison to NA pervious concrete.

The infiltration characteristics of pervious concrete were affected by the curing factor. The infiltration rate during the 7th curing day is slightly higher than the infiltration rate at the 28th curing day. On the other hand, the infiltration rate of the NA pervious concrete is slower than RCA pervious concrete regardless of having cement-aggregate ratio of 1:3 and 1:5 respectively. In this study, the focus is more on the RCA pervious concrete. Hence, RCA pervious concrete with cement aggregate ratio 1:5 was higher infiltration rate than RCA pervious concrete with cement aggregate ratio 1:3.

As for the conclusion, the RCA pervious concrete designed using 1:3 cement-aggregate ratio would be more proportionate if the compressive strength is emphasised for the intended use without neglecting the fact that NA pervious concrete will give a better result whereas for infiltration characteristic, it is preferably designed by using 1:5 cement-aggregate ratio in order to achieve higher infiltration rate.

6. References

[1] Rajasekhar K and Spandana K 2016 Strength Properties of Pervious Concrete Compared with Conventional Concrete *Journal of Mechanical and Civil Engineering* **13**(4) 97-103

[2] Katrina McNeil and Thomas H K K 2013 Recycled Concrete Aggregate: A Review *International Journal of Concrete Structures and Materials* 7(1) 61-69

[3] Evangelista L, Guedes M, De Brito J, Ferro A C and Pereira M F 2015 Physical, chemical and mineralogical properties of fine recycled aggregates made from concrete waste. *Construction and Building Material* **86** 178-188

[4] George Wardeh, Ethan G and Hector G 2014 Mix Design and Properties of Recycled Aggregate Concretes: Applicability of Eurocode 2 *International Journal of Concrete Structures and Materials* **9**(1) 1-20

[5] Cosic K, Korat L, Ducman V and Netinger I 2015 Influence of Aggregate Type and Size on Properties of Pervious concrete *Construction And Building Materials*, **78**, 69-76

[6] Ming-Gin L, Yi-Shuo H, Tao-Kuang C and Chun-Hua P 2011 Experimental Study of Pervious Concrete Pavement, *Geotechnical Special Publication No. 218* ASCE 93-99

[7] Ajamu S O, Jimoh A A and Oluremi J R 2012 Evaluation of Structural Performance of Pervious Concrete in Construction *International Journal of Engineering & Technology* **2**(5),829-236

[8] Sriravindrarajah R, Neo D H W and Lai J W E 2012 Mix Design for Pervious Recycled Aggregate Concrete *International Journal of Concrete Structures and Materials* **6**(24) 239-246

[9] Xiao J, Li L, Shen L and Poon C S 2015 Compressive behaviour of recycled aggregate concrete under impact loading *Cement and Concrete Research* **71**, 46-55

[10] Ismail A R, Hasrudin H and Mujahid A A Z 2009 Assessment of Recycled Aggregate Concrete *Modern Applied Science* **3**(10) 47-54

Acknowledgment

This research was supported by Universiti Teknologi MARA Pahang and Universiti Teknologi Malaysia.

IOP Conf. Series: Materials Science and Engineering **601** (2019) 012032 doi:10.1088/1757-899X/601/1/012032

Identification of the Activity-based Hazards/risks Involved in the IBS Construction Process: Case Study of Project That Uses Prefabricated Steel Framing System and Prefabricated Timber Framing System

Mohamad Ariff Mohd Amin[1], Nor Haslinda Abas[1*], Azreen Harina Azman[2] and Siti Khatijah Mohamad[2]

[1]Jamilus Research Centre, Faculty of Civil and Environment Engineering, Universiti Tun Hussein Onn Malaysia, 86400 Parit Raja, Batu Pahat, Johor, Malaysia
[2]Civil Engineering Department, Politeknik Sultan Mizan Zainal Abidin, KM 8, Jalan Paka, 23000 Dungun, Terengganu
*Corresponding author: nhaslin@uthm.edu.my

Abstract. Construction industry is known for its hazardous physical working environments and high accident risks. In reducing risk at the workplace, Occupational Safety and Health (OSH) is an important aspect. It is a standard which are set in legislation with the purpose to eliminate and reduce hazards at workplace. The objectives of this paper are to identify the major activities of the construction process in IBS residential and the OSH risks associated throughout the construction. This study involved a field observation to two different methods and two factory that use IBS as construction method which are 'prefabricated timber framing system', and 'prefabricated steel framing system'. This involved interview to the executive staff, operatives and supervisors, and observing and documenting operations. In achieving this, four case study were descriptively analysed and availing a revelation exposure to UV and mobile plant risks as the major risk in IBS construction site. This paper intends to contribute in providing a way for designers to integrate construction process knowledge into design to eliminate or reduce hazards during construction. It will effectively address the designer's role in making decisions in their designs and further illuminate the level of OSH risk their designs pose.

1. Introduction

The construction industry is known for its hazardous physical working conditions and high accident risks. It is one of the most hazardous industry compared to other industries due to the nature of activities at the worksites [1-2]. The construction industry is one of the four major contributing industries in the workplace. Despite, the average five years of death for this industry in Malaysia is 99 employees per year from 2011 to 2015. The death rate for 100,000 construction workers in 2015 was 10.94, higher than the average mortality rate for five years 2011-2015 of 8.17. The trend of fatal injuries to construction workers has risen since 2012, and 140 construction workers have died in 2015, the highest since 2001 and the highest in the 21st century.

As shown in Figure 1, the number of fatal injuries to workers in all industries, overall showed a decline, but the number of fatal injuries of workers in the construction industry showed an increase. The five-year average from 1999 to 2003, number of fatal injured workers in all industries was 898, while the average five years from 2011 to 2015 was 639. The five-year average for the deaths of all industry workers showed a 28.8% decline. Besides, the five-year average from 1999 to 2003, the number of fatal injured workers in the construction industry was 115, while the average five years from 2011 to 2015 was 99. Five-year average for the fatal injuries of all industry workers showed a decrease, but at lower rates at 13.9%.

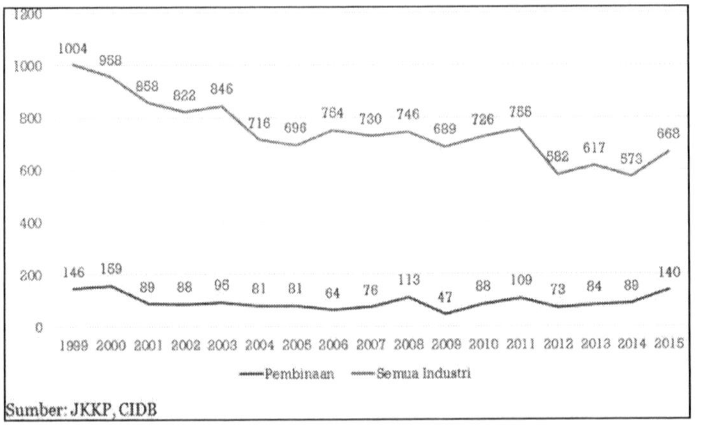

Figure 1. Accident statistics for construction and overall industry from 1999-2015 [3].

Accidents do regularly occur on construction site, though some of the accident causation or prevention models have failed to mention the types of accident on building site, whereas it is domineering to expose the various types of accident occurring on building sites[4]. Building site accidents range from fall-related accidents [5], fire outbreak [6], explosion [7], electrocution/electrical incidents [8], vehicle accidents [9], roof construction falls [10], fall of heavy objects during lifting, and contact with electric current [11]. Additionally, different types of accident on construction sites among many others to be: falls from roof, scaffold, stairs, ladder, ramp, and elevator shaft; crane accident; struck by moving vehicle, struck by an operating machine or defective machines (boilers, bulldozers, backhoes), run over by operating equipment; equipment/tools-related accident; slips and trips; electrocutions/electricity; and explosions [12].

Due to frequent accidents, IBS construction has been suggested as a replacement to traditional construction methods of the many initiatives that could be implemented to improve OSH performance [13-14]. The Malaysian government is actively promoting the adoption of IBS and encouraging a paradigm shift in the construction process, from a traditional to an industrialized approach [15]. This is demonstrated by the promulgation of the Construction Industry Master Plan (CIMP) 2006-2015 [13], which specifically mentions IBS and its implementation through IBS Roadmaps.

IBS is contrary from that traditional processes in term of the nature of activities that consist of four stages which manufacturing, delivery, component installation and in situ works [2]. IBS is an industrialized process in which components of a building are conceived, planned, and fabricated, and then transported to and erected on site [15][16]. Compared to the traditional method, IBS or more specifically offsite construction are claims can reduce site accidents [17-20][15].

This paper presents the findings of the risk involved in a construction process that use IBS as the construction method. This finding of the study is presented by showing the list of the hazards involve in the construction building wall for several IBS method which pre-fabricated steel framing and pre-fabricated timber framing.

2. Methodology

This study involved a field observation on two different IBS system construction sites and two factory of IBS component. Two sites have been visited, which represent IBS types which are 'prefabricated timber framing system' and 'prefabricated steel framing system'. This involved talking to the managerial staff, operatives and supervisors, and observing and documenting operations.

This study focused on the major activities of the construction process in IBS residential construction that only cover for structure and building envelope. Data has been collected on on-going construction projects, which used the selected construction approaches to observe the process involved and to identify the associated OSH risks.

3. Result

This study was completed through a combination of field observation, interviews and document analysis. These served to identify the construction process in terms of activities and the risks associated with the process. This section summarizes the data from field observation and the interviews.

3.1. Prefabricated steel framing system

This type of IBS classification is commonly used with pre-cast concrete slabs, steel columns and beams. Steel framing systems have always been the popular choice, used extensively in the fast-track construction of skyscrapers. Recent development in this type of IBS includes the increased usage of light steel trusses consisting of cost-effective profiled cold-formed channels and steel portal frame systems as alternatives to the heavier traditional hot-rolled sections [21].

The project was a government project with the contract worth about RM 430,000.00. The system used for the construction of the building uses pre-fabricated steel framing system, by integrating assembling components of a structure in a factory or other manufacturing site and transporting complete assemblies or sub-assemblies to the construction site where the structure is to be located. The connection system used for this project is a combination of bolt and screw and welding joint.

The main risks were associated with the lifting component of materials to the required position, for example site conditions and the stability of materials. In addition, radiation and thermal also highlighted during installation of H-column and I-beam. The risks were then categorized into damaging energies for evaluation. Table 1 provides a clearer view of the risks involved in each activity.

Table 1. Activities and risks on prefabricated steel framing system [15].

Process	Activities	Risk/Hazard and description
Manufacturing	Material preparation	MSDs
	Forcing pliable metal through large rollers;	MSDs Exposure to UV Electrical shock Dermatitis
	Heat hot-rolled beam above the recrystallization point	Heat from the burning point MSD, RSI Mechanical handling Inappropriate material handling method
	Cutting of metal	MSD Excessive noise
	Assemble of component	Mechanical handling
	Transferring the components to stockyard	Inappropriate stacking method
Delivery	Load component onto truck	Mobile plant risks (vehicle injury) Musculoskeletal Disorder (MSD), Repetitive Strain Injury (RSI)

		Mobile plant risks
	Transport and deliver materials	Road traffic risks Instability of materials
	Off-load materials on-site	Mobile plant risks (vehicle injury) MSD, RSI Mobile plant risks
	Stacking component	Collapse of elements/materials
Pre-in situ	Installation of BRC for slab	MSD, RSI Exposure to UV
	Casting and Vibrating	MSD, RSI Dermatitis Hand-arm vibration (HAVs) Exposure to UV
Installation	Lifting component	Excessive noise from mobile plant Exposure to UV Craneage risk
	Installation of H-column	Craneage risk, working at height, Due to inappropriate installation Vehicle injury MSD, Hand injury Excessive noise from mobile plant Exposure to UV Exposure to welding heat
	Installation of I- Beam	Craneage risk, working at height, Due to inappropriate installation Vehicle injury MSD, Hand injury Excessive noise from mobile plant Exposure to UV Exposure to welding heat
	Adjusting component to required position	Craneage risk Vehicle injury MSD, hand injury Exposure to UV
	Installation of bracing or propping	HAVs Electrical shock From machine Due to inappropriate propping Exposure to UV
In-situ	Prepare for bricklaying	
	Transfer brick	MSD, RSI Mobile plant risk Lifting equipment risk Exposure to UV
	Moving brick to workface	MSD
	Mix mortar	Dermatitis MSD, RSI Exposure to UV
	Cut brick	Electrical shock

		Exposure to excessive noise from cutting machine Struck by cut brick from saw
	Laying mortar	Dermatitis MSD, RSI Working at height Exposure to UV
	Laying of each course of brick	Working at height MSD, RSI Dermatitis Exposure to UV
	Fix wall ties;	MSD, RSI Working at height
	Plastering and finishing	Exposure to chemical compound Working at height RSI Exposure to UV

3.2. Timber framing system

Prefabricated timber framing systems consist of timber building frames and timber roof trusses. While the latter are more popular, timber building frame systems also have niche markets; offering interesting designs from simple dwelling units to buildings requiring high aesthetic values such as chalets for resorts [21].

The project was a private project with the cost about RM 150,000.00. The system used for the construction of the building uses timber framing system, by integrating assembling components of a structure in a factory or other manufacturing site and transporting complete assemblies or sub-assemblies to the construction site where the structure is to be located. The connection system used for this project is a combination of bolt and screw.

The OSH risks highlighted were mainly associated with the ergonomic-related risks, and exposure to UV. The cutting timber involved RSI due to repetitive works. The use of chemicals was identified in the activities planning works. In mortise and tenon cutting, the main OSH risks highlighted were MSD and manual handling. Mobile plant risks were identified in moving and storing the panels. The risks associated with the activities are depicted in Table 2.

Table 2. Activities and risks on timber framing system method [15].

Process	Activities	Risk/Hazard and description
Manufacturing	Material preparation	-
	Wooden log split	Cuts Excessive noise from cutting machine Electrical shock RSI
	Timber cutting	Cuts Excessive noise from cutting machine Electrical shock RSI
	Planing the timber	Cuts Excessive noise from cutting machine

		Electrical shock RSI Dermatitis
	Transfer to mortise place	Road traffic risk Stability of material
	Mortise and tenon cutting	Cuts Excessive noise from cutting machine Electrical shock RSI
	Drilling hole for connection	Cuts Excessive noise from cutting machine Electrical shock RSI
Delivery	Load component onto truck	Mobile plant risks (vehicle injury) Musculoskeletal Disorder (MSD), Repetitive Strain Injury (RSI) Mobile plant risks
	Transport and deliver materials	Road traffic risks Instability of materials
	Off-load materials on-site	Mobile plant risks (vehicle injury) MSD, RSI Mobile plant risks
	Stacking component	Collapse of elements/materials
Pre-in situ	Installation of BRC for slab	MSD, RSI Exposure to UV
	Casting and Vibrating	MSD, RSI Dermatitis Hand-arm vibration (HAVs) Exposure to UV
Installation	Lifting component	Excessive noise from mobile plant Exposure to UV MSD
	Installation of timber column	Working at height, MSD, Hand injury Excessive noise Exposure to UV
	Adjust the position of the column	MSD, hand injury Exposure to UV
	Installation of timber beam	Working at height, MSD, Hand injury Excessive noise Exposure to UV
	Adjusting component to required position	MSD, hand injury Exposure to UV

	Installation of bracing or propping	HAVs Electrical shock From machine Due to inappropriate propping Exposure to UV
	Bolt and screw the connection between beam and column	MSD
	Installation of timber plank	MSD
	Nails the end of the plank	Excessive noise from hammer Exposure to UV Working at height
	Finishing	Exposure to chemical compound Working at height

4. Results and Discussions

From the table above, the delivery stage provides likenesses in the process for most construction approaches; pre-fabricated steel framing system; and pre-fabricated timber framing system. This likeness provides general of the processes. However, there are differences in element nature between the approaches, whether it is timber component or steel framing. The difference is designed by the way the process is done and the physical feature of the element which will eventually give a different risk. Besides, manufacturing stage, pre-fabricated steel framing shown that MSD is the major risk due to were manual handling activities. Pre-fabricated timber framing shown RSI is the major risk due to repetitive work on manufacturing stages. For installation of component, exposure to UV were the major risk for both constructions approaches due to exposed to sun during the installation activities. For instance, 'lifting component' activities for both construction approaches shown that exposure to UV and excessive noise from the mobile plant were the risk which provides likenesses in the process. The in-situ stages provide similarities in the activities which will eventually give a same risk for both construction approaches.

5. Conclusion

This paper had identified and discussed the activities and risk associated in IBS construction in Malaysian construction industry. The level of safety in IBS industry in the construction industry was still low even though the government had seriously promoted and encouraged the use of IBS to have a better practice in the construction industry. The risk on IBS construction site faced by the workers had been discussed and mainly revolves around UV exposure, MSD, working at height, excessive noise and mobile plant risk. Despite the negative perspective concerning IBS method, this does not mean the future for adoption of IBS method by the prospects would be low. Future research is sought for more investigation to be done for the qualitative stakeholders' perspectives on the factors that impede the IBS adoption in the construction industry.

6. References

[1] Hassanein A A G and Hanna R S 2008 Safety Performance in the Egyptian construction industry *J. of Construction Engineering and Management* **134** 451–456

[2] Rahman R A 2015 Managing safety at work issues in construction works in Malaysia : A proposal for legislative reform *Moder Applied Science* **9**(13) 108–121

[3] Department of Occupational Safety and Health 2014 *Accident Statistics* (Kuala Lumpur: Ministry of Human Resources)

[4] Ng K, Laurlund A, Howell G and Lancos G 2012 Lean safety: Using leading indicators of safety incidents to improve construction safety. *Proc. for the 20th Annual Conf. of the Int. Group for Lean Construction: Are We Near a Tipping Point?* (San Diego: Int. Group on Lean Construction)
[5] Orji S E 2016 Hazards in building construction sites and safety precautions in Enugu Metropolis, Enugu State *Imperial J. of Interdisciplinary Research* **2**(1) 282–289
[6] Health and Safety Executive 2006 *Health and Safety in Construction* (London: Health and Safety Executive)
[7] Hovden J 2010 Is there a need for new theories, models and approaches to occupational accident prevention? *Safety Science* **48**(8) 950–956
[8] Nkem A N 2015 Relationship between unsafe acts/ condition and accidents in construction company in Nigeria *J. Teknologi* **75**(6) 73–77
[9] Edwards D J and Nicholas J 2002 The state of health and safety in the UK construction industry with a focus on plant operators *Structural Survey* **20**(2) 78–87
[10] Weeks J L 2011 *Health and Safety Hazards in the Construction Industry* Retrieved from http://www.iloencyclopaedia.org/component/k2/item/518
[11] Umeokafor N, Evaggelinos K, Lundy S, Isaac D, Allan S, Igwegbe O, Umeokafor K and Umeadi B 2014 The pattern of occupational accidents, injuries, accident causal factors and intervention in Nigerian factories *Developing Country Studies* **4**(15) 119–127
[12] Asanka W A and Ranasinghe M 2015 Study on impact of accident on construction project *Proc. 6th Int Conf. on Structural Engineering and Construction Management* (Kandy, Sri Lanka) p 58–67
[13] Construction Industry Develoment Board 2008 *Master Plan for Occupational Safety and Health in the Construction Industry 2005-2010* Retrieved from https://www.cidb.gov.my/cidbv4/images/pdf/OHSAS2006v2.pdf
[14] Construction Industry Develoment Board 2007 *Malaysian Construction Outlook 2007* (Kuala Lumpur: Construction Industry Develoment Board)
[15] Abas N, Blismas N and Lingard H 2011 Development of a knowledge-based energy damage model to assess occupational health and safety (OSH) construction risks in Malaysia *Proc. of Prevention: Means to the End of Construction Injuries, illnesses, and Fatalities* (Virginia Tech: Washington) p 1–16
[16] Junid S M S 1986 Industrialized building systems. *Proc. of a UNESCO/FEISEAP Regional Workshop* (Serdang: Universiti Pertanian Malaysia)
[17] Alistair G F G 1999 *Off-site Fabrication: Prefabrication, Pre-assembly and Modularisation* (United Kingdom: Whittles Publishing Services)
[18] Toole T M and Gambatese J 2008 The trajectories of prevention through design in construction *J. of Safety Research* **39**(2) 225–230
[19] Gangolells M, Casals M, Forcada N, Roca X and Fuertes A 2010 Mitigating construction safety risks using prevention through design *J. of Safety Research* **41** 107–122
[20] Mckay L J 2010 *The Effect of Offsite Construction on Occupational Health and Safety* (United Kingdom: Loughborough University)
[21] Construction Industry Develoment Board 2003 *Survey on the Usage of Industrialised Building Systems (IBS) in Malaysian Construction Industry* (Kuala Lumpur: Construction Industry Develoment Board)

Acknowledgement
This paper was partly sponsored by the Research Management Centre (RMC), Universiti Tun Hussein Onn Malaysia (UTHM). (Vot No. U784)

IOP Conf. Series: Materials Science and Engineering **601** (2019) 012033 doi:10.1088/1757-899X/601/1/012033

Performance of concrete containing mussel shell (Perna viridis) ash under effect of sodium chloride curing

Muhammad Shabery Sainudin[1*], Nor Hazurina Othman[1,2*] and Shahiron Shahidan [1,2]

[1] Department of Structural and Material Engineering, Faculty of Civil and Environmental Engineering, Universiti Tun Hussien Onn Ma-laysia, Batu Pahat 86400, MALAYSIA.
[2] Cluster of Advanced Construction Materials, Jamilus Research Center, Universiti Tun Hussien Onn Malaysia, Batu Pahat 86400, MALAYSIA.
*Corresponding author: hazurina@uthm.edu.my

Abstract. Mussel shell is a type of waste that consist large amounts of calcium content ($CaCO_3$) (>95%). It is potentially to be used as concrete filler which can reduce porosity in concrete micro-structure (compacted concrete) and indirectly reduce quantity of water used to maintain concrete workability. This paper investigates the performance of mussel shell waste as an admixture in normal concrete. These waste were crushed using ball mill for 5 hour then sieved with 75μm sieve sizes. Five different percentages of mussel shell ash (MSA) were incorporated in concrete as an admixture 0%, 1%, 2%, 3% and 4% (S0, S1, S2, S3 and S4) base on w/c 0.5 according to DoE method with total of 90 samples. Specimens then were cured using 2.37% of sodium chloride (NaCl) solution for 7 and 28 days. Data were analysed according to physical, mechanical and its durability properties of materials and its specimens. Particle size analysis (PSD), specific gravity, SEM image and setting time are measured on MSA, OPC and samples MSA concrete included slump analysis for concrete workability. While it's mechanical and durability properties of MSA concrete were analyses through compressive, split tensile and water capillary absorption. The result showed that increment percentages of MSA could effect on the strength of both compressive and split tensile. However, minimum percentages of MSA of 1% gained higher strength compared to control specimen. The differences between MSA percentages and its curing duration incorporated indicating an improvement while S1 and S3 indicates lower rate of capillary absorption for 7 and 28 days each. Overall, the results to establish concrete strength improvement of MSA in concrete should be limited to 1% uses. While S2 and S4 for concrete permeability and workability. With this percentages, concrete display an improvement behaviours according to its strength and durability towards sodium chloride exposure.

1. Introduction

Waste management had becomes one challenging issue in Asian region included other developing country since before. The waste generated seen to give negative impact due to its increment rate per year and it's prevail the ability of related parties in controlling its production [1]. According to a study by Badgie *et al.*, [2], it was expected that average waste generated in Asian will increased to 1.8 million tonnes which is 5.2 million meter square each days if it was not controlled. The key factor of this issue is caused by socioeconomic by community accordance with economic upturn in terms of business, manufacturing and others [3].

Mussel production is one of fishery sector that contributed in economic resource to national income in Malaysia. According to statistic by Malaysia Department of Fisheries, it was estimated about 1,827.27 metric tonnes green mussel were produced in year 2016 and increased about 154.14 tonnes compared to previous year [4]. However, this mussel production had contributed to waste increment [3] which 33% from mussel product is consist of it shell [5] that will be discarded after mussel fill been collected. This shells could lead to excessive quantity of mussel waste that generates from mussel production. Thus, it was confirmed that the fisheries industry could generates over 1 million tonnes of waste in worldwide which was implied a significant waste problem on a global scale [5].

One of the simply way to overcome and reduce waste increment issue by applying recycle method towards material which can minimize waste quantity [3]. Further research had developed in recent years to study the use of mussel shell waste as construction material. Recently, mussel shell as an addition in cementitious material that used in concrete mixing and its effect on concrete properties was studied. Various studies against mussel shell has been undertaken by researchers in identifying the potential of material. It was applied in many departments course to reduce mussel shell waste quantity and it clearly giving positive feedback among researcher. Xiong [6] found that mussel shell could become filtration media in reducing the amount of phosphate up to 55% in water. Based on construction materials sector, seashell such as mussel species is one natural occurrence that consist of calcium source in form of calcium carbonate which were applied in cements industries other than lime stone, lime and dolomite [7]. Martínez-García *et al.* used recycle materials consist of mussel shell as aggregates partial replacement could improve strength development of harden concrete [5]. In a previous experimental investigation by Oliviaa *et al.*, Kuo *et al.* and Othman *et al.*, it also found that the utilisation of seashell in concrete giving positive reaction toward quality of concrete based on its quantity to cement mixture within 4% to 5% respectively [8]-[10].

Durability is one of important properties in concrete that should be stressed which will affect towards its strength and serviceability. Concrete consist of low resistance towards ion penetration due to its pores structures [11]. Exposure of concrete towards aggressive environment such chloride, sulphate, carbon dioxide and etc. that came naturally or industries waste could disturbances its durability properties [12]. Exposure of concrete to sodium chloride is one an example of aggressive environment that came from sea-water due to it highly amount of sodium (Na) and chloride (Cl) ions from its salinity [7]. NaCl is classified as salt type that has been found could damage concrete strength and durability through its chemical interactions [13]. A study by Qiao [14] state that penetration of NaCl into concrete could lead to the formation of Friedel's salt forms in its cementitious materials due to reaction of Cl ion and C_3A. This could cause to loose of concrete flexural strength due to dissolution and leaching of $Ca(OH)_2$ and C-S-H and also crystallization pressure of Friedel's salt. [14-15]. While the present of other ion such sulphate could cause exceeded amount of gypsum and ettringite due to its reaction between $Ca(OH)_2$ and C_3A which can cause concrete deterioration [16]. Thus, the penetration of aggressive ions should avoided and solved in order to reduce its impact towards concrete strength and quality.

Mussel shell is one of the material with $CaCO_3$ (>90%) composition which exceed $CaCO_3$ percentage contained in limestone (>75%) [12][17] According to Sun & Chen, (2018) calcium carbonate possess filler properties which consists of small particle sizes compared to Portland cement [18]. It also potentially to increased concrete workability and reduce quantity of water used [12]. Since mussel shell consist of filler properties due to its higher $CaCO_3$ content, it was expected to reduce porosity and permeability of concrete due to its incorporated between cementitious product into concretes. Thus, the penetration of ion and aggressive environment toward concrete could be difficult due to low permeability in term of capillary pores and maintained properties of concrete. A study by Higgins [19], addition calcium carbonate that is small in concrete which contains ground granulated blast furnace slag (GGBS) could increase potential towards sulphate resistance. The analysis was observed according to the concrete with addition and absence of $CaCO_3$ then were cured in magnesium sulphate ($MgSO_4$) and sodium sulphate (Na_2SO_4). It was explained that addition of $CaCO_3$ had increased compressive strength compared to normal concrete. It because $CaCO_3$ has filling effect that potentially fill small pores in concrete causing difficulty Na_2SO_4 and $MgSO_4$ penetrates into concrete specimens. Overall, through application of mussel shell either as addition or partially replacements with Portland cements was significantly to be used and complied awareness towards sustainability development.

The aim of this research is to assess effect of utilisation of mussel shell as an admixture on the concrete properties towards aggressive environment. Thus, mussel shell ash (MSA) were added to concrete mixture then cured using sodium chloride solution. The physical mechanical properties of MSA were studied to evaluate its behaviour. The rationality and accuracy of mixed design developed are verified through comparison of two different factor on experimental observation. Finally, the influences of differences percentage and duration curing on the compressive and tensile strengths and water capillary absorption are analysed and discussed to determined optimum percentages of MSA in concrete based on its mechanical and durability properties for each specimens.

2. Experimental and physical properties of materials

2.1. Raw material

The specimens were analyzed through the preparation of raw materials that used in this study included water, coarse aggregate, fine aggregate, ordinary Portland cement and mussel shell. Mussel shell that had originally dumped from mussel production process were collected, cleaned and dried. Then the MSA product were obtained through crushing process of mussel shell for five hours using ball mill then were sieved to 75μm. The physical properties (specific gravity, setting time, PSA, slump test) were define in table 1.

Table 1. Physical properties of OPC and MSA as admixture for cementitous materials

Description	Chemical compound	OPC	MSA	Percentage total replacement (%)			
				S1	S2	S3	S4
Physical properties	Specific gravity	3.09	2.51	-	-	-	-
	PSD						
	10%	3.81	0.85	-	-	-	-
	30%	12.55	2.53	-	-	-	-
	50%	18.29	8.48	-	-	-	-
	60%	21.15	11.67	-	-	-	-
	70%	24.93	15.49	-	-	-	-
	Setting time (mins)						
	Initial	98	-	98	95	95	92
	Final	289	-	288	285	282	270
	Slump value(mm)						
		37	-	35	42	53	59

- = not determined

Cleaned mussel shell and its product of mussel shell ash (MSA) was shown as in figure 1 and figure 2. The raw of MSA and Ordinary Portland cements (OPC) properties were identified through its specific gravity analysis. OPC and MSA were complies with specification according to BS EN 197-1: 2000 with value of 3.09 while 2.51 respectively (10g sample of OPC and MSA) [20]. Neville state that there were relationship between specific gravity and setting time on concrete properties [16]. Thus, setting time analysis were obtained in this paper according to BS EN 196-3: 2005 for both ordinary OPC and MSA towards its initial and final setting time (400g of MSA, OPC and with addition MSA S1, S2, S3 and S4, 200mL water) [21].

Figure 1. Cleaned mussel shell

Figure 2. Mussel shell ash (MSA)

The result show that early setting time for MSA paste (S2, S3 and S4) developed faster early setting time (<96 minutes) compared to S1 and OPC (98 minutes). According to data observe from Table 1, the increment of percentages of MSA had lead to faster initial setting time same goes to its final setting time value (>289 minutes). According to Wen-Ten *et al.*, it's due to MSA properties that consist high rates of hydration to absorb water [22]. Apart from that, particle size distribution (PSD) of MSA were analysed according to standard method of BS ISO 13320:2009 using CILAS 1180 Liquid (0.04μm – 2500μm) [23]. Fineness of substances could accelerates strength of concrete due to faster reaction of hydration process [12]. Thus, PSD analysis were obtained in the study to identified fineness of MSA and OPC particle diameter which can contributes in promoting the development of concrete at early strength due to its surface contact to react with hydration products in concrete mixture.

PSD analysis for OPC and MSA was defined as in figure 3 and figure 4. According to both figure, MSA consist smaller diameter sizes compared to OPC. Data obtained for 10% cumulative, it shows the size of MSA were 0.85μm and 3.81μm for OPC. While for 50% and 60%, it was 8.48μm and 18.29μm, 1.67μm and 21.15μm each. According to 70%, the cumulative was 15.49μm and 24.93μm. Based on maximum diameter of materials from both graph and cumulative value for each percentages, MSA indicates lower diameter compared to OPC which were 20μm for MSA while 25μm for OPC. The value from each cumulative percentages in both graph proved that particle diameter of MSA was significantly suitable materials to be used as concrete admixture due to its small diameter particle sizes compared to OPC.

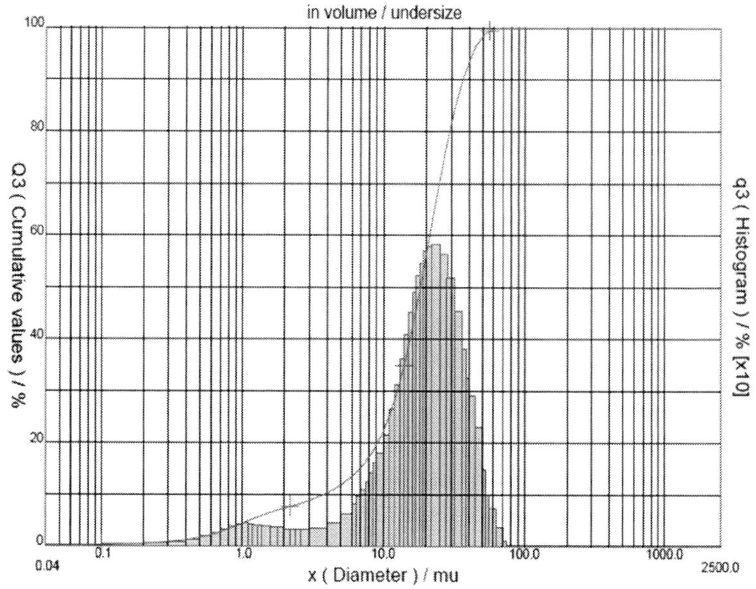

Figure 3. PSD analysis of OPC

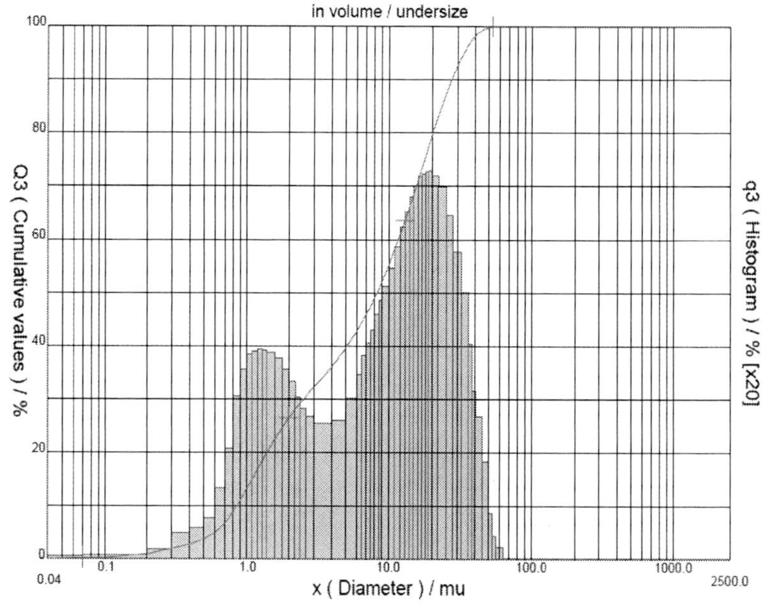

Figure 4. PSD analysis of MSA

SEM image was identified according as in figure 5 and figure 6. Based on both figures, it showed that 1.00KX zoomed MSA consist of fine particle compared to OPC which suitable to be used as filler in concrete. Particle characteristics of the MSA agglomerates showed that the shapes and size are consistent and almost similar to each other compared to OPC which formed in different sizes and particle shapes. It is due to OPC comprised from four type of minerals compounds such tri-calcium silicate (C_3S), di-calcium silicate (C_2S), tri-calcium aluminate (C_3A) tetra-calcium aluminoferrite (C_4AF) and oxide composition such CaO, SiO_2, Al_2O_3, Fe_2O_3, MgO, SO_3 and etc. [5, 7, 24] compared to MSA which major oxide composition consist of 95% calcium carbonate ($CaCO_3$) [17]. According to Hawlett, OPC compound and oxide orientated with various size fraction such coarser than 90μm and in between 45μm to 90μm [17]. Due to in variety sizes, it microscopy sizes and shape would be varies compared to MSA.

Figure 5. SEM image of OPC

Figure 6. SEM image of MSA

2.2. Concrete mixture

In this study, specimens were separated in classes which were mixed in 5 separate batch of concrete mixing (S0, S1, S2, S3 and S4 MSA). It was calculated and designed according to Design of Engineering (DoE) method with 0.50 (w/c). The concrete mixing process was carried out using mechanical method (mixer). The details of materials used for each specimens of water, fine and course aggregates were define as in table 2 and table 3. Whereas, admixture value of concrete design to ash materials (S1, S2, S3 and S4) were 0.024kg, 0.048kg, 0.072kg and 0.096kg each for 0.001m^3 mould and (S1, S2, S3 and S4) 0.048kg, 0.096kg, 0.145kg and 0.193kg for 0.002m^3 mould.

Table 2. Details of cement replacement in mixtures

Code	Percentages admixture
S0	100% concrete
S1	100% concrete + MSA 1%
S2	100% concrete + MSA 2%
S3	100% concrete + MSA 3%
S4	100% concrete + MSA 4%

Table 3. Mixing proportions of concrete design (kg)

Materials	Cements	Course Aggregates	Fine Aggregates	Water
Cube	0.42	1.21	0.57	0.21
Cylinder	0.84	2.42	1.14	0.42

According to concrete workability summarized in table 1, the slump value of each 1%, 2%, 3% and 4% MSA specimens were 35mm, 42mm, 53mm and 59mm respectively. Overall, it showed that the concretes slump value increases with the increases of MSA percentages. S4 indicates the highest slump value followed with S3 and S2. It proved that, $CaCO_3$ were able in improving concrete workability and handling purposes [12]. On the other hand, percentage value of S1 record lowest slump value compared to plain concrete. This is due hot weather during mixing process of S1 mixture carried out had accelerate further process cements hydration and reaction towards formation of C-S-H gel [25]. Although it is not a preferred choice of the interferences arising from the presence including the formation of interfaces during hot weather, however S1 slump value according to DoE calculation were in the range of 30-60mm and it was considered accepted due to its medium workability level with range of 35-75mm according to Neville [12, 16]. Further examination of the literature revealed that the variability observed in the appraisal of literature appears to the assessed variability of the effect of MSA towards concrete properties.

2.3. Solution preparation

Sodium chloride will be used in this paper to study its reaction and effect towards MSA concrete. Type of NaCl used were C1119 Bendosen 3301231. While the percentages of NaCl that will be used is 2.73% which been mixed with water for curing purposes. The percentages was selected according to its salinity content 3.5% from sea water salinity level, 78% of them is consist of NaCl [12]. Hence, to obtain similar impact towards NaCl composition contained in sea water, utilisation of 2.73% NaCl solvent were used in this study. Measurement method for quantity of NaCl uses are based on the total weight of water that been used for curing purposes. While, the temperature of cured water was maintained with 20 °C to 29 °C according to minimum and maximum temperature of water been measured using thermometer.

2.4. Compressive strength test

Compressive strength is one of mechanical analysis which done to concrete specimens to obtain the maximum limits of load been ex-posed until its failed [24]. It was normal analysis that been done in order to identified concrete strength. The amount and strength percentages are depends on numerous factor such type of cements, type of mixes, admixture uses and etc. [17]. Thus, the effect of MSA in concrete were study in this paper. Normally, there are two type of specimens that will be used in this analysis either cube or cylinder specimens. However, cube (100mm x 100mm x 100mm) specimens were selected in this research due to handling purposes and reduce quantity of materials used. The compressive strength were analysed according to each percentages of MSA and also curing duration using NaCl. The procedure of these analysis ac-cording to BS EN 12390-3:2009 [26].

2.5. Split tensile test

In general, concrete were designed to identify maximum resistance towards its compression limit when exposed to load. However, some relevant knowledge about resistance towards tension also need to be considered to obtain its maximum load which can cause cracks on concrete [12, 16]. Tensile analysis can be divide into three categories which are direct tensile strength analysis, bending analysis and split tensile analysis [12]. On the other hand, split tensile strength analysis were selected in this study which accordance to EN BS standard 12390-6:2009 [27]. Concrete cylindrical with 100mm x 150mm sizes were obtained using compression machine where each sample placed between saucer in line with horizontal axis and then load imposed until sample breach form separatist along diameter vertical axis sample.

2.6. Water capillary absorption test

Water absorption through its higher capillary content are cause by the concrete consist of high porosity [28]. Water molecule will penetrate through its capillary pores when attraction force between pores exceed the attraction force between water with water. Thus, the rates of pressure absorption are depending on its diameter of porous concrete structure [29]. According to this study, water capillary absorption were analysed to identify the rates of water absorption through its capillary which been measured according to its weight specimens when contact with bottom surface of specimens only. This is because tape will be wrap on surrounding surface of specimens to prevent absorption happened on other surface. This experiment was carried out according to RILEM CPC 11.2 standard [30].

3. Mechanical and durability properties of concrete

3.1. Compressive strength.

Compressive strength is usual analysis that be analysed for harden concrete. It meet the criteria in strength properties of concrete itself. Compressive strength of MSA concrete were measured through difference MSA used (0%, 1%, 2%, 3% and 4%) and curing period (7 and 28 days) to evaluate the features and performance for each sample. The result of this analysis are shown in table 4, where each value were taken according to average data for three cube. It show that MSA used in these study is superior to MSA properties as filler which can enhance strength of concrete. The result shows that the higher compressive value for 7 days curing is S0 with 34.15MPa while for 28 days is S1 with 41.7MPa. While S4 indicates lower compressive value in 7 and 28 days which were 18.03MPa and 20.9MPa each. As in figure 7, it shows the graph pattern plotted obtained relationship of compression strength against curing duration. From the graph, concrete with MSA mixture recorded decline. Among of each of different MSA percentages used, MSA with S1 showed the best performed which result the highest compressive strength value compared to control and other specimens.

Table 4. Compressive strength of control and MSA concretes.

Percentages of MSA (%)	Average compressive strength (MPa)	
	Curing period (days)	
	7	28
S0	34.15	38.15
S1	33.4	41.7
S2	30.5	32.2
S3	22.8	29.1
S4	18.03	20.9

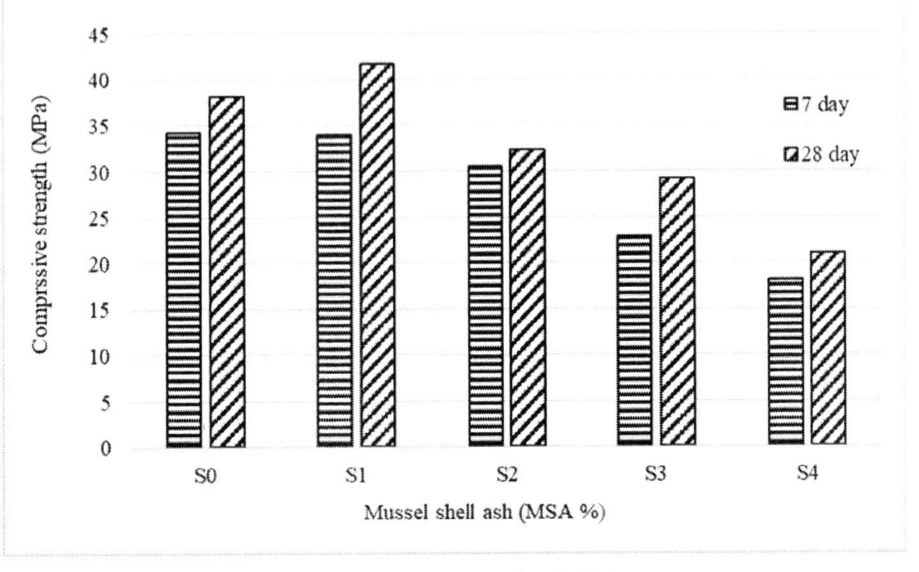

Figure 7. Compressive strength of MSA concrete.

However, it seem that the increments of MSA could decrease its compressive strength, although it increased the strength at the age of 28 days compared to 7 days curing. It can also be seen from Table 4 that at the age of both days, the addition of MSA in concrete resulted in a higher concrete strength with a minimum MSA percentages. A study by Lee state that, utilization of $CaCO_3$ could lead to the decreases strength of concrete [31]. The reaction of OPC with $CaCO_3$ could effect on the binding properties of concrete which resultant to strength loses [12][18].

3.2. Split tensile.

Tensile strength one of major parameter in analyses concrete strength in this research towards cracking value. Figure 8 and table 5 indicates the results of the split tensile strength, where each value was taken from three average specimens. In 7 and 28 days curing, it's showed that S0 records the highest tensile strength opposed S1, S2, S3 and S4 which is 2.86 MPa and 3.33MPa each. Meanwhile, MSA expansion recorded reduction in split tensile strength. It shows that S1, S2 and S3 in 7 and 28 curing period day were 2.72MPa and 2.77MPa, 2.55MPa and 2.76MPa, 2.34MPa respectively. Based on data obtained, S4 records the lowest split tensile strength value in 7 and 28 curing day. From both graph in figure 7 and figure 8, it shows that both split tensile and compressive strength giving almost similar reduction pattern with the increment of MSA percentages. According to Yan, compressive and split tensile strength are significantly related which split tensile strength can be predicted based on it compressive strength value [32]. Thus, the graph pattern produced for compressive strength and split tensile almost the same in accordance with increase of MSA.

Table 5. Split tensile strength of control and MSA concretes.

Percentages of MSA (%)	Average split tensile strength (MPa)	
	Curing period (days)	
	7	28
S0	2.86	3.33
S1	2.72	2.92
S2	2.55	2.76
S3	2.34	2.43
S4	1.71	2.02

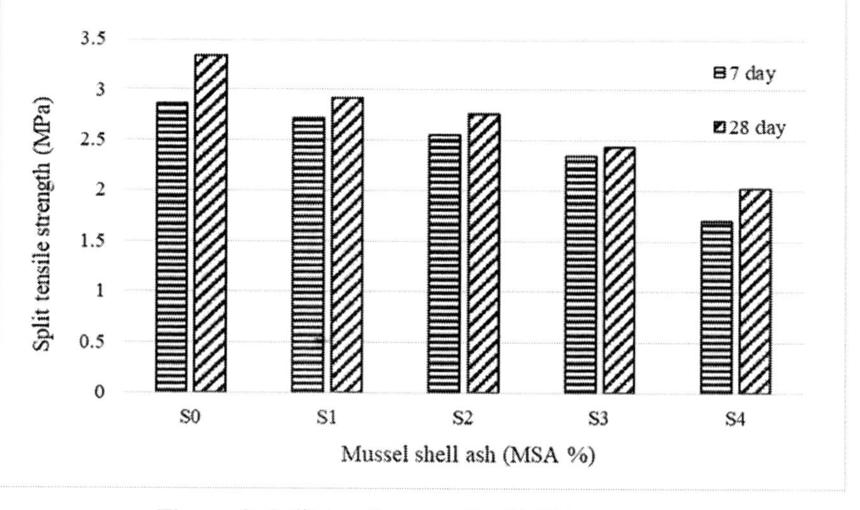

Figure 8. Split tensile strength of MSA concrete.

3.3. Water capillary absorption.

Permeability of concrete were controlled by its capillary porosity which the coefficient of permeability are rely with capillary pores [12]. Thus, the efficiency of the absorption force depends on the nature of the pores surfaces and diameters of concrete itself [33-34]. Capillary absorption was measured based on 7 and 28 days curing in NaCl solution and analysed for 24 hour maximum duration. The capillary absorption coefficient (k-value) were obtained in Table 6 and graph were plotted according to 7 and 28 days curing specimens presented as in figure 9.

Table 6. Results of water capillary absorption coefficient (k-value).

MSA percentages (%) / days	k-value ($x10^{-3}$)(cm/s)							
	Time (s)							
	300	600	1200	1800	3600	10800	21600	86400
S0 7 days	10.007	2.041	1.443	1.021	1.833	1.636	0.953	0.964
S0 28 days	6.351	1.089	0.962	0.629	0.833	0.866	1.497	0.374
S1 7 days	10.392	1.837	1.010	1.061	1.750	1.636	0.987	0.966
S1 28 days	3.464	0.953	0.577	0.471	0.556	0.321	0.544	0.339
S2 7 days	2.021	0.408	0.674	0.314	0.556	0.642	0.408	0.476
S2 28 days	4.234	0.408	0.722	0.629	0.444	0.289	0.431	0.318
S3 7 days	2.502	0.408	0.674	0.314	0.500	0.577	0.295	0.635
S3 28 days	4.041	0.816	0.866	0.629	0.333	0.609	0.227	0.306
S4 7 days	3.175	1.225	0.481	0.471	0.556	0.738	0.340	0.459
S4 28 days	3.657	0.544	0.577	0.314	0.611	0.577	0.249	0.373

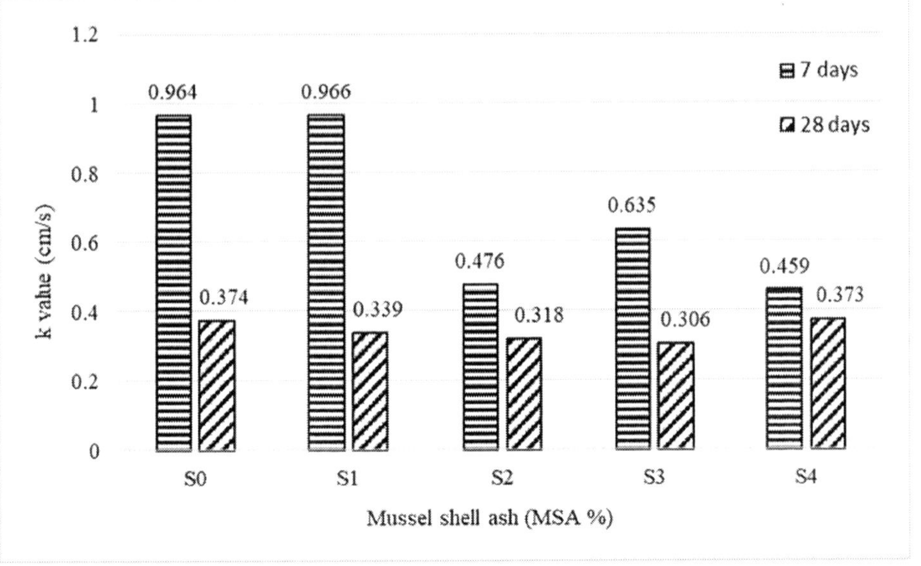

Figure 9. Water capillary absorption k-value versus time.

The result show that higher water absorption through capillary had decreased by the increases amount of MSA used in concrete admixture. According to both value and plotted graph in figure 10, the capillary absorption through 7 days curing of control concrete 10.007cm/s for initial five minutes. Increases of MSA to 2%, 3% and 4% lead to capillary absorption to 2.021cm/s, 2.502cm/s and 3.175cm/s each during first five minutes. Even tough, absorption rate for 1% value is higher compared

0% value and the increases of MSA had increased water absorption through its capillary, however the absorption value had decreased accordingly with the time. During the final time (24 hours), the absorption of S1, S2, S3 and S4 are 0.966cm/s, 0.476cm/s, 0.635cm/s and 0.459cm/s which decreasing compared to previous time and it was lower compared S0 specimens.

Based on 28 days curing, it shows that the absorption had decreases compared to 7 days duration for S0 and S1 in five minutes intervals. However, S2, S3 and S4 indicates higher capillary absorption in 7 days curing which are from 2.021cm/s to 4.234cm/s, 2.502cm/s to 4.041cm/s and 3.175cm/s to 3.657cm/s respectively in the first five minutes. As the increases of duration time to 24 hours, the rate of capillary absorption start decreasing to 0.318cm/s, 0.306cm/s and 0.373cm/s. It can be seen that the water proofing effects of MSA tend to appear more obviously in the concrete mixtures containing MSA admixture compared to control specimens. From both value in 7 and 28 days specimens, its shows that the water capillary absorbing more water at early time compared to final time. According to Jalal et al., its due to fully dried sample are more tendency and consist higher rates absorbing water at early times compared to sample which were immersed for long times in water, however longer time resulting more realistic absorption value [35]. Thus, the value were taken only according to 24 hours duration.

According to data obtained, the results obviously show descending trends by increase in time of coefficients k-value. Its shows that MSA obviously tend in giving effects towards water proofing in concrete compared to normal specimens. It may due to MSA itself which contained filler properties due high $CaCO_3$ which reduce capillary pores in concrete specimens. Less porosity could lead to less permeability concrete which can reduce the ions from penetrates into the concrete which can give impact to its strength [12, 17, 16]. Thus, the addition of MSA with micro particles tend in improving pore structure of concrete which can reduce the rate of capillary absorption significantly.

4. Conclusion

In this study, utilisation of mussel shell ash that generates from mussel waste as an admixture in concrete has been studied. Then, its materials, concrete specimens were analysed according to data collected above. Based on results in this study on the performance of mussel shell ash concrete under curing effect on sodium chloride solution. The physical properties related to MSA (specific gravity, setting time, PSA, slump value) remain unimpaired with OPC which MSA consist small particle diameter of 10%, 30%, 50%, 60% and 70% cumulative and lower specific gravity value compared to OPC. It was analysed that small particle suitable to a filler in concrete. According to workability analysis, it seen that the increment of MSA could improve fresh concrete workability with the higher slump by using 4% MSA. The setting time for MSA indicates faster initial and final value due to its properties in hydration reaction with cements. The MSA addition percentages in term of performance of fresh properties were S4 as in (Table 1).

Addition of MSA into concrete contributed in reducing its mechanical properties. It shows in Table 4 and Table 5 which the increment of MSA percentages had reduce it compressive and split tensile strength. S4 indicates the lower mechanical strength in 7 and 28 days curing in NaCl solution with 18.03MPa, 20.9MP for compressive value and 1.71MPa, 2.02MPa for tensile value. Split ten-sile strength indicates the same patterned as compressive value with the increment of MSA. It due to reaction sequences in MSA (>95% $CaCO_3$) [17] as filler which later stages causes formation of ettringite leading first then to gypsum deposition and finally to thaumasite formation due to C-S-H breakdown forming from present of sulphate [31, 36-39]. However, MSA still can be applied in concrete mixture but should permitted in minimum quantity.

According to water capillary absorption analysis, it seen that MSA could reduce concrete permeability. Data were obtained according based on final times of analysis due to its realistic value compared to early value. Result as in Table 6 indicates that rate capillary absorption of S0 (0.964cm/s and 0.374cm/s) is higher compared to S2, S3 and S4 during final time (24 hours) for 7 and 28 days curing with 0.476cm/s and 0.318cm/s, 0.635cm/s and 0.306cm/s, 0.459cm/s and 0.373cm/s respectively. Is seen that the lowest rate of absorption was S3 for 28 days curing and S4 for 7 days curing. Overall, the significant k-value for water capillary absorption is S2.

Although the addition of MSA to concrete does not affect the relationship between it permeability of the each specimens to its strength, the optimum percentages on MSA content for concrete strength was stressed by S1, S2 for permeability and durability and S4 for workability and early strength development. The utilisation of mussel shell might be in small amount and percentages, however if it was applied in larger scope in construction materials, it could involve in large quantities which was potentially improving concrete properties and also reduce the mussel shell waste generated.

5. References

[1] Malakahmad A, Abualqumboz, M S, Kutty S R M and Abunama J 2017 Assessment of carbon footprint emissions and environmental concerns of solid waste treatment and disposal techniques; case study of Malaysia *Waste Management* **70** 282-292
[2] Badgie D, Mohd Armi Abu Samah, Manaf L A and Muda A 2011 Assessment of municipal solid waste composition in Malaysia: Management, practice, and challenges. *Pol. J. Environ. Stud.* **21** (3) 539-547
[3] Ali H, Dermawan D, Ali N, Ibrahim M and Yaacob S 2012 Masyarakat dan amalan pengurusan sisa pepejal ke arah kelestarian komuniti: Kes isi rumah wanita di Bandar Baru Bangi, Malaysia. *Geografia: Malaysian Journal of Society and Space.* **8** (5) 64 - 75
[4] Jabatan Perikanan Malaysia 2017 Data Statistik Tahunan [Internet]. Diakses pada 29 Jun, 2017 daripada laman sesawang: http://www.dof.gov.my/dof2/ resources/Perangkaan
[5] Martínez-García, C, González-Fonteboa B, Martínez-Abella F, and Carro- López, D 2017 Performance of mussel shell as aggregate in plain concrete *Construction and Building Materials* **139** 570–583
[6] Xiong J, Qin Y, Islam E, Yue M. and Wang W 2011 Phosphate removal from solution using powdered freshwater mussel shells. *Desalination* **276** 317–321
[7] Mehta P K and Monteiro P J M 2014 *Concrete: Microstructure, Properties and Materials. United States of America. Fouth Edition* (United States of America: McGraw Hill Professional) pp 85-86
[8] Oliviaa, M, Mifshella A A and Darmayanti L 2015 Mechanical properties of seashell concrete Procedia Engineering **125** 760 – 764
[9] Kuo W T, Wang H Y, Shu C Y and Su D S 2013 Engineering properties of controlled low-strength materials containing waste oyster shells *Construction and Building Materials* **46** 128–133
[10] Othman N H, Abu Bakar B H, Don M M and Johari M A M 2013 Cockle shell ash replacement for cement and filler in concrete. *Malaysian Journal of Civil Engineering* **25**(2) 201-211
[11] Yang J, Tang Q, Wu Q, Li X and Sun Z 2017 The effect of seawater curing on the properties of magnesium potassium phosphate cements *Construction and Building Materials* **141** 470-478
[12] Neville A M 2011 *Properties of concrete. 5th Edition* (England: Prentice Hall) pp 120-121
[13] Farnam Y, Villani C, Washington T, Spence M, Jain J and Weiss W J 2016 Performance of carbonated calcium silicate based cement pastes and mortars exposed to NaCl and $MgCl_2$ de-icing salt *Construction and Building Materials* **111** 63–71
[14] Qiao C, Suraneni P and Weiss J 2018 Damage in cement pastes exposed to NaCl solutions.*Construction and Building Materials* **171** 120–127
[15] Shi Z, Geiker M R, Lothenbach B, De Weerdt K, Garzón S F, Enemark-Rasmussen K and Skibsted J 2017 Friedel's salt profiles from thermogravimetric analysis and thermodynamic modelling of Portland cement-based mortars exposed to sodium chloride solution *Cem. Concr. Compos.* **7** 73–83
[16] Neville A M and Brooks J J 1987 *Concrete Technology.* (England: Longman Scientific & Technical Press) pp 57-58

[17] Hewlett C P 1998 *Lea's Chemistry of Cement and Concrete Fourth Edition.* (Butterworth-Heinemann: Elsevier) pp 250-255
[18] Sun J and Chen Z 2018 Influences of limestone powder on the resistance of concretes to the chloride ion penetration and sulfate attack *Powder Technology* **338** 725–733
[19] Higgins D D 2003 Increased sulfate resistance of GGBS concrete in the presence of carbonate *Cement & Concrete Composites* **25** 913–919
[20] British Standards Institution 2000 Cement composition, specifications and conformity criteria for common cements : British Standard European Norm: BS EN 197-1: (2000) London
[21] British Standards Institution 2005 Methods of testing cement. Determination of setting times and soundness: British Standard European Norm: BS EN 196-3: (2005) London
[22] Wen-Ten K, Wang H Y, Shu C Y and Su D S 2013 Engineering properties of controlled low-strength materials containing waste oyster shells *Construction and Building Materials* **46** 128-133
[23] British International Standards 2009 Particle size analysis. Laser diffraction methods:British International Standards: BS ISO 13320: (2009) London
[24] Gambhir M L 2013 *Concrete Technology: Theory and Practice. Fifth Edition* (New Delhi: Tata McGraw-Hill Education) pp 75-351
[25] Day K W 1999 Concrete Mix Design Quality Control and Specification. Second Edition. (Victoria, Australia: E & FN SPON) pp 53-54
[26] British Standards Institution 2009 Testing hardened concrete–Part 3: Compressive strength of test specimens: British Standard European Norm: BS EN 12390-3: (2009) London
[27] British Standard Institution 2009 Testing Hardened Concrete. Tensile Splitting Strength of Test Specimens: British Standard European Norm: BS EN 12390-6: (2009) London.
[28] Karagiannis N, Karoglou M, Bakolas A and Moropoulou, A 2016 Building materials capillary rise coefficient: concepts, determination and parameters involved *J. Delgado (Ed.), New Approaches to Building Pathology and Durability. Building Pathology and Rehabilitation, Springer Singapore* **6** 27-44
[29] Al-Naddaf M A 2018 new automatic method for continuous measurement of the capillary water absorption of building materials *Construction and Building Materials* **160** 639–643
[30] Rilem 1982 Absorption of water by concrete by capillarity : CPC 11.2 (1982)
[31] Lee S T, Hooton R D, Ho-Seop Jung, Du-Hee Park and Chang Sik Choi. 2008 Effect of limestone filler on the deterioration of mortars and pastes exposed to sulfate solutions at ambient temperature *Cement and Concrete Research* **38** 68–76
[32] Yan K, Xu H, Shen G and Liu P 2013 *Prediction of Splitting Tensile Strength from Cylinder Compressive Strength of Concrete by Support Vector Machine. Advances in Materials Science and Engineering.* Retrieved on April 20, 2018 from https://www.researchgate.net/publication/258395854_Prediction_of_Splitting_Tensile_Strength_from_Cylinder_Compressive_Strength_of_Concrete_by_Support_Vector_Machine
[33] Debouchaa W, Lekloua N, Khelidja A and Oudjit M N 2017 Natural pozzolana addition effect on compressive strength and capillary water absorption of Mortar *Energy Procedia* **139** 689-695
[34] Habeeb G A and Mahmud H B 2010 Study on properties of rice husk ash and its use as cement replacement material *J. Material Research.* **13** 185-190
[35] Jalal M, Reza A, Pouladkhan, Norouzi H and Choubdar G 2012 Chloride penetration, water absorption and electrical resistivity of high performance concrete containing nano silica and silica fume *Journal of American Science* **8** (4) 278-284
[36] Irassar E F, Bonavetti V L, Trezza M A and González M A 2005 Thaumasite formation in limestone filler cements exposed to sodium sulphate solution at 20 ºC.*Cement & Concrete Composites.* **27** 77–84

[37] Ramezanianpour A M and Douglas Hooton R D 2013 Thaumasite sulfate attack in Portland and Portland-limestone cement mortars exposed to sulfate solution *Construction and Building Materials* **40** 162–173
[38] Hanaa Y Ghorab, Fouad S. Zahran, Mohamed Kamal and Amr Said Meawad 2018 On the durability of Portland limestone cement: Effect of pH on the thaumasite formation *Housing and Building National Research Center. HBRC Journal* **14** (3) 340-344
[39] Justnes H 2003 Thaumasite formed by sulfate attack on mortar with limestone filler *Cement & Concrete Composites* **25** 955–959

Acknowledgement
The authors would like to thank the Faculty of Civil and Environmental Engineering, UTHM and Cluster of Advanced Construction Materials, Jamilus Research Centre, UTHM for the opportunities, facilities and support in conducting this study.

IOP Conf. Series: Materials Science and Engineering **601** (2019) 012034 doi:10.1088/1757-899X/601/1/012034

Physical and Mechanical Properties of Concrete Containing Green Mussel (*Perna viridis*) Shell Ash as an Admixture

Nurul Nabilah Ismail[1], Nor Hazurina Othman[1,2*] Goh Wan Inn[1,2] and Muhammad Shabery Sainudin[1]

[1] Faculty of Civil and Environmental Engineering, Universiti Tun Hussein Onn Malaysia, 86400 Parit Raja, Batu Pahat, Johor, Malaysia
2 Jamilus Research Centre, Universiti Tun Hussein Onn Malaysia, Batu Pahat, Johor, Malaysia *Corresponding author: hazurina@uthm.edu.my

Abstract. In line with the aim of applying green technology in construction for a sustainable living and environmental preservation, this study focused on the utilization of green mussel shells, which come from unmanaged aquaculture residues. The shell ashes is used as an admixture material in the concrete mix. Green mussel shell, which had undergone drying, crushing, grinding and sieving processes were added into the concrete mix at different percentages of 1%, 2%, 3% and 4%. The concrete specimens were water-cured for 7, 28 and 60 days with the target strength of 30MPa in the 28 days. This research also focused on material characterization in term of physical properties of the mussel shell ash. The physical properties of the mussel shell ash were determine by the specific gravity test, setting time analysis and the particle size distribution. Furthermore, workability of the wet concrete was verified by conducting the slump test. Testing of the compressive strength, tensile strength, and capillary absorption have been conducted on the hardened concrete that contained the mussel shell ash mix right after the 7, 28, and 60 curing days. The high calcium carbonate (CaCO3) content in the mussel shell served as filler within the pores of the hardened concrete. Furthermore, the concrete specimens containing the mussel shell ash (GMSA2) demonstrated higher compressive strengths when compared to the control specimens. Apart from supporting the sustainability concept, this research also concluded that the addition of the mussel shell ash has improved the mechanical properties of the hardened concrete.

1. Introduction

The Eleventh Malaysia Plan (11MP) has made green technology growth has an importance in the socio-economic development of the country. This is made through capacity and capabilities enhancement of the green technology innovation in the construction industry. Advantages of applying green technology concepts in construction is the reduction of the negative environmental impact and improving human health by reducing waste disposal, controlling pollution and minimizing environmental degradation. As energy sector, which is as the thrust of the 11th Malaysia Plan, building sector that pursuing in applying green technology in construction also among advancements that need to be put into practice. Usage of waste materials as research in construction materials has been widely adopted. There have been many reported waste materials that have the potential to be used as a replacement or admixtures in building materials. Some had focused on the waste from rice husk ash as partial cement replacement [1], while another focused on fly ash and coconut shell fiber as admixtures in the concrete mix [2].

The importance of the green mussel in Malaysia's economic field is undeniable, as its production have reached 1,827.27 tonnes in the past two years [3]. The green mussel, which is a contributor to the agricultural sector under the aquaculture category, contributed 5.9% of the Gross Domestic Product in the second quarter of 2017 [4]. Each 100g of the green mussel meat contains about 11.90g of protein [5]. Therefore, the mussels are suitable for dietary purposes because of the high protein content and the low fat content. Due to health awareness, adults today are keen to take more seafood in view of their nutritional benefit and one of the cheapest protein [5]. A survey made in the vicinity of the Danga River of Johor showed that the green mussel by product which is the shell itself, simply discarded in the farming process without systematic waste management and over time, it caused rotting smell which attracted all sort of flies.

The extremely slow decaying process of the shells can be a source of environmental pollution such as air pollution [6]. Previous research showed the suitability of many types of seashells to be used as filler materials [6] and cement replacements [6-9]. There are also other researches demonstrated the potential use of periwinkle shells as the fine aggregate substitution [10-11]. The use of these shells waste can enhance the microstructure, mechanical properties and durability of the concrete which is quite difficult to achieve with the use of ordinary Portland cement alone [12]. In this regard, there was a preliminary study found that the high calcium carbonate, CaCO3 contained in the chemical composition of green mussel shell (> 90%) that almost matched with the calcium content in the ordinary Portland cement and it almost equal to the limestone that is used to produce the Portland cement [13]. Calcium carbonate (CaCO3) was found to be the highest amount of compound in the green mussel shell with 95.25%, 95.6% and 95.7 % by weight [14-13][16].

The crystalline structure in the green mussel is made up of the araganite and the calcite that contribute to its higher strength and density than the typical limestone [13]. The main layer of seashell was outer layer made of calcite whereas the inner layer or known as nacre indicated that is calcium carbonate content. [16] In addition, green mussel shell ash can be a filler in the concrete pores to increase the concrete strength [6]. Garcia et al [14] reported micro-structure in green mussel shell show outer part have prismatic layer and periostracum layer, while inner layer content has nacre layer that rich with CaCO3.This study focuses on the potential use of the green mussel shell ash as an admixture in concrete.

2. Material and mix proportion

Ordinary Portland cement Type 1 used is in accordance to the requirements in the BS EN 197-1:2000 and supplied by Tasek Group.

Coarse and fine aggregate passed the requirement set by the BS 882: 1992. They were sieved to ensure the aggregate used passed the 20 mm mesh and retained on the 10 mm mesh for the coarse aggregate and passed the 5 mm mesh for the fine aggregate. Apart from determining the particle size distribution, sieving was also used to segregate impurities.

The green mussel shells were collected from a mussel processing farm in the Sungai Danga, Johor. The green mussel shells went under several processes involving cleaning, drying, crushing, grinding and sieving before turning into ash. Cleaning process was done to remove impurities attached to the mussels by immersing in a clean water reservoir. Drying was carried out by exposing the shells under direct sunlight to ensure that the scalp are fully dried. Then, green mussel shell was crushed using Jaw crusher machine and grinded by using steels ball that rotated in ball mill for 4 hours. Next, the usable mussel shell ash was obtained by collecting the ashes that passed through the 75μm mesh.

Water used in mixing and curing must be free of impurities such as organic matter, silt, clay, acid, alkali, salt and wastewater. Generally, the permissible pH value of the water is between 6 and 8 [17]

The concrete mix design in this research was implemented in an accordance to the Design of Experiment (DoE) method. The concrete mix design process was repeated until the target mixture ratio that meets the pre-set concrete workability has been achieved. The green mussel shell ash admixture percentage levels selected for the study were 0% (C), 1% (CMA1), 2% (CMA2), 3% (CMA3) and 4% (CMA4). Specimen were prepared in two types of mould which were the cube (100mm x 100mm x 100mm) and the cylinder (ø100 x 200mm). The concrete were designed to achieve an equivalent of grade 30 MPa concrete after 28 days curing. All the concrete specimen were subjected to 7, 28 and 60 days curing.

3. Methodology

3.1. Material characterization

Several material characterization were determined by the physical properties of the green mussel shell ash. The physical properties interest were specific gravity (BS EN 1097-6:2000), setting time (BS EN 196-3:2005) and particle size distribution (BS ISO13320:2009). Specific gravity formula is as pointed in equation (1).

$$Sg= (m_2-m_1)/ (m_4-m_1)-(m_3-m_2) \quad (1)$$

where;
m_1= weight of empty flask
m_2 = weight with stopper
m_3 = weight with sample and half kerosene
m_4 = weight with sample and fully kerosene

3.2. Fresh state

Concrete workability was measured immediately after mixing by using slump test that is in reference with the BS EN 12350-2:2009. In this research, the slump test was carried out in each every mixing process to ensure the quality of each batch is maintained.

3.3. Hardened properties

There were two types of mechanical properties test for the concrete. The first one was carried out to assess the compressive strength of the control concrete and concrete with different level of admixture percentage according to the standard procedure of the BS EN 12390-3: 2009. Next, split tensile test was performed in accordance to the BS EN 12390-6:2009 for the cylindrical sample. The capillary water absorption test was conducted using the RILEM CPS 11.2 as reference as per shown Figure 1. All the hardened properties test were conducted after the curing processes of 7days, 28days and 60days. Sample was seal and stopper was added under the specimen to ensuring water uptake from below surface. Reading from 3 samples were averaged. The capillary water absorption coefficient (K) was measured using eq (2).

$$Q/A = k\, t^{0.5} \quad (2)$$

where:-
k = capillary absorption coefficient (cm/s).
Q = quantity of water absorb (cm^3)
A = area of concrete surface that water passes through it (cm^2)
t = time in second (s)

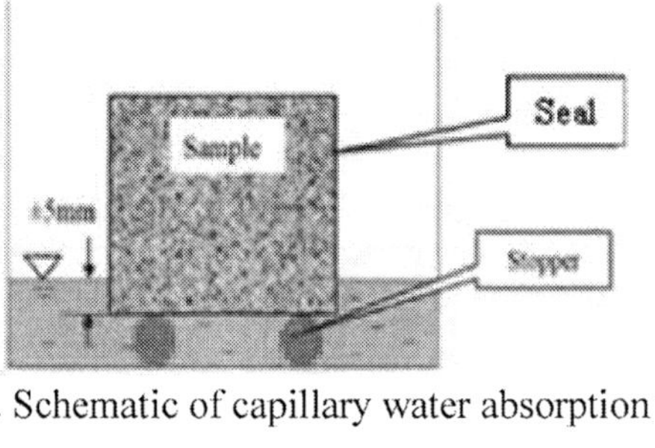

Figure 1. Schematic of capillary water absorption test

IOP Conf. Series: Materials Science and Engineering **601** (2019) 012034 doi:10.1088/1757-899X/601/1/012034

4. Results and Discussions

This part of the paper presents the results and related discussion on the exploration of the green mussel shell ash as admixture in the concrete mixture.

4.1. Material characterization

Physical properties of the green mussel shell ash (GMSA) and the ordinary Portland cement (OPC) were analyzed based on specific gravity, setting time and particle size distribution. Setting time was assessed for cement and cement containing different percentage levels of the green mussel shell ash.

4.1.1 Specific gravity

Table 1 show the specific gravity result of the OPC and the GMSA. On the basis of Table, the specific gravity of the OPC is 3.14. This is quite similar with previously reported value of 3.11 and 3.15 [15-16]. Meanwhile, the specific gravity of the GMSA is 2.52 which is lower than the OPC. Garcia et al. [14] found a slightly higher specific gravity value for the GMSA which is 2.62. Another researcher Jayaraman et al. [18] found that the specific gravity of lime is 2.64. Although, the difference between three values of green mussel shell ash are quite significant, green mussel shell ash still show it is lighter than ordinary Portland cement. Admixture that are finer and lightweight will give advantage in concrete in terms of workability, density, capillaries and segregation [20]. Hence, specific gravity of green mussel shell ash indicates that it is light and have potential as admixture and filler in concretes.

4.1.2 Setting time

The initial time and final setting time analysis were conducted on the OPC and the OPC containing 1%, 2%, 3% and 4% of the GMSA. This test give behaviour of the OPC and different percentage of the CMA when cement paste will be stiffening. Actually, initial time was started during that time. Table 1 show the outcome of the analysis. Overall, both initial and final setting time of all samples decreased, as the percentage of the GMSA increased. Based on the results obtained, adding the GMSA to the concrete as an admixture can reduce the hardening period. Adding 1% the GMSA did not indicate any change in the initial setting time (98 minutes), but a very minimal change be observed in the final setting time (289 minutes versus 288 minutes). The initial setting time result for the 2% and 3% GMSA decreased to 95 minutes from 98 minutes for both cases. When 4% of the GMSA was added, the final setting time decreased to 270 minutes. As discussed in Taruk et al. [19], usage of material with same properties as in the limestone will reduce the initial and final setting time which occur due to the formation of carboaluminate in the cement particles. Considering that the GMSA is mainly made up of lime, then the statement made by [19] support the addition of the GMSA can reduce the hardening duration of the paste. Hardening (setting) time important in concrete mixing, transporting and placing of concrete process.

4.1.3. Particle Size Distribution

Particle size distribution of the OPC and the GMSA are shown in Figure 2 while Table 1 shows the particle sizes of samples according to their respective cumulative percentages. Comparison done between the OPC and the GMSA clearly showed that the GMSA has finer particle size compared to the OPC. The particle size D50 was a diameter of the sample particle size show whereas sample mass half is finer and half is coarse particle size.It show the D50 for the OPC was 18.187μm and the GMSA 8.284μm. Finer size will provide more specific surface area for a faster reaction of the hydration process. As stated by Neville [20], accelerating the hydration process, will resulted in an accelerated development of the concrete strength.

Table 1. Material characterization of ordinary Portland cement and green mussel shell ash

Physical characteristic		
Specific Gravity		
Type	**Value**	
GMSA	2.52	
OPC	3.14	
Setting time		
Type of mix	**Initial time (min)**	**Final time (min)**
OPC	98	289
GMSA1	98	288
GMSA2	95	285
GMSA3	95	282
GMSA4	92	270
Particle size distribution		
Cumulative percentage (%)	**GMSA**	**OPC**
20	0.642	5.283
25	8.233	19.697
50	8.284	18.187
75	24.68	33.536
90	27.374	36.318

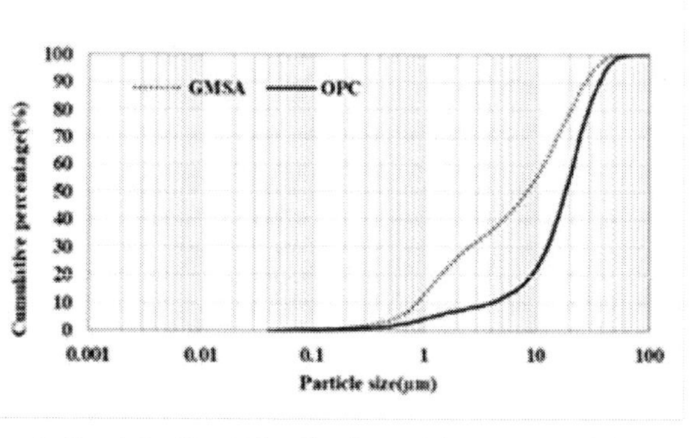

Figure 2. Particle Size Distribution of the OPC and the GMSA

4.2. Workability

Based on Figure 3, the control concrete show the highest workability than the concrete containing the GMSA. As the percentage of the GMSA is increased, the slump value will subsequently decreased. The control concrete slump was rise gradually compared to others concrete containing the GMSA. True slump has been observed in the entire workability test. Particle that are finer size absorb more water and will reduce the workability of the concrete [19]. Othman et al.[6] mentioned that the usage of seashell ash in concrete will fill the concrete pores during the compaction process due to the fineness of the particle size.

4.3 Compressive Strength

As depicted in Figure 4, the compression strength result of the control and the experimental concrete. The results indicated that the compression strength of the concrete containing 2% GMSA of increased significantly from the control concrete at 7 and 28 days. At later age (60 days curing), the GMS2 concrete showed some reduction in strength but still roughly on par with the control concrete's strength. However, the GMSA4 concrete expressed significant reductions compared to every other concretes. Overall, most concretes with different GMSA percentages achieve the concrete target strength of 30 MPa at 28 days saves for the GMSA4 concrete. The optimum percentage of the GMSA as ad-mixture in this study is found to be at 2% GMSA with compressive strength 42.75 MPa, 52.6 MPa and 54 MPa for 7, 28 and 60 days respectively.

4.4. Tensile strength

The average value of the tensile strength containing different percentages of the GMSA was presented in Figure 5. For 7 and 28 days, we can clearly see that the tensile strength of the GMSA2 rose sharply compared to any other concrete including the control concrete. The control concrete at 60 days tensile strength indicated an increment compared to GMSA2 concrete but the difference is only barely evident. As the compression strength results, GMSA4 has been expected to have consequently the lowest tensile strengths across all variant of concrete. However, the GSMA1 and GMSA3 concretes yielded tensile strengths that are approaching to the control concrete strength. At this point, this concluded that 2% of GMSA is suitable dosing admixture in concrete.

4.5. Capillary water absorption

As can be seen in the Figure 6, capillary water absorption coefficient of the concrete containing the GMSA have slightly decreased when the percentage of the GMSA increased. GMSA4 concrete recorded the lowest K value compared to the other specimen at 7 and 60 days ($1.81x10^{-6}$ and $1.21x10^{-6}$). The control concrete at 28 days curing indicated lower capillary water absorption coefficient compared to the other specimen which is $1.58x10^{-6}$. The result of 60 day for the GMSA3 and the GMSA4 coefficient sharply decreased from $1.86x10^{-6}$ the control concrete value to the concrete containing the GMSA $1.85x10^{-6}$ and $1.81x10^{-6}$ respectively.

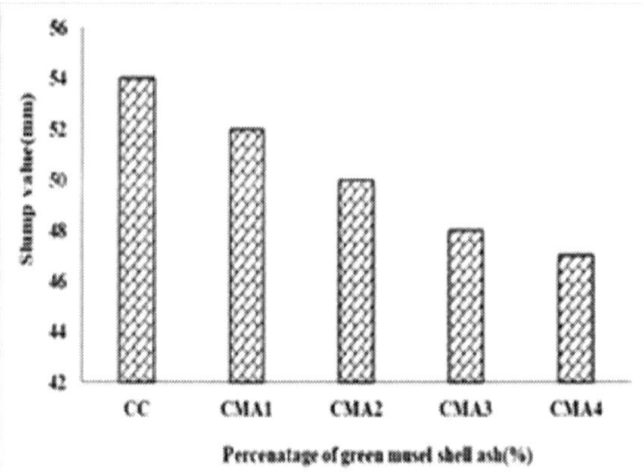

Figure 3. Slump value of the control concrete and concrete containing the GMSA

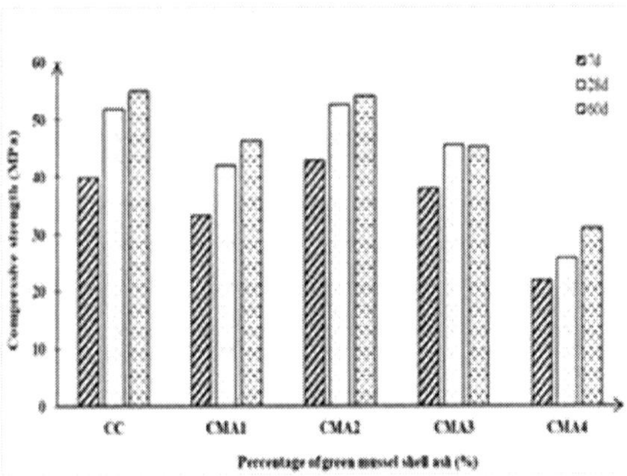

Figure 4. Compressive strength of the control concrete and the concrete containing the GMSA

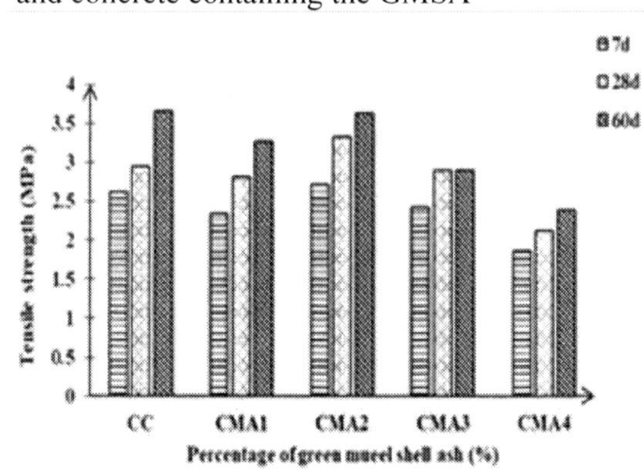

Figure 5. Tensile strength of the control concrete and the concrete containing the GMSA

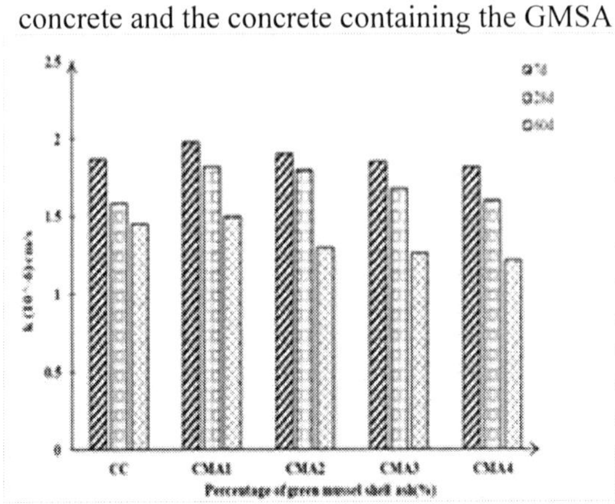

Figure 6. Result of capillary water absorption for the control concrete and the concrete containing the GMSA

IOP Conf. Series: Materials Science and Engineering **601** (2019) 012034 doi:10.1088/1757-899X/601/1/012034

5. Conclusion

In summary, this study show that GMSA has an efficiency as an admixture in concrete. The results of this study may differ with other researchers regarding of the raw material features, components of concrete, proportion and mixing constituents, curing process and among other parameters.

1. Physical properties indicated that high fineness of GMSA particle gives advantages to the hardening time and the workability of the concrete.
2. Slump values decreased as the percentage of the GMSA increased. This means that the concretes have become cohesive and even required a lot of water to maintain the pre-set workability.
3. The optimum percentage of the GMSA for the compressive and tensile strength is 2% of the GMSA.

Overall, the GMSA can produce high strength concrete compared to the control concrete whilst ensuring the concept of sustainable living and environmental preservation are adopted.

6. References

[1] Krishna N K, Sandeep S and Mini K M 2016, Study on concrete with partial replacement of cement by rice husk *ash IOP Conference series: Material and Science Engineering* Vol 148

[2] Ahirwar S K, Chandraul P K and Singh P M K 2017 Experimental Study on Concrete Using Fly Ash and Coconut *International Research Journal of Engineering and Technology (IRJET)* **4**(6) 2381–2385

[3] Statistik hasil pengeluaran akuakultur Malaysia (2016) Retrieve on October 5, 2018 from https://www.dof.gov.my/dof2/resources/user_29/Documents/Perangkaan Perikanan/2016/Kumulatif. pdf. Revised March 16, 2017

[4] Keluaran Dalam Negeri Kasar Suku Tahun Kedua 2017 Retrieved on November 29, 2018 from https://www.dosm.gov.my/v1/index.php?r=column/pdfPrev&id=dlI0RUxjVDFrQW1 Revised August 18, 2017

[5] Teerawut S, Arjpong R and Kwan-On P 2017 Microbiological and physical assessment of Green Mussel (Perna viridis) Coated with Green Tea and Ascorbic Acid Stored Under Modified Atmosphere Packaging *Thammasat International Journal of Science and Technology* **22**(2) 1-9

[6] Othman N H, Bakar B H A, Don MM and Johari M A M 2013 Cockle shell ash replacement for cement and filler in concrete *Malaysia Journal of Civil Engineering* **25**(2) 201–211

[7] Olivia M, Mifshella A A and Darmayanti L 2015, Mechanical properties of seashell concrete *Procedia Eng*. **125** (2015) 760–764

[8] Kuo W, Wang H, Shu C and Su D 2013 Engineering properties of controlled low strength materials containing waste oyster shells *Constr. Build. Materials* **46** 128–133

[9] Yang E, Yi S and Leem Y 2005 Effect of oyster shell substituted for fine aggregate on concrete characteristics: Part I. Fundamental properties *Cement and Concrete Research* **35** (11) 2175–2182

[10] Umoh A A and Ujene A O 2015 Improving the strength performance of high volume Periwinkle shell ash blended cement concrete with sodium *Journal of Civil Engineering, Science and* Technology **6**(2) 18–22

[11] Gagan J and Lejano B 2016 Evaluation of the effects of combining pig-hair fiber as fiber reinforcement and green mussel shells as partial cement substitute to the properties of concrete Proceeding of *DLSU Research Congress(Manila)* vol 4(2012) pp5–10

[12] Siddique R 2008 *Waste material and by-products in concrete*. (Switzerland: Springer Velag Berlin Heidelberg)

[13] Lertwattanaruk L, Makul N and Siripattarapravat C 2012 Utilization of ground waste seashells in cement mortars for masonry and plastering *J. Environ. Manage* **111** 133–141

[14] Garcia M C, Fonteboa B G, Abella F M and Lopez D C 2017 Performance of mussel shell as aggregate in plain concrete *Construction and Building Materials* **139** 570–583

[15] Hamester M R R, Balzer P S, & Becker D 2012 Characterization of calcium carbonate obtained from oyster and mussel shells and incorporation in polypropylene *Materials Research* **15**(2) 204–208
[16] Mageswari M, Manoj C R, Siddarthan M, Saravanan, T P and Princepatwa G 2016 To increase the strength of concrete by adding seashell as admixture *International Journal of Advanced Research in Civil, Structural, environmental and Infrastructure Engineering and Developing* **2**(2) 165–174
[17] Laskar A I 2015 *Concrete Technology Practices* (Oxford, United Kingdom: Alpha Science International Ltd)
[18] Jayaraman A, Senthilkumar V and Saravanan M 2014 Compressive and tensile strength of concrete using lateritic sand and lime stone filler as fine aggregate. *International Research Journal of Engineering and Technology* (IRJET) **3**(1) 79–84
[19] Tarun R N, Canpolat F and Yoon-moon C 2003 Limestone powder use in cement and concrete. Report No. CBU-2003-31, (The University of Wisconsin, Milwaukee: Center for By-Products Utilization).
[20] Neville A M and Brook J J 2011 *Properties of Concrete* (Harlow, United Kingdom: Pearson Education Limited)

Acknowledgement
The authors would like to thank the Faculty of Civil and Environmental Engineering, UTHM and Cluster of Advanced Construction Materials, Jamilus Research Centre, UTHM for the opportunities, facilities and support in conducting this study.

The Effect of Prefabricated Steel Framing System Towards Construction Occupational Safety and Health (OSH)

Mohamad Ariff Mohd Amin[1], Nor Haslinda Abas[1*] and Rafikullah Deraman[1]

[1] Faculty of Civil and Environmental Engineering, Universiti Tun Hussein Onn Malaysia, 86400 Parit Raja, Batu Pahat, Johor, Malaysia
*Corresponding author:nhaslin@uthm.edu.my

Abstract. Construction industry is known for its hazardous physical working environments and high accident risks. In reducing risk at the workplace, Occupational Safety and Health (OSH) is an important aspect. It is a standard which are set in legislation with the purpose to eliminate and reduce hazards at workplace. The objectives of this paper are to identify the major activities of the construction process in IBS residential and the OSH risks associated throughout the construction and to compare the impact of IBS construction with conventional method. This study involved field observation to one (1) construction sites that use IBS construction which are 'prefabricated steel framing system'. Observations were also conducted at the manufacturing facilities of the prefabricated steel framing system. This also involved interview to the executive staff, operatives and supervisors and documenting analysis. The analysis is done by comparing the effect of IBS method with conventional method. In achieving this, one (1) case study was descriptively analyzed and availing a revelation exposure to UV and mobile plant risks as the main risk in IBS construction site. This paper intends to contribute in providing a way to integrate construction process knowledge to eliminate or reduce hazards during construction.

1. Introduction

The construction industry is known for its hazardous physical working conditions and high accident risks. It is one of the most hazardous industry compared to other industries due to the nature of activities at the worksites [1-2]. The construction industry is one of the four major contributing industries in the workplace. Despite, the average five years of death for this industry in Malaysia is 99 employees per year from 2011 to 2015. The death rate for 100,000 construction workers in 2015 was 10.94, higher than the average mortality rate for five years 2011-2015 of 8.17. The trend of fatal injuries to construction workers has risen since 2012, and 140 construction workers have died in 2015, the highest since 2001 and the highest in the 21st century.

As shown in Figure 1, the number of fatal injuries to workers in all industries, overall showed a decline, but the number of fatal injuries of workers in the construction industry showed an increase. The five-year average from 1999 to 2003, number of fatal injured workers in all industries was 898, while the average five years from 2011 to 2015 was 639. The five-year average for the deaths of all industry workers showed a 28.8% decline. Besides, the five-year average from 1999 to 2003, the number of fatal injured workers in the construction industry was 115, while the average five years from 2011 to 2015 was 99. Five-year average for the fatal injuries of all industry workers showed a decrease, but at lower rates at 13.9%.

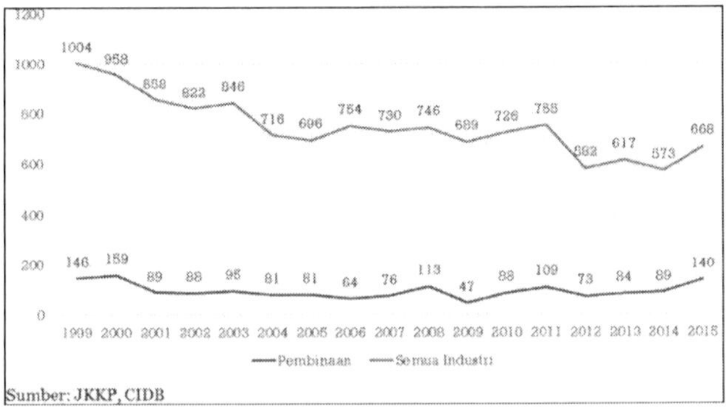

Figure 1. Accident statistics for the construction and others industry from 1999-2015 [3]

Accidents do regularly occur on construction site, though some of the accident causation or prevention models have failed to mention the types of accident on building site, whereas it is domineering to expose the various types of accident occurring on building sites[4].

Due to frequent accidents, IBS construction has been suggested as a replacement to traditional construction methods of the many initiatives that could be implemented to improve OSH performance [5-6]. The Malaysian government is actively promoting the adoption of IBS and encouraging a paradigm shift in the construction process, from a traditional to an industrialized approach [7]. This is demonstrated by the promulgation of the Construction Industry Master Plan (CIMP) 2006-2015 [5], which specifically mentions IBS and its implementation through IBS Roadmaps To date, the importance of IBS in improving construction industry is highlighted under Construction Industry Transformation Programme (CITP) 2016-2020. IBS offers better productivity, quality and safety, and can further help towards a better construction industry, as well as enhance the global competitiveness of Malaysia construction industry [2].

The implementation of IBS changes the nature of activities, which are different from traditional processes. In IBS, the process is industrialized by which components of a building are conceived, planned, fabricated, transported to and then erected on site [8]. Even though there are several studies indicating IBS can significantly reduce OHS risks in traditional construction [9], [10] extent of IBS impact upon safety and health in construction is still unclear as there is no current system to assess OHS risk in the construction process. UK government reports including identified offsite as a strategy for health and safety performance improvement, however, there is an ongoing need to address the lack of knowledge regarding offsite and its affects. Knowledge relating to the identification of the occupational safety and health risks in the offsite factory is required. This study aims to investigate the effect of IBS methods towards construction OSH risk.

The findings are presented by showing the comparison of the hazards/risks associated with the activities of the construction process in IBS and conventional building construction. Due to the limitation of time, this study only covers three IBS methods, which are: pre-fabricated steel framing, pre-fabricated timber framing and formwork system.

2. Literature Review

2.1. Overview of construction industry

The new interest in Industrialized Building Systems (IBS) to replace the traditional in-situ construction method has been fostered as the huge demand for construction projects especially in building construction. The importance of IBS implementation is highlighted in the *Construction Industry Master Plan* (CIMP 2006-2015), under the *Strategic Thrust 5* [6]. The Government of Malaysia has emphasized the full utilization of IBS for government projects by the inclusion of not less than a 70% IBS component [5]. Further, the IBS Roadmap 2011-2015 aims to raise the existing IBS score from 70% to 80% by 2015 for government projects above RM10 million.

The Construction Industry Transformation Programme (CITP) 2016-2020 was initiated to streamline the construction industry, with the vision to create an advanced, highly productive construction industry that will be a major contributor towards Malaysia's ambition of becoming a high-income nation by 2020. The CITP, which sets important strategic goals and milestones to bring Malaysia's construction industry to the next level, aims to transform the construction industry, encompassing four strategic thrusts: 1. Raising the overall productivity level of the industry, 2. Environmental sustainability being incorporated in the design, construction, and subsequent maintenance of buildings and infrastructure, 3. Focusing on improving competitiveness in the capability and capacity of industry players to foray internationally, and 4. Improving the overall quality, safety and professionalism of the industry [11].

2.2. Types of IBS

In Malaysian context, there are six (6) types of construction methods classified under IBS categories [23], which are: i) precast concrete framing, panel and box system; ii) prefabricated steel framing; iii) prefabricated timber framing; iv) formwork system; v) blockwork system; vi) innovative system. In this study, only one (1) types of IBS are covered which is prefabricated steel framing system. The details of selected type of IBS method are described below:

2.2.1. Pre-fabricated steel framing system

This type of IBS classification is commonly used with pre-cast concrete slabs, steel columns and beams. Steel framing systems have always been the popular choice, used extensively in the fast-track construction of skyscrapers. Recent development in this type of IBS includes the increased usage of light steel trusses consisting of cost-effective profiled cold-formed channels and steel portal frame systems as alternatives to the heavier traditional hot-rolled sections [11].

2.3. The phases of IBS construction compare to conventional method

There are differences between construction process between IBS and conventional methods. Commonly, conventional construction only involved delivery of construction materials and components to site, before massive in-situ construction activities begin. The process is different from IBS, in which IBS requires the manufacturing of the building components either in factory or onsite, before transporting to- and installation of the components at the site. The different stages between IBS and conventional is shown in Figure 2. In each stage, there are several activities involved in which these activities create hazards and risk to the workers involved.

The different situation for IBS (off-site) and traditional (in-situ) construction is depicted in Figure 2. For traditional construction, the activities require materials and components (fixtures and fittings) to construct building on site. In contrast, IBS involves materials and components to assemble building elements in a factory, which are delivered and installed on site. Within these phases, there are several activities involved within them.

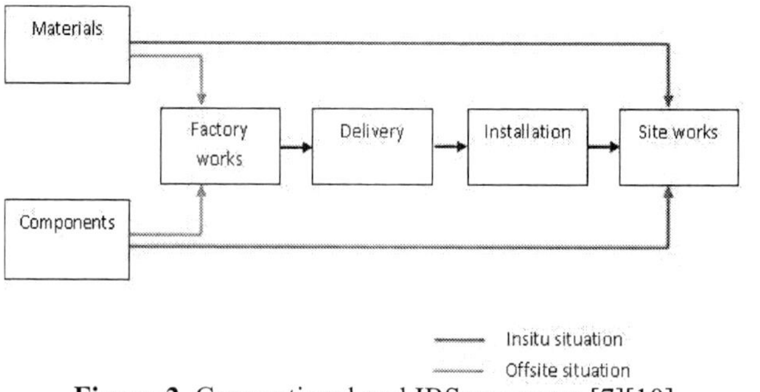

Figure 2. Conventional and IBS processes [7][10]

3. Methodology

This study involved a field observation on one IBS system construction site and one factory of IBS component of prefabricated steel framing system. One site has been visited, which represent IBS types which is 'prefabricated steel framing system'. This involved interview to the managerial staff, operatives and supervisors, and site observing and documenting operations.

Observation can complement interviews as a valuable source of additional data. During the site observation, a significant amount of time has been spent directly observing the activities and process operations of each case.

For the purpose of this research, semi-structured interviews have been used which offer open responses from participants to specific questions. Several artefacts and documentary sources have been collected during the empirical data gathering activities. These included case study organizations' corporate publications, and case study organizations' public website and presentation material. This documentary evidence acts as a method to cross-validate information gathered from interviews and observations and provide guidelines in assisting the author with the inquiry during interview. Documents and records pertaining to the construction process observed are analyses. The corroboration of multiple qualitative techniques for these case studies enhances the validity and reliability of findings. This method also used to support the knowledge obtained from field observation and interview methods.

In addition, this study used the data gathered by Abas [7] when obtaining the information about activities and associated hazards/risks involved in the construction process of conventional method. Next, the data is analyzed by comparing the risk changes for IBS method with conventional in each construction stage. The comparison is made by the authors based on referred guidelines, legislations, journals, and other published materials.

4. Result and Discussion

4.1. Result

This study was completed through a combination of field observation, interviews and document analysis. These served to identify the construction process in terms of activities and the risks associated with the process. This section summarizes the data from field observation and the interviews.

4.1.1. Prefabricated steel framing system

The project was a government project with the contract worth about RM 430,000.00. The system used for the construction of the building uses pre-fabricated steel framing system, by integrating assembling components of a structure in a factory or other manufacturing site and transporting complete assemblies or sub-assemblies to the construction site where the structure is to be located. The connection system used for this project is a combination of bolt and screw and welding joint.

The main risks were associated with the lifting component of materials to the required position, for example site conditions and the stability of materials. In addition, radiation and thermal also highlighted during installation of H-column and I-beam. The risks were then categorized into damaging energies for evaluation.

4.2. Comparison of IBS method and conventional method

4.2.1. Comparison between conventional and pre-fabricated steel framing system

The summary of comparison is shown in Table 1 (Appendix 1). The risks change is described:

Risks removed by IBS

There are several risks that eliminated with the usage of IBS as compared to conventional method. One of the risks that being emphasized with the use of pre-fabricated steel framing system is the risk of manual handling within the activity of installation of column and beam formwork. Besides, the risk that also can be eliminated is the exposure to chemical compound during curing process. The system eliminated the usage of human energy during installation of component, a major benefit in eliminating this serious risk type using offsite.

Risk changes – amelioration of risks

Furthermore, three risks were analyzed to be more controllable and of a less serious nature in the transition from in-situ to IBS, this helps to explain the changes in risk which the take up of offsite has had on the manufacture of products/units. The installation of column and beam were less likely to incur exposure to chemical compound, repetitive tasks and excessive vibration.

Activities with similar risks

From the Table 2, the specific activities; the delivery of materials; load component, delivery of element, off-loading the element and stacking of element was analyzed to be the same for both in-situ construction and IBS. The suggestion is that there are certain activities where the risk remains unchanged regardless of the construction technique or technology used.

Additional risks

Several risks that were apparent as same extent between the IBS techniques and in-situ activities. Welding of the component during component installation were expose to the UV where it led to the exposure to radiation and eye irritating which have a more serious consequence.

4.3. Discussion

From the table above, the delivery stage provides likenesses in the process for most construction approaches; conventional method and pre-fabricated steel framing system This likeness provides general of the processes. However, there are differences in element nature between the approaches, whether it is timber component, formwork system or steel framing. The difference is designed by the way the process is done and the physical feature of the element which will eventually give a different risk.

The main risk associated with the conventional method were to be associated with the transport of material, road traffic, site access, inappropriate method of handling and mobile-plant risk. The main risk associated with in situ site work were associated with dermatitis, repetitive task, manual handling and MSD.

For IBS method, the main risk associated with the pre-fabricated steel framing system at Table 1 were MSD, mobile plant risks, manual handling and exposure. For installation of component, exposure to UV were the major risk for all constructions approaches due to exposed to sun during the installation activities. For instance, 'lifting component' activities for both construction approaches shown that exposure to UV and excessive noise from the mobile plant were the risk which provides likenesses in the process. The in-situ stages provide similarities in the activities which will eventually give a same risk for all construction approaches.

The results also revealed that, for prefabricated steel framing construction method, risks keep on in several activities. For example, the risk in laying mortar during in situ works is similar to that involved in the in-situ concrete and masonry method. Moreover, it consists of several activities during in situ works that are like those of the traditional method, e.g. bricklaying. This is in agreement with [15] that suggest that designers cannot assume that IBS will remove all occupational health and safety risks. Rather, the risk profiles showed that for certain damaging energies, they may be absent or relatively low in one phase but exist in other phases.

5. Conclusion

This paper had identified and discussed the activities and risk associated in IBS construction in Malaysian construction industry. The risk on IBS construction site faced by the workers had been discussed and mainly revolves around UV exposure, MSD, working at height, excessive noise and mobile plant risk. Future research is sought for more investigation to be done for the qualitative stakeholders' perspectives on the factors that impede the IBS adoption in the construction industry.

6. References

[1] Hassanein A A G and Hanna R S 2008 Safety Performance in the Egyptian Construction Industry *Journal of Construction engineering and Management* **134**(6) 451–455
[2] Rahman R A 2015 Managing Safety at Work Issues in Construction Works in Malaysia : A Proposal for Legislative Reform *Modern Applied Science* **9(13)** 108–121.
[3] DOSH 2014 Accident Statistics. Department of Occupational Safety and Health, Ministry of Human Resources. Kuala Lumpur, Malaysia: Department of Occupational Safety and Health (DOSH)
[4] Ng K, Laurlund A, Howell G and Lancos G 2012 Lean safety: Using leading indicators of safety incidents to improve construction safety *In I. D. Tommelein & C. L. Pasquire (Eds.), Proceedings for the 20th Annual Conference of the International Group for Lean Construction: Are We Near a Tipping Point?*
[5] CIDB 2008 Master Plan for Occupational Safety and Health in the Construction Industry 2005-2010. Kuala Lumpur, Malaysia: Construction Industry Development Board (CIDB) *CIDB*
[6] CIDB 2007 Malaysian Construction Outlook 2007. Kuala Lumpur
[7] Abas NH 2015 Development of a Knowledge-Based Energy Damage Model to Assess Occupational Health and Safety (OSH) Construction Risks in Malaysia Development of a Knowledge-Based Energy Damage Model to Assess Occupational Health and Safety (OSH) Construction Risks
[8] Junid SMS 1986 Industrialised Building Systems", *Proceedings of a UNESCO/FEISEAP Regional workshop*, Serdang.
[9] Mckay L J 2010 *The Effect Of Offsite Construction On Occupational Health and Safety.* Loughborough University
[10] Gangolells M., Casals M., Forcada N., Roca X and Fuertes A. 2010 Mitigating construction safety risks using prevention through design *Journal of Safety Research* **41** 107–122
[11] CIDB 2003 *Survey on the Usage of Industrialised Building Systems (IBS) in Malaysian Construction Industry*. Kuala Lumpur

Acknowledgement

The author also would like to acknowledge to Research Management Center (RMC), Universiti Tun Hussein Onn Malaysia (UTHM) for the financial and technical support (Vot No. U784).

IOP Conf. Series: Materials Science and Engineering **601** (2019) 012035 doi:10.1088/1757-899X/601/1/012035

APPENDIX 1

Table 1. Comparison between in situ and masonry method and pre-fabricated steel framing system

In Situ and masonry method				Pre-fabricated steel framing system			
Main Delivery Activities				Main Delivery Activities			
cf IBS	Main Risks	Main Hazards	Main Off-Site Activities	Main Off-Site Activities	Main Hazards	Main risks	cf insitu
S	Vehicle (mobile-plant) risk Musculoskeletal Disorder (MSDs) Repetitive Strain Injury (RSI)	Mobile plant Manual handling	Load onto transport	Transfer to stockyard	Mobile-plant Caught in between object Repetitive work, manual handling Exposure to UV	Mobile plant risk (death) MSD Skin cancer, cornea	S
S	Road traffic risks, Musculoskeletal Disorder (MSDs) Repetitive Strain Injury (RSI)	Mobile plant Inappropriate method of bracing the element	Delivery of element (transportation)	Load onto transport	Manual handling Mobile Plants	MSD, RSI Mobile plant risks (death)	S
S	Mobile plant risks, vehicle (mobile plant) risks	Manual handling,	Off-loading the element at site	Delivery of element (transportation)	Road traffic Inappropriate method	Mobile plant MSD, RSI	S
S	Musculoskeletal Disorder (MSDs) Repetitive Strain Injury (RSI)	Inappropriate method of bracing the element	Stacking of element	Off-loading the element at site	Crane Manual handling Mobile plant risks	Creneage risk RSI, MSD Mobile plant	S
				Stacking of element	Inappropriate method of stacking element	MSD, RSI	S
Main in-situ work activities				Main Component Installation Activities			
cf IBS	Main Risks	Main Hazards	Main On-Site Activities	Main On-Site Activities	Main Hazards	Main Risks	cf insitu
S	Chemical burn from wet concrete, Respiratory tract irritation RSI, MSD Skin cancer, cornea	Exposure to chemical compound Manual Handling Exposure to UV	Mix concrete	Install the BRC for slab	Manual handling Exposure to UV Inappropriate method	MSD, RSI Skin cancer, cornea	LL
S	RSI, MSD Chemical burn from wet concrete, Respiratory tract irritation Hand-arm vibration (HAVs)	Manual handling, repetitive tasks Exposure to chemical compound Excessive vibration Exposure to UV	Concreting and vibrating	Casting slab and vibration	Excessive vibration Electricity	HAVs Electric shock	S

IOP Conf. Series: Materials Science and Engineering **601** (2019) 012035 doi:10.1088/1757-899X/601/1/012035

	Skin cancer, cornea						
ML	RSI, MSD Skin cancer, cornea	Manual handling, repetitive tasks Exposure to UV	Fix rebar/steel bar for column	Column stump construction	Manual handling Exposure to sun, heat	MSD, RSI Heat fatigue	LL
LL	RSI, MSD	Manual handling	Install column formwork, propping	Drill the stump	Repetitious drilling Electricity Manual handling Exposure to spark Exposure to noise Excessive vibration	RSI. HAVs Electric shock MSD, RSI Skin burns Excessive noise, loss of hearing HAVs	A
A	Chemical burn from wet concrete, Respiratory tract irritation Hand-arm vibration (HAVs) Skin cancer, cornea	Exposure to chemical compound Excessive Vibration Exposure to UV	Pour concrete into formwork & vibrating	Anchor bolt setting	Repetitive works Exposure to UV	RSI Skin cancer, cornea	LL
LL	Chemical burn from wet concrete, Respiratory tract irritation	Exposure to Chemical compound	Curing compound	Preparation of column and beam installation	Exposure to UV Repetitive works Muscle injury Manual handling working at height	Skin cancer, cornea RSI MSD MSD, RSI Falling from height	A
A	RSI, MSD	Manual handling	Formwork removal	Install the H-column and I-beam	Manual handling, Excessive noise, Exposure to UV Inappropriate methodworking at height	MSD, RSI Loss of hearing Skin cancer, cornea MSD	A
S	Vehicle (mobile plant) risks Lifting risks, RSI, MSD Skin cancer, cornea	Mobile plant Manual handling, repetitive tasks Exposure to UV	Transport bricks (from storage to working areas)	Welding the H-column and I-beam	Exposure to chemical compound Exposure to UV Electricity	Chemical burn from wet concrete, Respiratory tract irritation Skin cancer, cornea Electric shock	A
				Main In-situ Work Activities			
S	RSI, MSD	Manual handling	Carriage/moving bricks to workface	Transport brick	Manual handling, repetitive tasks Vehicle (mobile plant) Exposure to UV	MSD, RSI Vehicle (mobile plant) risks Skin cancer, cornea	S
A	RSI, MSD Falling from height	Manual handling, overhead work Working at height	Erecting work platforms	Carriage of brick to workplace	Manual handling	MSD, RSI	S
S	Chemical burn from wet concrete, Respiratory tract irritation Skin cancer, cornea	Exposure to chemical compound exposure to UV	Mix mortar	Mix mortar	Exposure to chemical compound exposure to UV	Chemical burn from wet concrete, Respiratory tract irritation Skin cancer, cornea	S
S	Loss of hearing MSD, hand injury	Excessive noise from cutting machine Struck by the saw	Cut bricks	Cut brick	Excessive noise from cutting machine	Loss of hearing Hand injury, MSD	S

					Struck by the saw		
S	Chemical burn from wet concrete, Respiratory tract irritation MSD, RSI Skin cancer, cornea Falling from height	Exposure to chemical compound Manual Handling Exposure to UV Working at height	Laying mortar	Laying mortar	Exposure to chemical compound Manual Handling Exposure to UV Working at height	Chemical burn from wet concrete, Respiratory tract irritation MSD, RSI Skin cancer, cornea Falling from height	S
S	Falling from height, RSI, MSD Chemical burn from wet concrete, Respiratory tract irritation Skin cancer, cornea	Working at height Manual handling, repetitive tasks Exposure to chemical compound Exposure to UV	Laying of each course of bricks	Laying of each course of brick	Working at height Manual handling, repetitive tasks Exposure to chemical compound Exposure to UV	Falling from height MSD, RSI Chemical burn from wet concrete, Respiratory tract irritation Skin cancer, cornea	S
S	RSI, MSD	Manual handling, repetitive tasks	Fix wall ties; fix mesh reinforcement; place lintels and DPC	Fix wall tiles	Manual handling, repetitive tasks	MSD, RSI	S
S	Exposure to toxic chemical Falling from height MSD, RSI Skin cancer, cornea	Exposure to chemical compound Working at height Repetitive task, exposure to UV	Plastering/finishing	Painting and finishing	Exposure to chemical compound Working at height Repetitive task, exposure to UV	Exposure to toxic chemical Falling from height MSD, RSI Skin cancer, cornea	S

Risk change code: **S= Same (no change), A = Additional (new risk), LL = Less likely, ML = More Likely, LC = Less serious consequence, MC = more serious consequence, C = more controllable**

IOP Conf. Series: Materials Science and Engineering **601** (2019) 012036 doi:10.1088/1757-899X/601/1/012036

Identification of Construction Waste Generated at Precast Concrete Plants: Case study

Nurul Ashikin Suhaini[1], Nor Haslinda Abas[1*], Sasitharan Nagapan[1] and Kageishieny Nadarason[1]

[1] Faculty of Civil and Environmental Engineering, Universiti Tun Hussein Onn Malaysia, 86400 Parit Raja, Batu Pahat, Johor, Malaysia
*Corresponding author: nhaslin@uthm.edu.my

Abstract. The government of Malaysia has introduced Industrialised Building System (IBS) or commonly termed offsite construction in other countries to replace the conventional method of construction, as one of the initiatives to overcome the problems related to waste generation and negative environmental impact in the construction industry. There are few studies of the construction waste generation focus on the offsite, particularly at precast production plant as compared to the construction site. This study aims to investigate the significant type of construction waste generated at precast concrete plants in Johor. In order to identify the type of waste generated in the process of manufacturing the precast concrete component, this study involved field observation to three precast concrete plants in Johor, together with analysing the selected organisation's document on the waste generation and management record. Interviews were also conducted with the production man-ager/engineer to support the data collection process. The findings revealed that, the types of construction waste that commonly found in the precast concrete plants are concrete and steel. The most significant type of waste generated at precast concrete factory is concrete (total 1,298.56 tonnes for 3 factories, followed by 820 tonnes, 350 tonnes and 128.56 tonnes). This study provides useful information concerning waste assessment data to achieve a better understanding of construction waste in the precast concrete manufacturing industry.

1. Introduction

Malaysian construction industry is associated with a serious problem regarding the environmental sustainability that produces a large amount of construction waste quantities and negative environmental impacts [1]. In Malaysia, the rapid development of the construction industry led to a lot of waste generated due to increasing of infrastructure projects, commercial and housing buildings [1-2]. In 2016, about 35,335 tons per day of waste disposed in landfill and waste expected to be generated by Malaysian in the year 2020 is approximately 16.76 million tons [3]. Thus, the generation of construction waste is an important issue in Malaysia [4].

According to Augustine [5], construction wastes are resulting from the process of construction structures, including building structure and infrastructure projects. Construction waste can be divided into two groups namely the physical and non-physical waste [6]. The physical waste commonly consists complete material loss which can noticeable on site and used to treat as final disposal. These waste may include timber, concrete, brick, sand, ceramic, plastic, packaging, metal, plastic and cement [6–8]. Meanwhile, the non-physical waste normally as delay time and cost overrun including in inefficiency use of equipment, mishandling materials and poor workmanship during the construction [5], [9-10].

In a study by Lu and Yuan [11] in Hong Kong, construction waste is the main source of urban waste and typically ranging from 10-30% of the total waste disposed of at landfill sites. Lai et al. [12] report that there are about 1.2-1.9 million tons per year generation of construction waste in Taiwan. Whereas, in a study by Jereme et al. [6] revealed that about 18,000 tons of wastes are generated in Malaysia per day and 75% of the wastes generated are collected and disposed to dumpsite or landfills that are not sanitary. The government of Malaysia has introduced prefabricated technology method or known as IBS to replace the conventional method of construction, as one of the initiatives to overcome the problems related to waste generation and negative environmental impact in the construction industry.

Prefabricated method enables manufacturing building components in a controlled environment and then assemble it on site which could produce better building quality and faster project completion. However, Malaysia is still lagging behind in terms of IBS adoption in the local construction industry compared to the developed countries such as Australia, United State, United Kingdom and Japan. For example in Japan, the degree of industrialization for prefabrication and construction industry has becoming advanced where the use of robotics in building construction is now adopted, whereas in Malaysia the degree of industrialization is in mechanization stage (the use of mechanized machines is common)[13].

As indicated by Azmi et al.[14], the use of IBS components for building constructions is an effective waste minimization solution for the general project and commercial project in the construction industry. The most important benefits of this system are can reduce the overall cost, time, manpower, produces better quality buildings and improve the environmental performance for overall site conditions [15-16]. However, there is still weakness in the understanding of these benefits among players in the construction sector. In the past, researchers have highlighted numerous studies the general of construction waste generated on sites, however construction waste also could be produced in a plant environment which are normally generated during manufacturing processes and the causes of waste generated is limitedly reported.

Regarding to this matter, this study seeks to identify the type of waste generated throughout the process of manufacturing of precast concrete components at the precast concrete plant and further clarify the significant types of waste generated. This study is focusing on the waste generated during manufacturing of precast concrete components at the precast concrete plant which does not include the design stage and assembly on sites.

2. Literature review

Manufacturing is different things with construction. The manufacturing described as the process is carried out in the factory by a factor of a controlled environment [16]. In contrast, construction is typically work outsides which refers to the creation of physical structures such as buildings, bridges or roadways [17]. According to Azman et al.[18], IBS concept is similar to Modern Method of Construction (MMC) in the UK and Off-Site Manufacturing (OSM) is used in Australia and the UK construction industry. Meanwhile, prefabrication is used both in Hong Kong and Singapore.

Precast concrete is the one of IBS which made in a factory, where a dedicated concrete batching plant; equipment that combines various ingredients to form concrete and thus to produce a specially designed concrete for building precast products such as structural beams, columns and wall systems and so on. Considering the precast elements have huge volume and heavyweights, thus the mobile crews are required for the manufacturing process [17]. The main materials used in precast elements are concrete and steel bars[11]. In term of the type of elements, precast concrete has two main elements type namely precast reinforced concrete element and prestressed concrete element [20].

Precast reinforced concrete components commonly manufactured in the factory by using pre-assembled form and rebar or wire mesh reinforcement. After casting and curing, these components can be shipped directly to the construction site for installation. For instance, facade or parapet walls, beams, columns, slab, refuse chutes and staircases. Meanwhile, prestressed concrete commonly casting concrete into pre-assembled form in combination with rebar and strand reinforcement. The strand reinforcement would be tensioned to approximately three-quarters of their ultimate strength. Once the concrete cures to its required strength, the tensioning is released. For example, hollow core slabs, beams and planks.

It is evidenced that precast concrete manufacturing plant also could produce waste during the manufacturing process of the component. For example, in Hong Kong, Lu & Yuan [11] investigated three prefabrication plants by taking into account its upstream processes of manufacturing and transportation of precast concrete components. The findings revealed that, the waste generation rate of the precast concrete components are around 2% by weight or lower. In the UK, one of the precast concrete manufacturing plant organisations has come out with it's own plant's waste generation, and revealed that concrete and timber were the most significant type of waste generated at the plant (2% and 1% of waste generated, respectively)[21]. This particular organisation has successfully limited the production wastes (by disposing them to landfill) down to less than 1% of the material used. This was achieved through decreasing the demand on natural resources, efficiently reuse and recycle materials in the production chain.

Meanwhile, Wimalaya et al. [8] were conducted the surveys a few of the precast concrete production plants in Japan, the amount precast concrete production waste varies with different types of products. These types of products were classified into popular (infrastructure work), large-sized (big size and/or heavy weight product) and small-sized products (a product that can easily be carried like the one that is usually sold at a home center). Their findings were found that plants of small-sized products generated in average the highest amount of concrete wastes which was 6.07% of their total production, followed by plants of popular products with 4.45% and plants of large-sized products with only 2.96%.

3. Methodology

The methodologies adopted are discussed in this sub-topic. It includes discussion on the method of data collection process, the approach used, whereby research considerations such as the field observation, interviews and document analysis are also taken into account. Before that pilot study was done to validate the actual processes of precast concrete factory by field observation and discussion interviews with selected industry professionals. Then, the selection of case study for four precast concrete factories were carried out in order to collect actual data for this study.

3.1. Pilot Study

Before conducting a real data collection at precast concrete manufacturing plants, a pilot study was executed by visiting one precast concrete plant in Johor. This pilot study has provided a fundamental representation of the common waste generated in plant. The aim of the pilot study is to test research protocol and data collection instruments in preparation for a real study [22]. As a result, from the observations of manufacturing process at this plant, several common waste were identified such as concrete, steel and timber waste. Hence, this pilot study can assessing the feasibility of a (full-scale) study.

3.2. Selection of Case Study

According to Yin[24], there are four types of case study: single-case (holistic) designs (Type l), single-case (embedded) designs (Type 2), multiple-case (holistic) designs (Type 3), and multiple-case (embedded) designs (Type 4). Type 4- the multiple-case (embedded) approaches have been chosen for this study in order to make a comparison of the type of waste generated between case studies and expected reasons or either similar result from the studies[23]; as cited in [24]

This study involved 3 case studies of precast concrete plants. The selected precast concrete plants are situated in Johore, Malaysia. In this paper, the author code the selected precast concrete plants as A, B and C. The selection of plants took into account the location of the companies (easy access by route transportation) and the availability of participant of interviewees in this study. The methods used in this study are field observation and interviews with production's coordinator of precast concrete plants. Site observation and personal interview were made regarding the current production processes and situation in the plants started from material procurement, manufacturing, storage and delivery precast components process in order to identify types of construction waste and what the most of construction wastes were generated in precast concrete plants. The research tool used and the expected outcome of this study is shown in Table 1.

The participants selected for interview ware the individuals who involved directly in the precast plant and capable to provide appropriate answer and justifications to the interview questions. The details of the interviewees are shown in Table 2.

Table 1. Research tools used and expected outcomes of the study

Objectives	Research tools	Expected outcomes
To identify the types of construction waste produced at precast concrete plant.	Field observation, interview	Collect information regarding common waste generated and waste management practices.
To determine the amount of construction waste produced at precast concrete plant.	Document analysis with record of amount production purchased and supplied, interview	Calculate and tabulate the data of total amount of waste produced.

Table 2. Details of the interviewees

Precast concrete plant	Participants	Position	Working experience
A	Interviewee 1	Quality Assurance & Control (QA&QC) Engineer	4 years
B	Interviewee 2	Head of Production (Hollow core components)	5 years
C	Interviewee 3	Production Engineer	4 years

4. Results and Discussions

4.1. Precast concrete plant A

The precast plant is located in Pekan Nenas, Pontian. This company has had manufacturing division that is one of the leading manufacturers in Industrialised Building System ("IBS"), which offers a full range of precast concrete products for various projects across the Asian region. The manufacturing operation offers all types of building components such as precast reinforced concrete beams, columns, staircases, façade walls, precast prestressed double tee slabs, and hollow core slabs. The company also manufacturing of precast concrete products for infrastructure such as bridge crossings, tunnel linings, railway sleepers and so on. A production facility was set up to produce and supply 15,000 tonnes of concrete products and applied 'Design and Build' option using proven integrated fast track industrialized building systems based on the precast concept for residential buildings, multi-storey carparks, commercial buildings, schools and light industrial buildings.

The processes observed at this plant are precast concrete hollow core slabs manufacturing process. The activities involved in the precast plant during field observation are:

- Receipt of concrete form batching plant
- Cleaning moulds
- Fixing of moulds or casting beds
- Fixing of rebar cages or pulling wire strands
- Concreting
- Curing
- Demoulding or cut off wire strands
- Stripping
- Drilling drainage holes
- Repair or finishing
- Storage

The type of construction waste that identified from field observations in this precast concrete plant were concrete, steel and wood. According to Interviewee 1, the main materials used for building components of precast concrete are cement, sand, aggregates, water and steels. Interviewee 1 also showed the location of construction waste were generated were found from the manufacturing process and sorted out (see Figure 1). Regarding the amount of waste generated, this plant did not possess the proper record of waste generated throughout the process. However, through interviews with interviewee 1, the most significant type of wastes produced at this plant is concrete, followed by steel and wood. Concrete wastes were generally found from concrete pouring activities and losses during the transportation and placing activities in moulds or casting beds and drilling drainage holes for hollow core slab components. The steel waste was generated during the cutting and bending activities.

When asked about the causes of the wastes generated, interviewee 1 was stated that the common causes were mainly due to labour's mistakes and the condition of equipment/machine. Example of labour's mistakes include improper concrete finishing and improper drilling drainage holes. However, the construction waste generated remains low at an estimated 1% as raw material quality is strictly controlled and the machine will be well maintained for production of precast concrete. The quantities of concrete waste were produced solely based on rough estimation by the selected interviewees by reason of they are reluctant to reveal the information of the organisation's documents. The estimated production precast concrete waste quantities are tabulated in table 3.

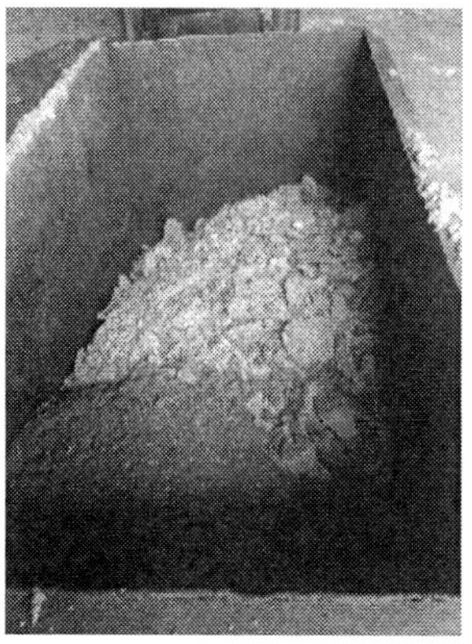

(a)

(b)

Figure 1. (a) Concrete waste generated; (b) Steel waste

4.2. Precast concrete plant B

The precast plant is located in Ulu Choh, Pontian. The precast is specializing in precast concrete business was established in 2002. The company has permanent sheltered factories that have quality management standard to supply precast concrete components all over Malaysia and overseas. The factory's facilities are fully mechanized and automated with a production capacity which able to produce precast products supply 6,215 tonnes per year. The production in precast concrete plant offers robust structural with faster & cleaner construction environment for Industrialised Building System (IBS) products such as precast reinforced beams, columns, staircases, precast concrete wall panels, bay windows, bathroom units, precast prestressed planks and hollow core slabs.

The processes observed at this plant are precast concrete hollow core slabs and beam components. The activities involved in the precast plant during field observation are:

- Receipt of concrete form batching plant
- Cleaning moulds
- Preparing mould/ casting bed
- Fixing reinforcement cage/ pulling wire strands
- Concreting / forming slab around the reinforcement
- Marking slab
- Curing precast concrete elements/ slabs to required length
- Demoulding
- Drilling drainage holes
- Repair
- Lifting components/ hollow core slabs to storage yard

During the field observation, it was noticed that the type of material waste that had found were concrete and steel. The participant also showed the location of construction waste were generated from the manufacturing process and sorted out (see Figure 2). Through interviews with interviewee 2, concrete wastes often occur after the manufacturing process of hollow core slabs components. However, the steel waste generated in this plant is slightly than concrete waste because most reinforcement steel is delivered to the factory on a just-in-time basis in the form of welded mesh that has been manufactured to the right dimensions. There are also several causes of waste generation identified and mentioned by him such as poor attitudes of workers and poor materials handling either in pour concrete activities or lifting components to storage yard. The quantities of concrete waste were produced solely based on estimation by the selected organisation's document. The estimation method used by the organisation; waste = quantity of material bought – quantity of material used. The production precast concrete waste quantities are tabulated in table 3.

(a) (b)

Figure 2. (a) Hollow core concrete waste; (b) Steel waste

4.3. Precast concrete plant C

The precast plant is located in Seelong, Senai. There are major precast elements produced which are box culvert, precast concrete pipes, U-shape drain and tunnel precast pipe. The production of the infrastructure of the pipe is quite varied from the manufacturing process compared to building components. All pipe products are manufactured under controlled factory environment and occupy an area of about 45 hectares, has full casting facility necessary for mass production of precast components. It is demarcated into a few areas, namely the batch plant, casting and molding area, demolding and finishing area, and the storage yard. Site offices are located in proximately to the casting area. A production capacity which able to produce precast concrete products approximately 25,000 tonnes per year. The production processes were observed are tunnel precast pipe.

The processes observed at this plant are manufacturing process of tunnel precast concrete. The activities involved in the precast plant during field observation are:

- Receipt of concrete form batching plant
- Cleaning moulds
- Preparing moulds
- Fixing reinforcement cage
- Concreting
- Curing precast concrete elements
- Demoulding
- Repair
- Lifting components/ hollow core slabs to storage yard

During the field observation, it was noticed that the type of material waste that had found were concrete and steel. However, through interviews with interviewees 3, the most significant type of wastes produced at this plant is concrete, followed by steel. The interviewee was showed the location of construction waste generated from the manufacturing process and it were sorted out to the temporary area (see Figure 3). Through interviews with interviewee 3, concrete wastes were generally found causes of losses during the transportation and placing concrete activities in moulds. Meanwhile, the steel waste was generated due to the cutting and bending activities.

There are also several causes of waste generation identified and mentioned by her such as mishandling machines. For example, the component was fall during lifting to storage yard. Besides that, poor condition of machines and poor workmanship also can contributes the precast concrete had badly damage and rejected. The quantities of concrete waste were produced solely based on rough estimation by the selected interviewees. The reasons could be that either they are reluctant to reveal the information of the organisation's documents. The production precast concrete waste quantities are tabulated in table 3.

(a)

(b)

Figure 3. (a) Tunnel precast pipe concrete waste; (b) Steel waste

5. Discussion

Concrete is the most widely used man-made construction material which can shape the built environment after water in the world [25]. However, the demand in global market for current housing, the construction industry has shifted its focus to mass production assembly and standardising product development[20]. In Malaysia, for example reports that the products of IBS precast concrete is the popular IBS component's producer with a number of 207 plants.

Even so, these study have few studies or have not focused on the construction waste generation in offsite, particularly at precast production plant as compared to the construction site. In this study, we investigated the significant type of construction waste generated at precast concrete plants in order to identify the type of waste generated in the process of manufacturing the precast concrete component.

We found that in the construction waste for existing case studies, it was noticed that the types of construction wastes found on each of the precast concrete plants were almost similar in nature. The type of construction wastes were found is concrete and steel waste. Based on data collections, concrete waste are the highest construction waste generated in the precast concrete plants, followed by steel and wood waste. These findings extend the most type of construction waste were significant produced within manufacturing process which started from fixing reinforcement cage to lifting precast concrete components processes. In addition, the construction waste generated in the precast concrete plants still remains low as estimated 1-3% of waste. It is because of the organisations are strictly controlling materials in the precast concrete plants and the component were made in time as scheduled to deliver to the clients.

From the survey in the precast plants, it is observed that the total amount of precast concrete production was reaching a number of 294 701.44 tonnes in 2018. From the table, the biggest total amount of precast concrete wastes was produced from precast concrete plant C, representing 8,200 tonnes followed by precast concrete plant A and B which are 350 tonnes and 128.56 tonnes, respectively.

Table 3. Generation of construction wastes within precast concrete plants

No.	Type of plant	Amount of production (t/year)	Amount of waste (t/year)	% of waste
1	Precast concrete plant A	15,000	350	2.33
2	Precast concrete plant B	6,215	128.56	2.06
3	Precast concrete plant C	25,000	820	3.28

As discussed with personal individual during field observations, the majority of interviewees stated that they have experienced concrete waste are mostly produced in precast concrete plants. As a result, the causes of construction waste were generated because of workers' lack of awareness and poor performance of machines. However, from the view of a production manager, the manufacturing process relies on materials being delivered on a just- in-time basis and the quality and sustainability of the materials bought from supplier is also strictly supervised and controlled to avoid unnecessary waste of time, resources and, ultimately, materials.

6. Conclusion

In summary, we have presented identified type of construction waste generated that arising in manufacturing process for existing case studies. We have also shown that this is possible for causes of construction waste generated per manufacturing activities. The main kinds of waste, including concrete, steel and wood are investigated individually. We find that the construction wastes generated in Malaysian precast concrete plants are mainly caused by worker and condition of machines or equipment.

Hence, in order to minimise worker's mistake from generation waste, the organisation's management needs to play a role by establishing waste production checklists on material waste for each projects involved in prefabricated of precast concrete components. Thus, the sources of materials that have been used can be identified to determine the amount of waste generated and to be reviewed based on recorded data. Other than that, for the initial phase, the client must provide the specific objectives of the project by setting waste management policy to provide an effective waste management plan with the organisation's precast concrete plants. Thus, the organisations can hire consultants who can provide guidance on waste streams and treatment options in the precast concrete plants that are appropriate to meet the needs of the client project.

Most notably, this is the first study to our knowledge to identify the type of construction waste generated throughout the process of manufacturing at the precast concrete plants and clarify the significant types of waste generated in order to provide useful information concerning waste assessment data in the precast concrete manufacturing industry. However, some limitations are worth nothing. As noted earlier, only construction waste was available in this study. Further work is needed to determine another type of waste that happens in factory such as waste of overproduction, waiting time, transportation, inventory, motion and defect product and where its exists in the factory.

7. References

[1] Rahim M H I A., Kasim N, Moham I, Zainal R, Sarpin N and Saikah M 2017 Construction waste generation in Malaysia construction industry: Illegal dumping activities *IOP Conf. Ser. Mater. Sci. Eng.* 271 012040

[2] Begum R A, Satari S K, and Pereira J J 2010, Waste Generation and Recycling : Comparison of Conventional and Industrialized Building Systems *American Journal of Environmental Sciences* **6** (4) 383-388

[3] Taha M P 2016 Intergrated Solid Waste Management: Challenge And Future *Natl. Environ. Heal. Action Plan Conf.* pp. 1–53

[4] Begum R A, Siwar C, Jacqueline J and Jaafar A H 2007 Factors and values of willingness to pay for improved construction waste management – A perspective of Malaysian contractors, *Waste management* **27** 1902–1909

[5] Augustine S 2011 *Managing Waste and Cost in the Construction Industry : A Case Study of the Road Construction*, Master Thesis (Institute of Distance Learning: Kwame Nkrumah University of Science and Technology) pp 1–118

[6] Jereme I A, Siwar C and Alam M 2014 Waste Recycling in Malaysia : Transition from Developing to Developed Country *Indian Journal of Education and Information Management* **4** (1) 1-14

[7] Mtech Group 2017 *Waste Reduction Potential of Precast concrete Manufactured Offsite Precast concrete manufactured off-site negates the need for temporary* Project code: WAS 003.003 pp 1-10

[8] Mia W, Akihiro F and Kenji K 2017 Environmental Impact of Waste Concrete Treatment in Precast Concrete Production *JCI Member* (Japan: Hiroshima University) pp 2-7

[9] Nagapan S, Rahman I A, Asmi A, Memon A H and Latif I. 2012 Issues on Construction Waste : The Need for Sustainable Waste Management Issues on Construction Waste : The Need for Sustainable Waste Management *IEEE Colloq. Humanit. Sci. Eng. Res. (CHUSER 2012)* pp. 325–330

[10] European Commission 2005 What a waste; SWM in Asia urban development sector unit East Asia and pacific region *Solid Waste Management* pp. 170–194

[11] Lu W and Yuan H 2013, Investigating waste reduction potential in the upstream processes of offshore prefabrication construction, *Renew. Sustain. Energy Rev.* **28** 804–811

[12] Lai Y, Yeh L, Chen P, Sung P and Lee Y 2016 Management and Recycling of Construction Waste in Taiwan, *Procedia Environ. Sci.* **35** 723–730

[13] Malaysian Industrial Development Finance [MIDF] 2014 Construction IBS Practical solution to rising costs *MIDF Res.* pp. 1–24

[14] Bari N A A, Abdullah N A, Yusuff R, Ismail N and Jaapar A 2012 Environmental Awareness and Benefits of Industrialized Building Systems (IBS) *Procedia - Social and Behavioral Sciences* **50** (December) 392–404

[15] Kermanshahi E, Hossein M, Isaabadi Z and Harirchian E 2015 Material Waste Reduction between IBS and Conventional Construction *IOSR J. Mech. Civ. Eng. Ver. II* **12**(3) 2320–334

[16] Azman M N A, Kamar K A M and Nawi M N M 2013 Industrialised Building System in Reducing Waste of Construction Industry *JSTE* **2** 96-103

[17] Ko C H 2015 Arranging Precast Production Sheduled Using Demand Variability, *Proceedings for the 20th Annual Conference of the International Group for Lean Construction* pp 5-8

[18] Neil K 2019 The Difference Between Construction & Manufacturing. Retrieve on March 14,

2019 from http://smallbusiness.chron.com/difference-between-construction-manufacturing-20748.html

[19] Azman M N A, Ahamad M S S, Majid T A and Shah M N S A 2012, A study of precast concrete in Malaysia, *Concrete* **46**(10) 50–52

[20] Building and Construction Authority 2016 Retrieved on March 14, 2019 from https://www.bca.gov.sg/Professionals/IQUAS/others/precastfabrication.pdf.

[21] Mtech Consult Group, Waste reduction potential of offsite manufactured pods, pp. 1–10, 2012.

[22] Zailinawati A H, Peter S 2006 Doing A Pilot Study: Why is it essential? *Malaysian Fam. Physician* **1**(2) 70–73

[23] Yin R K 2006 Case Study Reserach - Design and Methods, *Clin. Res.* **2** 8–13

[24] Gustafsson J 2017 *Single case studies vs. multiple case studies: A comparative study* Dissertation Retrieved on March 14, 2019 from http://urn.kb.se/resolve?urn=urn:nbn:se:hh:diva-33017

[25] CIDB 2017 Country Report Malaysia (Construction Industry Development Board Malaysia (CIDB) Kuala Lumpur, Malaysia) pp. 5–6

IOP Conf. Series: Materials Science and Engineering **601** (2019) 012037 doi:10.1088/1757-899X/601/1/012037

The selection of optimum water-cement ratio for production of low thermal conductivity cement sand brick with Oil Palm Mesocarp Fibre as admixture

Nurul Fatina Md Sali[1] and Rafikullah Deraman[1*]

[1] Deparment of Building and Construction Engineering, Faculty of Civil and Environmental Engineering, Universiti Tun Hussein Onn Malaysia, 86400 Parit Raja, Batu Pahat, Johor, Malaysia

*Corresponding author: rafikullah@uthm.edu.my

Abstract. The development of housing construction project in Malaysia has rapidly increasing. However, the development of the projects only focusing on the exterior look without giving consideration on the thermal properties of materials used for building envelopes which may influence to external gain and thermal comfort for the occupants inside. As way to encounter the problem, one of the methods is by passive design which through utilized a low thermal conductivity of brick wall in creating good thermal insulation of the building envelops. This study investigated the characteristics of the cement sand brick (CSB) with incorporated of Oil Palm Mesocarp Fibre (OPMF) as admixture by using different water-cement ratio ranging from 0.45 to 0.65. All bricks were tested for compressive strength, water absorption and thermal conductivity. The result reveals that the higher water-cement ratio increased the compressive strength and water absorption, while decreased in thermal conductivity of bricks. It was found that, by incorporating of 0.65 water-cement ratio enhances the production of low thermal conductivity brick with value of 0.891 W/m°C, compressive strength of 12.4 MPa and water absorption of 1.99% that meets the standard requirement. Foremost, the incorporation of OPMF with water-cement ratio of 0.65 provides alternative solution for energy efficiency and improve the comfort in buildings by improving CSB insulation properties.

1. Introduction

Malaysia had generated in excess of 15,000 tons of solid waste per day in the form of biomass that consists of forest and mill residues, agricultural waste, wood wastes and municipal wastes [1]. The increase in the quantity of agricultural industries in Malaysia supports economic growth positively, but it affects the environment contrarily by generating large amounts of wastes. Internationally, 998 million tons of agricultural waste were produced in a year and in Malaysia, 1.2 million tons of agricultural waste is disposed into landfills annually [2].

In 2015, Malaysia Palm Oil Council (MPOC) reported that approximately 5.1 million hectares of land crosswise over Peninsular Malaysia, Sabah and Sarawak were planted with oil palm, with more than 200 palm oil mills currently operating [3]. The wastes such as Empty Fruit Bunch (EFB), Palm Kernel Shell (PKS), Oil Palm Mesocarp Fibre (OPMF), Palm Oil Mill Effluent (POME), Oil Palm Trunk (OPT), Oil Palm Leaves (OPL) and Oil Palm Fronds (OPF) are produced after the oil palm fruits harvesting, palm oil processing or during oil palm trees replantation [4]. These wastes are simply disposed of without any economic return [5].

Oil Palm Mesocarp Fibre (OPMF) otherwise known as palm pressed fibre (PPF) is the biomass residue acquired subsequent to pressing the palm fruits for palm oil extraction. Approximately, 11% of OPMF is created from the palm fruits after the oil extraction [6]. Being a lignocellulosic material, OPMF has grabbed the attention of researchers due to its potential utilization for biocomposite production [7-8], whereby the fibre can be used to strengthen polymer materials such as thermoplastics.

Hence, utilization of agricultural wastes in the bricks production seems to be a solution not exclusively to environmental pollution yet additionally to the design of green building [9]. Besides that, the use of wastes as materials to increase thermal insulation, turn into a solution which offers a reduction in resource use, advance reusing of the wastes, less dependent on toxic substances composes in wood or cellulose-based insulators, and reducing energy consumed by altering indoor air temperature conditions [10].

Modern housing buildings became more dependent on artificial means to provide comfortable thermal environment with high energy consumption. The increase in energy consumption by the housing buildings is mainly because of the growing demand for air-conditioning systems to provide thermal comfort for building's occupants. This high energy consumption is mostly related to poor thermal performance of building envelope [11]. This situation happened due to modern housing buildings are fully exposed to solar radiation, which will increase the ambient temperature inside and affects the occupants comfort.

Natural fibre bricks offer a competitive alternative to conventional building materials since they utilize local resources that can be both costs effective and energy efficient, and closely follow existing masonry construction practices. By adding natural fibres in cement sand bricks has been reported can improve the mechanical properties and flexibility of bricks [12-13]. There are numerous studies have been done on the development of thermal insulation materials from natural fibres used as environmental-friendly and renewable materials such as rice husk, bagasse and corncob [14], maize husk, maize cob, groundnut shell, coconut pith and paddy straw, eggplant stalks, tamarind hulls and cotton fibres [15].

Typically, there are many types of bricks manufactured in the factories all around the world. The most common brick is the traditional fired clay brick and a huge amount of energy is depleted throughout its production [16]. Sand cement brick (CSB) is a kind commonly used in low and medium cost housing development and other commercial constructions in Malaysia. CSB is easy to make and inexpensive to produce [17]. Thus, CSB will be a better way to replace clay brick which significantly less expensive.

Therefore, in this study, the utilization of OPMF as admixture material in CSB with different water-cement ratio was conducted. The aim of this study is to select the optimum water-cement ratio for the production cement sand brick with Oil Palm Mesocarp Fibre (OPMF) as admixture which meet the standard requirement and has a low thermal conductivity.

2. Materials and Method

2.1. Preparation of raw materials

The materials used in this study are Ordinary Portland Cement (OPC), sand, oil palm mesocarp fibre (OPMF) and water. The OPMF ware taken from Ladang Tereh Mill Kulim Plantation (M) Sdn. Bhd. located at Kluang, Johor, Malaysia. The fibre was treated by immersed with one portion of OPMF (g) and mixed with 20 portions (ml) of Sodium Hydroxide (NaOH) or 1% by weight for 30 minutes. The treated fibre then was washed with distilled water and oven dried at 50°C-60°C until constant weight [18]. The fibre was grinded into smaller pieces and surpasses sieved of 5 mm to ensure the particle size was uniform for brick manufacturing. Figure 1 shows the OPMF waste fibre with a treatment of NaOH.

a) NaOH treatment for OPMF before oven-dried.

b) NaOH treatment for OPMF after oven-dried.

Figure 1. NaOH treatment for MF waste

2.2. Manufacturing process of bricks

In this study, six different mix ratio (cement: sand: fibre) were considered as shown in Table 1. For each ratio, 15 samples were manufactured. The nominal dimension for brick size is 225 x 113 x 75 mm as followed JKR Standard Specification for Building Work (2005) [19]. To produce a control brick sample, cement and sand were weighed and mixed first by using a mechanical mixer at 10-litre capacity. The same step was repeated to make OPMF brick, with an incorporation of constant 1.5% waste into the mixture. However, for fibre bricks, sand, cement and fibres were mixed first so that the fibres were evenly distributed before water was added. Then, the samples were pressed into moulds according to the shape and compacted further. The newly-produced raw bricks were dried naturally for 24 hours at room temperature before the curing process. Basically, the bricks have been cured for 7 days and 28 days by using wet blanket curing. The manufactured bricks then underwent a series of test including physical, mechanical, chemical and thermal properties.

Table 1. Material composition and ratio

Sample	Cement: sand: fibre[a]	Water-cement ratio
1 (Control)	1:5:0	0.45
2	1:5:0.04	0.45
3	1:5:0.04	0.50
4	1:5:0.04	0.55
5	1:5:0.04	0.60
6	1:5:0.04	0.65

Note: 1 volume ratio of cement = 2.75 kg, 1 volume of sand = 3.25 kg.
Weighing was carried out using an electronic balance with a precision of 0.01 kg.
[a] Fibre ratio is in kg (corresponding to 1% of reference cement volume)

2.3. Properties of cement-sand brick

2.3.1. Physical and mechanical properties

All manufactured spceimens were tested for sieve analysis, specific gravity, compressive strength and water absorption test. For compressive strength test, these testing were according to ASTM C129-06 (2006) [20]. While, for water absorption followed ASTM C67-94 [21].

2.3.2. Chemical composition
X-ray fluorescence (XRF) test was conducted for OPMF in order to determine its chemical compound.

2.3.3. Thermal conductivity
Thermal conductivity was tested using a hot guarded plate method follow the ASTM C177 [22]. This device can measure the thermal conductivity of a material when a layer of materials of known thickness and area are heated form one side by an output. This test likewise measures the ability of a material to conduct heat. Then, following the procedure of BS EN ISO 8990 (1996) [23], the experiment was kept running for 100 minutes for each specimen and the information was recorded for each one minute.

3. Result and discussion

3.1. Sieve analysis test
Sieve analysis test was done to obtain the particle size of the raw material. The particle sizes can be calculated based on the weight of soils loss when passing sieve size. The test was conducted based on BS 1377: Part 2: 1990 [24]. Based on the result obtained, the sieve analysis of the sand sample is presented in Figure 2.

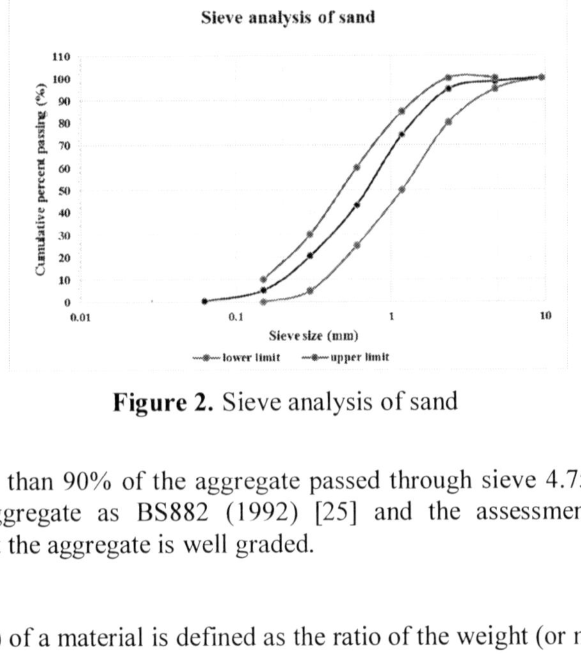

Figure 2. Sieve analysis of sand

From Figure 2, more than 90% of the aggregate passed through sieve 4.75 mm which places the aggregate as a fine aggregate as BS882 (1992) [25] and the assessment of the particle size distribution revealed that the aggregate is well graded.

3.2. Specific gravity
The specific gravity (Gs) of a material is defined as the ratio of the weight (or mass) of a given volume of the material to the weight (or mass) of an equal volume of water. Gs is useful for determining weight – volume relationships. The result for the specific gravity of sand and OPMF were 2.65 and 1.17 respectively.

3.3. X-ray fluorescent analysis (XRF)
This test was conducted to obtain the various chemical compounds present in OPMF. Based on Table 2, it shows that the chemical compound for Zirconium dioxide (ZrO_2) was the highest compound presence in OPMF waste. ZrO_2 is a white crystalline oxide of zirconium. Zirconia is produced by calcining zirconium compounds, exploiting its high thermal stability [26]. Hence, zirconia has to be compared with a material that is both aesthetic and has good mechanical properties for use as a restorative material [27] in brick manufacturing. Therefore, the presence of ZrO_2 tends to produce better properties of bricks.

Besides that, another compound such as Calcium oxide (CaO) contributes to the plasticity of the mix. It is additionally valuable in creating a very tenacious adhering power which helps the mix hold to forms and moulds.

Table 2. Chemical composition of OPMF.

Chemical compound	Formula	Concentration (%)
Zirconium dioxide	ZrO_2	12.20
Carbon	C	0.10
Hafnium dioxide	HfO_2	8.83
Silicon dioxide	SiO_2	7.62
Iron oxide	Fe_2O_3	4.60
Calcium oxide	CaO	3.47
Potasium oxide	K_2O	4.37
Sulfur trioxide	SO_3	1.03
Clorine	Cl	0.80
Aluminium oxide	Al_2O_3	0.97
Phosphorus pentoxide	P_2O_5	1.23
Sodium oxide	Na_2O	0.22
Magnesium Oxide	MgO	0.49

3.4. Water-cement ratio

Brick develops its strength by hydration of the cement and addition to form a complex series of hydrates. The initial hydration fixes the cement particles into a weak structure surrounded by a water-filled space. The higher the initial water content, the further will be the average separating between the cement grains [28]. Where the initial water-cement ratio is high, the subsequent pore structure within the hydrates is interconnected and the subsequent mixed has low strength, high penetrability and low durability. In practice, long-term strength gain will only occur in conditions where the mixed retains or gains adequate water for hydration to continue.

3.4.1 Effect of water-cement ratio on the compressive strength

The compressive strength of the bricks determines its ability to withstand loads before failure. The materials will crash when the limit of compressive strength is reached at the standard limit. According to ASTM C129-06 (2006), the minimum net area compressive strength requirements for a brick wall as for non-load bearing wall masonry is 3.45 MPa. Figure 3 shows that all the brick specimens carried the capacity above the minimum ASTM C129-06 (2006) requirement. The compressive strength of bricks measured at different curing times are given in Table 3.

Table 3. The average compressive strength of bricks.

OPMF content (%)	Water-cement ratio	Compressive strength (MPa)
0	0.45	6.6
1.5	0.45	4.2
1.5	0.50	5.0
1.5	0.55	5.9
1.5	0.60	7.8
1.5	0.65	12.4

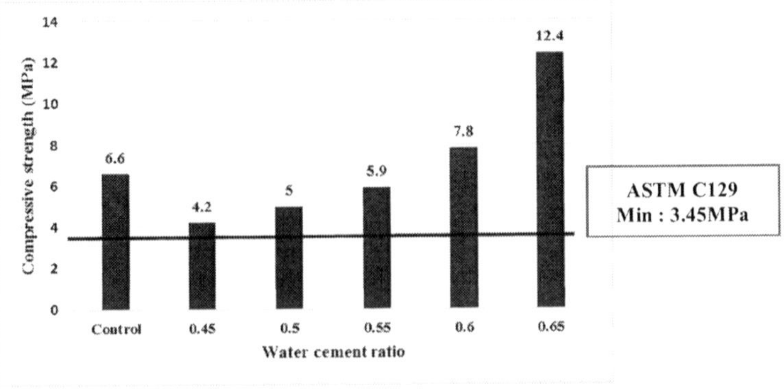

Figure 3. The compressive strength of specimen with different water-cement ratio

The content of water-cement ratio in the brick plays an important role in determining the compressive strength of the brick. Increasing of water-cement ratio ranging from 0.45 to 0.65 was found to impact the compressive strength of specimens significantly. From the graph, it shows that as the water-cement ratio increased, the compressive strength increased. This study is similar with [29] reveals that lower water-cement ratio reduces the workability and compressive strength due to the presence of air voids in bricks.

3.4.2. Effect of water-cement ratio to water absorption

Absorption measures the unit's total capacity to absorb moisture. The results of percentage water absorption with a different water-cement ratio at 28 days curing time are shown in Figure 4. It can be conclude that the absorption of bricks increased as water-cement ratio increased. This is due to the high amount of water that helps on bonding and reaction process of bricks. The present of OPMF fibre needs more water for hydration process and voids inside the bricks have the tendency to absorb more water during immersion of the brick. Overall, water absorption for the brick with 0.65 has the highest water absorption value (1.99%). However, it is meet the standard requirement that is lower than 18%.

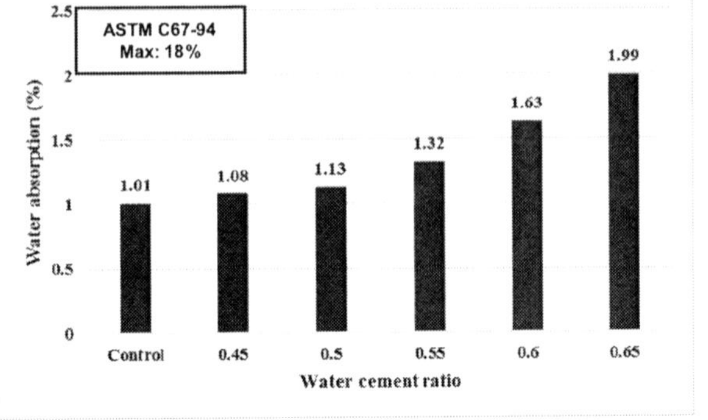

Figure 4. The average water absorption between different water-cement ratio

3.4.3. Effect of water-cement ratio to thermal conductivity

Figure 5 shows the result for thermal conductivity of bricks. The result shows that the thermal conductivity of bricks decreased as the water-cement ratio increased. That is due to the porosity of fibre brick that played an important role in determining the value of thermal conductivity. Porosity influences the value of thermal conductivity because of the existence of space and voids between brick particles. Therefore, from the comparison between a different water-cement ratio of bricks, it can be concluded that the water-cement ratio influenced the value of thermal conductivity. The results were also below the maximum requirement of ASTM C177 which are 1.16 W/m°C. The lower of thermal conductivity improves the insulation properties of a brick wall.

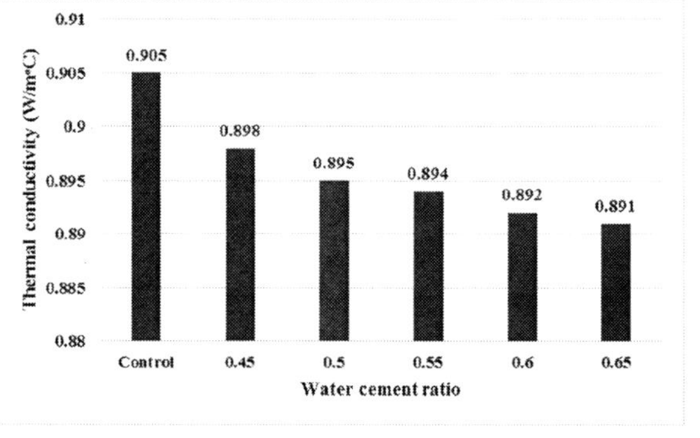

Figure 5. Effect of different water-cement ratio on thermal conductivity

4. Conclusion

The production of OPMF waste has been an issue in the palm oil industry. Often the accumulating wastes were not reused and further, disposed into landfills and caused another environmental issue. OPMF was worth being recycled as bricks as this could lessen the disposal problem and pollution apart from becoming a building material. The utilization of OPMF waste into bricks could act as a low thermal brick that provides an alternative solution for low consumption energy and improves the comfort in buildings. The OPMF waste could also be a potential for a low-cost waste admixture for the production of brick. In conclusion, it can be found that the different water-cement ratio gives a different compressive strength, water absorption and thermal conductivity to bricks. Meanwhile, air dry curing process brings a positive effect to the increasing strength of bricks. Therefore, water-cement ratio of 0.65 was selected as the optimum ratio for further manufacturing of OPMF brick as it fulfils the standard requirement.

5. References

[1] Manaf L A, Samah M A A and Zukki N I M 2009 Municipal solid waste management in Malaysia: Practices and challenges *Waste Management* **29** 2902-2906
[2] Agamuthu P and Fauziah S H 2009 Sustainable household organic waste management via vermicomposting *Malaysian Journal of Science* **28** 135-142
[3] Kadir A A, Mohd Zahari N A and Azizi Mardi N 2013 Utilization of palm oil waste into fired clay brick *Advances in Environmental Biology* **7** 3826–3834
[4] Mushtaq F, Abdullah T A T, Mat R and Ani F N 2015 Optimization and characterization of bio-oil produced by microwave assisted pyrolysis of oil palm shell waste biomass with microwave absorber, *Bioresource Technology* **190** 442-450
[5] Awal A S M A and Husin M W 2011 Effect of palm oil fuel ash in controlling heat of hydration of concrete *Procedia Engineering* **14** 2650-2657
[6] Riansa-ngawong W and Prasertsan P 2011 Optimization of furfural production from hemicellulose extracted from delignified palm pressed fibre using a two-stage process *Carbohydrate Research* **346** 103–110
[7] Wittaya T 2009 Microcomposites of rice starch film reinforced with microcrystalline cellulose from palm pressed fibre *International Food Research Journal* **16** 493–500
[8] Phattaraporn T, Waranyou S and Thawien W 2011 Effect of palm pressed fibre (PPF) surface treatment on the properties of rice starch films *International Food Research Journal* **18** 287–302

[9] Shakir A A, Naganathan S and Mustapha K N 2013 Development of Bricks from Waste Material : A Review Paper *Australian Journal of Basic and Applied Sciences* **7** 812-818

[10] Sadek D M 2012 Physico-mechanical properties of solid cement bricks containing recycled aggregates *Journal of Advanced Research* **3** 253–260

[11] Manioglu G and Yılmaz Z 2008 Energy efficient design strategies in the hot dry area of Turkey *Building and Environment* **43** 1301–1309

[12] Shafee M N H 2016 *A study of optimum corn cob ratio in cement sand brick* Undergraduate Thesis (Pahang: Universiti Malaysia Pahang)

[13] Aminudin E, Khalid A N H, Azman N A, Bakri K and Din M F M, Zakaria R, Zainuddin N A 2017 Utilization of Baggase Waste Based Materials as Improvement for Thermal Insulation of Cement Brick *MATEC Web of Conferences (Melaka)* vol 103 (Malaysia: MATEC Web of Conference) pp 1-8

[14] Akmal T, Fahmy M and El-Kadi A W 2011 Rice-straw based cement brick microclimatic thermal impact assessment in Cairo Egypt *World Renewable Energy Congress* (Sweden) vol 1 (Sweden: Low-Energy Architecture) pp 2094-2101

[15] Sadik C, El Amrani I and Albizane A 2012 Influence of chemical and mineralogical nature of clay and manufacturing process on the quality of ceramic tiles *MATEC Web of Conferences (Morocco)* vol 2 (Morocco: MATEC Web of Conference) pp 1-7

[16] Sadik C, El Amrani I and Albizane A 2013 Effect of Carbon Graphite on the Crystallization of Andalusite: Application to the Synthesis of Mullite and the Improvement of Refractory Quality *Materials Sciences and Applications* **4** 337-346

[17] Binici H, Aksogan O and Shah T 2005 Investigation of fibre reinforced mud brick as a building material *Construction and Building Materials* **19** 313-318

[18] Astimar A A, Das K, Husin M and Mokhtar A 2002 Effect of physical and chemical pre-treatments on xylose and glucose production from oil palm press fibre *Journal of Oil Palm Research* **14** 10-17

[19] Jabatan Kerja Raya Malaysia 2005 *Standard Specifications for Building Works* (Malaysia: Ketua Pengarah Kerja Raya Malaysia)

[20] ASTM International 2006 *ASTM C129 Standard Specification for Nonloading Concrete Masonry Unit* (West Conshohocken, PA)

[21] ASTM International 2007 *ASTM C67 -94 Standard Test Methods for Sampling and Testing Brick and Structural Clay Tile* (West Conshohocken, PA)

[22] ASTM International 1997 *ASTM C177 Standard Test Method for Steady-State Heat Flux Measurements and Thermal Transmission Properties by Means of the Guarded-Hot-Plate Apparatus* (West Conshohocken, PA)

[23] BS EN ISO 8990 1996 Thermal Insulation – Determination of Steady-State Thermal Transmission Properties – Calibrated and Guarded Hot Box (London: British Standard Institution)

[24] BS 1377-2 1990 *Methods of test for soils for civil engineering purposes* (London: British Standard Institution)

[25] BS 882 1992 *Specification for aggregates from natural sources for concrete* (London: British Standard Institution)

[26] Nielsen R *Zirconium and Zirconium Compounds" in Ullmann's Encyclopedia of Industrial Chemistry* Retrieved on October 20, 2018 from https://onlinelibrary.wiley.com

[27] Dalayan M, Jairaj A, Nagaraj K R and Savadi R C 2010 An evaluation of fracture strength of Zirconium Oxide posts fabricated using CAD-CAM technology compared with Prefabricated Glass Fibre Posts *The Journal of Indian Prosthodontic Society* **10** 213-218

[28] Killoh D, Parrott L J, Patel R and Gutteridge W A 1988 Influence of curing at different relative humidities upon compound reactions and porosity in Portland cement paste *Materials and Structures* **21** 192-197

[29] Ali N, Yusup N F M, Khalid F S, Shahidan S and Abdullah S R 2018 The effect of water-cement raio on cment brick containing High Density Polyethylene (HDPE) as sand replacement *MATEC Web of Conference (Melaka)* vol 150 (Malaysia: MATEC Web of Conference) pp 1-5

Institute of Physics
Dirac House, Temple Back
Bristol BS1 6BE UK

ISSN: 1757-8981
ISBN 978-1-5108-9479-2

CURRAN ASSOCIATES INC.
proceedings.com

4th International Scientific Conference SEA-CONF 2018

IOP Conference Series: Earth and Environmental Science
Volume 172

Constanta, Romania
17 – 19 May 2018

IOP
Institute of Physics